T0330856

ENGINEERING FOR SUSTAINABILITY

SUSTAINING THE MILITARY ENTERPRISE SERIES

Engineering for Sustainability
Dennis F.X. Mathaisel, Joel M. Manary, and Ned H. Criscimagna

Enterprise Sustainability: Enhancing the Military's Ability to Perform Its Mission
By Dennis F.X. Mathaisel, Joel M. Manary, and Clare L. Comm

Sustaining the Military Enterprise: An Architecture for a Lean Transformation (with CD)
By Dennis F.X. Mathaisel

ENGINEERING FOR SUSTAINABILITY

DENNIS F.X. MATHAISEL
JOEL M. MANARY
NED H. CRISCIMAGNA

CRC Press
Taylor & Francis Group
Boca Raton London New York

CRC Press is an imprint of the
Taylor & Francis Group, an **informa** business

CRC Press
Taylor & Francis Group
6000 Broken Sound Parkway NW, Suite 300
Boca Raton, FL 33487-2742

ISBN-13: 978-1-4398-5351-1 (hbk)

Library of Congress Cataloging-in-Publication Data

Mathaisel, Dennis F. X.
    Engineering for sustainability / Dennis F.X. Mathaisel, Joel M. Manary, Ned H. Criscimagna.
      p. cm. -- (Sustaining the military enterprise)
    Summary: "This book outlines a series of principles to help engineers design products and services to meet customer and societal needs with minimal impact on resources and the ecosystem. The third volume in a series on sustainable engineering, it provides up-to-date information on planning and implementing sustainable activities. Using examples and case studies from the government, military, academia, and commercial enterprises, the authors provide a set of tools for long-term sustainability and explain how an entire enterprise can be engineered to sustain itself"-- Provided by publisher.
    Includes bibliographical references and index.
    ISBN 978-1-4398-5351-1 (hardback)
    1. Environmental engineering. 2. Sustainable engineering. I. Manary, Joel. II. Criscimagna, Ned H. III. Title.

TA170.M38 2012
620.0028'6--dc23   2012021336

Visit the Taylor & Francis Web site at
http://www.taylorandfrancis.com

and the CRC Press Web site at
http://www.crcpress.com

Contents

Preface

Sustainability is an ability: the ability to endure. In ecology, sustainability describes how biological species survive. For the environment, it is assessing whether or not project outputs can be produced without permanent and unacceptable changes in the environment. For humans, it is our long-term physical and cultural well-being. For mechanical systems and structures, it is maximizing reliability while conserving required resources and reducing waste. For an entity or an enterprise, it is the ability of the enterprise, its products, and its systems to remain competitive and productive long term, without failure, while minimizing waste.

Sustainability and sustainable development have become popular goals. They have also become wide-ranging terms that can be applied to any entity or enterprise on a local or a global scale for long time periods. Sustainability has many interpretations. Recently, the term has been used more in the context of "green," which refers to having no negative impact on the environment, community, society, or economy (Bromley 2008). However, the traditional meaning centers on the words "to endure" or "to maintain" or "to survive," which is the context for sustainability used in this book. Here, sustainability means to adopt a strategy or prescription to maintain the ability of an entity or enterprise and its systems or services to survive with established performance requirements in the most effective and efficient manner possible over the entity's life cycle.

Engineering for Sustainability is the third volume in a series of manuscripts under the title *Sustaining the Military Enterprise*. The first volume, *An Architecture for a Lean Transformation* (Mathaisel 2007), focused on the various process improvement initiatives that are available to help sustain the military enterprise, and it presented a **Lean Enterprise Architecture** to accomplish that objective. The second volume, *Enterprise Sustainability* (Mathaisel, Manary, and Comm 2009), focused on five abilities (see Figure 0.1) that an enterprise must possess to be sustainable:

- Availability of required parts, facilities, tools, and manpower
- Dependability of the systems
- Capability of the enterprise to perform the mission
- Affordability and improving the life cycle cost (LCC) of a system or project
- Marketability of concepts and motivating decision makers

Engineering for Sustainability addresses the question: how does one engineer an enterprise or a product for sustainability? Sustainability engineering is a discipline that has become increasingly important as systems become more complex and development and support costs increase while budgets are being challenged. Achieving the high levels of sustainability needed in complex military and industrial systems is too often an elusive goal. Competing rules and regulations,

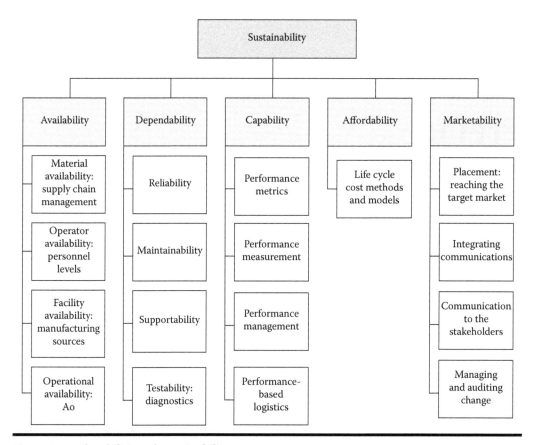

Figure 0.1   The abilities of sustainability.

conflicting goals and performance metrics, the desire to incorporate promising commercial off-the-shelf (COTS) technologies, and the pressures of maintenance schedules contribute to this elusiveness. To assist in countering this elusiveness, the authors wrote this book to provide common sense information for engineering, planning, and carrying out those tasks needed to sustain military products and services and, in turn, the entire enterprise.

Two factors pointed to the need for this third volume. First, military services are emphasizing the need to reduce costs and improve readiness. Second, increasing international competition and rising customer expectations are making economical and rapid maintenance critical product attributes. For example, noting the significant improvement in the quality of commercial products and the rapidity with which new technology is incorporated in commercial products, and facing a shrinking defense budget, the U.S. Department of Defense changed its acquisition policies to foster the evolution of a unified military and commercial industrial base. The objective is to capitalize on the "best practices" that American business has developed or adopted. The information in this book reflects both the move to incorporate commercial practices and the lessons learned over many years of sustainable systems.

When appropriate, commercial and military standards are cited for reference. These documents are familiar to both military and commercial companies, contain a wealth of valuable information, and often have no commercial counterpart. Although many of these documents emphasize what to do and how to do it, this book, in the spirit of the new policies regarding acquisition, focuses on the objectives of sound sustainability engineering and the tools available to meet these objectives. In particular, there are two handooks that contributed to some of the ideas presented in this book:

1. *Maintainability Toolkit* by Ned (Cris) H. Criscimagna (1999)
2. *Systems Engineering Process Guide: A Practical Framework for Acquirers and Providers of Systems Engineering Services* by Joel M. Manary (2004)

Maintainability, reliability, and systems engineering practices are implemented and referenced throughout this book. These principles and practices are key abilities that are necessary for an enterprise or an entity to be sustainable.

References

Bromley, Daniel W. 2008. "Sustainability." In *The New Palgrave Dictionary of Economics*, 2nd ed. Eds. Steven N. Durlauf and Lawrence E. Blume. Palgrave Macmillan. 293–318.

Criscimagna, Ned (Cris) H. 1999. *Maintainability Toolkit*. Alion System Reliability Center Publication SRC-HDBK-2000 (MKIT). Rome, NY: Reliability Analysis Center (RAC).

Manary, Joel M. 2004. *Systems Engineering Process Guide: A Practical Framework for Acquirers and Providers of Systems Engineering Services*. Rome, NY: Reliability Analysis Center (RAC). Alion System Reliability Center Publication SRC-HDBK-7010.

Mathaisel, Dennis F. X. 2007. *Sustaining the Military Enterprise: An Architecture for a Lean Transformation*. Boca Raton, FL: Auerbach Publications, Taylor & Francis Group.

Mathaisel, Dennis F. X., Joel M. Manary, and Clare L. Comm. 2009. *Sustaining the Military Enterprise: Enhancing its Ability to Perform the Mission*. Boca Raton, FL: CRC Press, Taylor & Francis Group.

Acknowledgments

The authors thank the Massachusetts Institute of Technology, the U.S. Air Force, the U.S. Army, and the U.S. Navy.

Criscimagna thanks the many civilian and military professionals with whom he has had the good fortune to work with throughout his career. Special thanks go to Patrick Hetherington for his encouragement and inspiration during Criscimagna's stint at the Reliability Analysis Center. Last, but certainly not least, he thanks his wife, Sandy, for her constant and steady support for more than 45 years of marriage.

Authors

Dennis F. X. Mathaisel is a professor of management science in the Department of Mathematics and Science at Babson College and received his PhD from the Massachusetts Institute of Technology (MIT). For 20 years, he was a research engineer at MIT. He was also cofounder and president of Scheduling Systems Incorporated, a computer software firm, and in the early 1970s he was a branch manager at the McDonnell Douglas Corporation.

Dr. Mathaisel's interests focus on the sustainability of complex and aging systems. He was an MIT colead for the Lean Sustainment Initiative for the U.S. Air Force. He has written two books and several papers on sustainability and enterprise transformation. Through his experience working with several government and commercial organizations, he has learned how an effectively designed and executed enterprise transformation plan can promote the vision, commitment, sense of urgency, senior leadership buy-in, and shared goals and objectives that are necessary for a successful adaptation of enterprise-wide lean sustainment.

Dr. Mathaisel has consulted for the Federal Aviation Administration, the National Aeronautics and Space Administration, the U.S. Air Force, and the U.S. Department of Transportation (Office of the Secretary); Pratt & Whitney; FedEx, and the Flying Tiger Line; and Continental Airlines, Garuda Indonesia Airline, Hughes Airwest, Iberia Airlines, Northwest Airlines, Olympic Airlines, Pan American World Airways, Trans World Airlines, U.S. Airways, among many other institutions. These assignments have focused on enterprise sustainability, the application of lean manufacturing to sustainability, decision support systems, maintenance and logistics, scheduling, fleet and route planning, and transport systems analysis and engineering.

Dr. Mathaisel is a private pilot and an owner of a Cessna 182 aircraft based at Hanscom Air Force Base, near Lexington, Massachusetts.

Joel M. Manary received his MS in logistics and systems acquisition management at the Air Force Institute of Technology. He is a research fellow at Massachusetts Institute of Technology (MIT) and has participated in several studies as part of the MIT Advanced Studies Program. He is a senior systems engineer for Ocean Systems Engineering Corporation and is homebased in Oceanside, California. He is the lead systems engineering subject matter expert in the Systems Engineering Process Office, a staff agency supporting SPAWAR (Space and Naval Warfare Systems Command) systems center Pacific in San Diego, California. He has over 20 years of experience in acquisition program management and systems engineering management. He was a program manager of an automated tool improvement project. He was a senior systems analyst, staff consultant, and advisor to senior managers in the Office of the Secretary of Defense, Department of Defense, and Naval Air Systems Command, where he provided and supervised direct support to project managers, senior systems analysts, senior systems engineers, scientists, and technical personnel. He also has over 20 years of experience in operational systems support, as an active duty Air Force maintenance officer. He performed as a senior systems analyst and maintenance officer at the organizational level, intermediate level, and military depot level. Joel is an avid classic car buff and has restored six Mustangs for his six children, as well as a 450 SL Mercedes for his wife and a Rolls Royce Silver Cloud for himself!

Ned H. Criscimagna is the owner of Criscimagna Consulting LLC, providing consulting services in reliability and maintainability (R&M). From June 1993 to the spring of 2006, he was a senior engineer with the System Reliability Center of Alion Science & Technology. Criscimagna received his bachelor's degree in mechanical engineering from the University of Nebraska–Lincoln, received his master's degree in systems engineering from the Air Force Institute of Technology, and did his postgraduate work in systems engineering and human factors at the University of Southern California. He completed the U.S. Air Force Squadron Officer School in residence, the U.S. Air Force Air Command and Staff College by seminar, and the Industrial College of the Armed Forces correspondence program in National Security Management. He is also a graduate of the Air Force Instructors Course and completed the ISO 9000 Assessor/Lead Assessor Training Course. Criscimagna is a former member of the American Society for Quality (ASQ) and a senior member of the Society of Logistics Engineers. He is a certified professional logistician, chaired the ASQ/ANSI Z-1 Dependability Subcommittee, was a member of the US TAG to IEC TC56, and secretary for the G-11 Division of the Society of Automotive Engineers. He has been involved in projects related to defense acquisition reform. These have included a project for the Department of Defense in which he led an effort to benchmark commercial reliability practices. He led the development of a handbook on maintainability to replace MIL-HDBK-470 and MIL-HDBK-471, and the update to MIL-HDBK-338, *Electronic Reliability Design Handbook*. Before joining Alion, he spent 7 years with ARINC Research Corporation and, prior to that, 20 years in the U.S. Air Force. He has over 32 years experience in project management, acquisition, logistics, R&M, and availability.

Chapter 1

Sustainability Engineering

1.1 The Concept of Sustainability Engineering

Sustainability engineering is a logically sequenced, consistent set of technical and management processes that translate a customer's needs and requirements into a successful product. This book provides guidance for accomplishing and improving this concept for any enterprise. The approach is based on proven systems engineering and management processes that would span the life cycle of the product. The objective is for the enterprise to endure. This means continually meeting customer requirements at manageable cost. An engineer committed to these objectives will consistently challenge the design of the enterprise and its products to uncover weaknesses and potential problems. The solution is to be preemptive and to design for sustainability. If this objective cannot be met and the product entity fails to meet sustainability objectives, corrective design changes will have to be made later in the entity's life cycle at significant expense. The primary emphasis should be to identify and correct problems early in the process when corrective procedures are relatively simple and inexpensive.

No product entity, and in turn the enterprise, can be sustainable if the enterprise does not possess the following abilities:

- Availability of the required parts, facilities, tools, and manpower
- Dependability of the products
- Capability of the enterprise to perform the mission
- Affordability and improving the life cycle cost (LCC) of the product
- Marketability to stakeholders and motivation of decision makers

Thus, to be sustainable, an enterprise must possess all of these abilities. It means that the enterprise must aim to maintain the readiness and operational capability of its systems or services through the adoption of a strategy that meets established performance requirements in the most effective, efficient manner over the entity's life cycle.

Sustainability engineering is concerned with the relative ease and economy of preventing failure (retaining an entity in a specified condition) or correcting failures (restoring an entity to a specified condition) through the necessary actions. So, sustainability is not simply the ability to keep an entity operating using prescribed procedures and resources. It is the ability to do so economically and efficiently. Thus, consolidating the idea that sustainability is the ability to endure with the economics of being sustainable yields the following definition of entity sustainability.

> Entity sustainability is the relative ease and economics of time and resources with which an entity (enterprise, product, or service) can be retained in, or at least restored to, a specified and satisfactory condition by personnel having the necessary skill levels and ability, using prescribed procedures and resources.

In this context, sustainability is a function of the entity's design. So, sustainability should be a design parameter. Although other influences, such as highly trained, motivated, and responsive people, can help keep the chance of failure to an absolute minimum, it is the inherent design that determines this minimum. Improving training or support cannot effectively compensate for the effect of a poorly designed entity on its operational performance and availability.

Sustainability can be measured in many different ways, quantitatively and qualitatively. Table 1.1 suggests some measures. These measures are based on the measures that are commonly applied to maintainability, one of the components of sustainability.

Table 1.1   Suggested Measures of Sustainability

Measure	Comment
Mean time to repair (MTTR), also called mean corrective maintenance time	A composite value representing the arithmetic average of the maintenance cycle times for the individual maintenance actions for a system (excludes preventive maintenance).
Mean preventive maintenance time	A composite value representing the arithmetic average of the maintenance cycle times for the individual preventive maintenance actions (periodic inspection, calibration, scheduled replacement, etc.) for a system.
Median active corrective maintenance time	That value of corrective maintenance time that divides all downtime (DT) values for corrective maintenance such that 50% are equal to or greater than the median and 50% are equal to or less than the median.
Mean active maintenance time	The mean or average elapsed time needed to perform maintenance (preventive and corrective), excluding logistic and administrative delays.
Maximum active corrective maintenance time	That value of maintenance DT below which one can expect a specified percent of all corrective maintenance actions to be completed. Must be stated at a given percentile point, usually the 90th or 95th. Primarily related to the lognormal distribution.
Mean time to restore system	For highly redundant systems, the time needed to switch to a redundant backup unit.

Table 1.1 Suggested Measures of Sustainability (*Continued*)

Measure	Comment
Mean downtime (MDT)	The mean or average time that a system is not operational due to repair or preventive maintenance. Includes logistics and administrative delays.
Maintenance labor hours per hour/cycle/ action/month	A labor hour factor based on operating or calendar time, maintenance actions, or operating cycles.

Table 1.2 Design Sustainability and Operational Sustainability Contrasted

Design Sustainability	Operational Sustainability
Used to define, measure, and evaluate supplier's program	Used to describe performance when operated in planned environment
Derived from operational needs	Not normally appropriate for contract requirements
Selected such that achieving them allows projected satisfaction of operational sustainability	Used to describe needed level of sustainability performance in actual use
Expressed in design parameters	Expressed in operational values
Includes only effects of design and manufacturing	Includes combined effects of item design, quality, installation environment, maintenance policy, repair, delays, and so on
Typical terms	Typical terms
MTTR (mean time to repair)	MDT (mean downtime)
A_i (inherent availability)	A_o (operational availability)

No matter how one may quantitatively measure sustainability, commercial and military users measure performance in their own ways, to suit their own needs. For example, an aircraft manufacturer is concerned with the cost of operation and its maintainability. An aircraft operator, on the other hand, is concerned with availability. Such measures may or may not include factors totally determined by the design. So, the way in which a customer measures the sustainability of an entity in use may not be meaningful to a designer, and a translation from the user's measures to measures more appropriate for design is needed. Table 1.2 shows how operational (the user's) sustainability and design sustainability might differ.

1.2 The Need for Sustainability Engineering

The level of sustainability designed into an entity or an enterprise is instrumental to its long-run success or failure. Each enterprise must determine what level is necessary. Factors to be considered include the characteristics of the market (e.g., market growth and competitors' strategies); cost (in terms of dollars as well as opportunities) of implementing or not implementing a sustainability program; and complete knowledge of the customer's expectations and use of the entity. Achieved sustainability may be relatively simple to quantify (i.e., product availability, success or failure, etc.), but it will also impact qualitative issues, which must be adequately addressed, such as

- Customer expectations
- Market competition
- Diverse needs

For military markets, a unique set of needs exists that must be adequately addressed in order to satisfy operational readiness requirements in an inherently hostile environment. These needs include availability, safety, support, operational factors, and the need for the entity to operate under a variety of adverse environmental conditions.

1.2.1 Customer Expectations

Regardless of the product or service being offered, or who the intended customer may be, it is reasonable to assume that the degree to which the product/service entity is successful directly depends on the ability of that product/service to meet or exceed customer expectations. The challenge is twofold:

- How to assess and define true customer expectations
- How to design, manufacture, and market products/services to best meet those expectations

Implicit in the second challenge are economic considerations involved with customer expectations:

- The cost of meeting some versus all customer expectations.
- The cost of "best-in-class" products/services—Is best in class needed to meet customer expectations?
- The cost of exceeding customer expectations.
  - Short-term costs versus long-term savings
  - Competitive advantage in marketplace (increased market share)
- The cost of not meeting customer expectations.
  - Short-term savings versus long-term costs
  - No competitive discriminators (decreased market share)

On the surface, it would appear that ascertaining the customers' expectations for a specific product or service would be a conceptually straightforward task—simply ask the customers what they want. Yet, cost or scheduling constraints may force many manufacturers to forego this direct approach. Instead, they must rely on

- Their instincts or perceptions of the customers' needs
- The traditional performance of similar products or services (without fully understanding whether or not customer expectations were met)
- Their ability to create customer expectations for their product/service, where they did not previously exist

A quality function deployment (QFD) matrix (Anthony and Dirik 1995; Dean 1993; Gillespi et al. 1990; Guinta and Praizler 1993; Reed, Jacobs, and Dean 1994; Schubert 1989) can be a useful mechanism to facilitate communications between a customer and his supplier to determine what the customer wants (customer expectations) and how the supplier can meet those expectations. Table 1.3 shows a basic QFD matrix representation between customer expectations and engineering requirements for an automobile.

Table 1.4 provides more general methods on how to gather information on customer needs and expectations.

1.2.2 Market Competition

Sustainability can be used as an effective strategic competitive tool and can be equally as important as price, quality, and features in defining product success. A variety of sustainability design, analysis, and test techniques can be used to achieve the desired level. However, the required level of sustainability for achieving "best-in-class" performance depends on the characteristics of the specific market. These characteristics include the competitive position of a company and its competitors, the definition and needs of the target market, the dynamics of the marketplace, and the changing perceptions/characteristics of the customer base.

A benchmarking study by the Reliability Analysis Center (RAC 1995) ascertained what commercial industry is doing with respect to reliability. Although the study originally was intended to address reliability and sustainability, the scope of the study was changed to focus only on reliability. However, some lessons for sustainability can be drawn from the study. The study results are based on the product characteristics shown in Table 1.5.

The study results provided insight into the reliability tasks that are important to the commercial sector and the point in the product life cycle when they are typically applied. What was discovered was that industry focuses not on specific tasks but on objectives. These objectives, focusing on reliability and sustainability, are shown in Table 1.6.

Table 1.7 characterizes competitive market factors.

Table 1.8 provides an overview of technological policies and generic competitive strategies that can be used by companies operating in an environment of product or process technological change.

Figure 1.1 provides a graphical backdrop for the product life cycle concept, suggesting points at which different competitive strategies may be advantageous.

1.2.3 Customer Needs

While there are numerous differences between the needs of a commercial customer and those of a military customer, the sustainability needs of the military focus primarily on operational requirements:

- Readiness (product performance on demand)
- Sustainability (high-tempo operations over an extended period)
- Supportability (effective, responsive, and economical maintenance)
- Robustness (maintenance in all environmental extremes)

Table 1.9 provides an overview of the generic differences between these two customers. One of the major considerations in determining the ability of a product to meet the operational sustainability needs of the military is an understanding of the environment in which the product is

Table 1.3 Example Relationship between Customer Expectations and Engineering Requirements

Customer Expectations	Engineering Requirements					
	Reliable Built-In Test	Good Accessibility	Standardization	Modularity	Reliability-Centered Maintenance	Human Factors
Repairs done right the first time	✓	✓				✓
Easily performed customer maintenance	✓	✓				✓
Quick accurate diagnostics	✓			✓		
Inexpensive dealer maintenance and repairs	✓	✓	✓	✓		✓
"Long" intervals between scheduled maintenance						

Table 1.4 Potential Methods for Determining Customer Needs and Expectations

Quantitative	
Questionnaires	Can be mailed, conducted via telephone, or completed via face-to-face interview.
Delphi approach	Solicit and quantitatively assess the opinions of experts in their field.
Qualitative	
Direct observation	Allows data collection in a natural environment (primarily for characterization of "internal" customers).
Document analysis	Applicable to customers not easily accessible, and well suited to assessing larger groups over longer periods of time.
Focus group	Brings customer cross section together for intensive and interactive discussions on their needs and expectations.
Partnering workshops	Similar to focus groups, except suppliers are also involved and consensus is reached on how needs and expectations will be met.

Table 1.5 Benchmarking Study Product Characteristics

Characteristics	*Description*
Types of products	Automobiles and automotive products; telephones and test equipment; heating, ventilation, and air conditioning equipment and systems; diesel engines; computer workstations; data communication products; aircraft fuel and speed controls; commercial and general aviation aircraft; diesel and electric locomotives; medical equipment
Product unit cost	Ranged from less than $100 per unit to several million dollars per unit
Production volume	Ranged from single unit per year (large, custom product) to millions of units per year
Product technologies	Ranged from proven, off-the-shelf technologies to state-of-the-art technologies for material, processes, and functional design
Product markets	Included U.S. mass markets; international markets; small, customer-niche markets; and all levels (i.e., industry, general public, etc.)

Table 1.6 Reliability and Maintainability Tasks Grouped by Objective

Design Objective	Contributing Tasks
For reliability	
Determine feasibility of meeting design goals	Predictions
Understand the impacts on design performance (single-point failures, key design parameters, predominant failure modes/ mechanisms)	Design reviews; failure modes and effects analysis (FMEA); failure reporting, analysis, and corrective action system (FRACAS); design of experiments (DOE); test, analyze, and fix (TAAF); thermal analysis
Use proper parts and apply correctly	Environmental Stress Screening (ESS), parts control
Address all sources of components, materials, and so on	Vendor/supplier control
Validate the design	Qualification testing
For maintainability	
Determine feasibility of meeting design goals	Predictions, comparability analysis
Understand the impacts on design performance	Design reviews, task analysis
Use proper parts and apply correctly	Parts control, standardization
Address all sources of components, materials, and so on	Vendor/supplier control
Validate the design	Demonstrations, simulation

Table 1.7 Characterizing Competitive Market Factors

Market Factors	Balance between Reliability and Economic Performance	
	Positive Impact	Negative Impact
Customer/manufacturer relationships	Trust and partnership	Antagonism and isolation
Decision process	Few decision makers, minimal oversight	Many decision makers, excessive oversight
Manufacturer/supplier relationships	Long term, partnership	Short lived, driven by statutory requirements
Economics	Value, fitness for use emphasized	Price, compliance emphasized
Risk/payoff	Risk high, but payoffs are greater	Low, but little improvement in the performance envelope
Customer requirements	Functional, stated as guidelines	Product related, mandatory procedures

Table 1.7   Characterizing Competitive Market Factors (*Continued*)

Market Factors	Balance between Reliability and Economic Performance	
	Positive Impact	Negative Impact
Documentation	Limited to that needed for design or minimizing liability	Extensive and expensive
Customer focus	Interested only in product performance	Extensive involvement in design details
Manufacturer focus	Primarily on meeting customer needs and being competitive	Primary focus is on meeting specification requirements
Product performance	Good performance can increase market share and competitive advantage	Bad performance can result in nonpayment, loss of business/markets, potential contractual issues

Table 1.8   Technological Policies and Generic Competitive Strategies

Attribute of Technological Change	Technological Policies Corresponding to Generic Corporate Strategy			
	Overall Cost Leadership Strategy	Overall Differentiation Strategy	Focus-Segment Cost Leadership Strategy	Focus-Segment Differentiation Strategy
Product technological change	Product development to reduce product cost by lowering materials content, facilitating ease of manufacture, simplifying logistic requirements, and so on.	Product development to enhance product quality, features, deliverability, or switching costs.	Product development to design in only enough performance for the segment's needs.	Product design to exactly meet the needs of particular business segment application.
Process technological change	Learning curve process improvement. Process development to enhance economies of scale.	Process development to support high tolerances, greater quality control, more reliable scheduling, faster response time to orders, and other dimensions that improve the ability to perform.	Process development to tune production and delivery system to segment needs in order to lower cost.	Process development to tune the production and delivery system to segment need in order to improve performance.

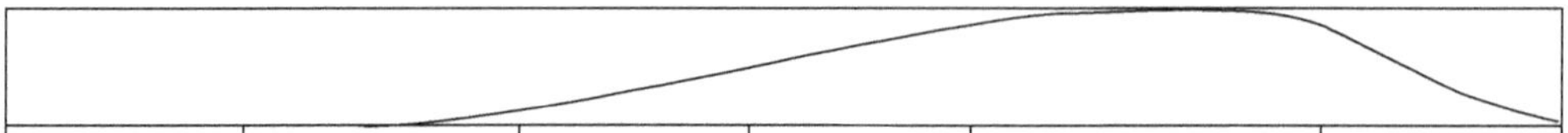

Life cycle phase	Market development (introductory period for high-learning products only)	Rapid growth (normal introductory pattern for very low-learning product)	Competitive turbulence	Saturation (maturity)	Decline
Strategy objective	Develop widespread awareness of product benefits	To establish a strong brand market and distribution niche	To maintain and strengthen the market niche	To define brand position and product category against other potential products, through constant attention to product improvement opportunities.	To milk the offering dry of all possible profit.
Outlook for competition	None in the early, unprofitable stages	Early entrance of aggressive emulators	Price and distribution squeezes on the industry	Market shares relatively stable except when a brand gains substantial added perceived value through product improvement or price repositioning.	Similar competition declining and dropping out.
Product design objective	Utmost attention to quality control and quick elimination of market-revealed defect in design	Modular design to facilitate flexible addition of variants	Intensified attention to product improvement	Constant attention to possibilities for product improvement and cost cutting. Reexamination of necessity of design compromises.	Constant pruning of line to eliminate any items not returning a direct profit.
Intelligence focus	Uncover any product weaknesses	Detailed attention to opportunities for market segmentation	Close attention to product improvement needs	Close analysis of competitors' strategies. Regular monitoring of trends in use patterns and possible product improvements.	Information helping to identify the point at which the product should be phased out.

Figure 1.1   Dimensions of the product life cycle concept important to competitive marketing. (Adapted from Burgelman, R. A., and Maidique, M. A, *Strategic Management of Technology and Innovation.* New York: McGraw-Hill Irwin, 1988.)

Table 1.9   Characteristics of Military Needs versus Commercial Needs

Market/Product Characteristics	Military Needs	Commercial Needs
Useful life	Typically 10–30 years	Variable
Safety factors	Low risk to personnel/equipment	Application dependent
Support factor	Full pipeline (100% availability)	Application dependent
Operational factors	Performance on demand is critical	Performance on demand is desirable
Purchase decision	"Best value" performance/price relationship	Consumer expectations met
Market need for product	Meet adversarial threat	Meet market expectations
Environmental factors	Product operation in extreme environments	Product operation in typical environments

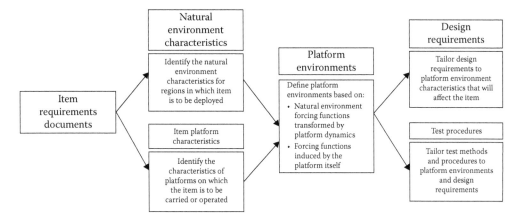

Notes:
1. Conventional meteorological data are not collected with military hardware in mind. Great care must be taken to ensure that the meteorological data used are relevant to the specific hardware items.
2. In this context, a platform is any vehicle, surface, or medium that carries the hardware. For example, an aircraft is the carrying platform for an avionics pod, the land itself for a ground radar, and a person for a hand-carried radio.

Figure 1.2 Environmental tailoring process for military hardware.

intended to operate. Figure 1.2 provides a graphical representation of an environmental tailoring process for military hardware. Figures 1.3a and b illustrate the types of natural and induced environments that can be expected over the operating and maintenance (O&M) portions of the military equipment life cycle.

The environmental stress events experienced by actual hardware may not always occur in the sequence shown in the profiles of Figures 1.3a and b. The generalized profile is intended to be a starting point for a tailored life cycle stress analysis and to provide confidence that all potentially significant environmental conditions have been considered. It provides only representative decision-making information. It does not impose a specific test order although it can aid in suggesting potentially useful environmental test stress combinations or sequences. Hardware may be subjected to any or all of the shipping/transportation modes shown. Therefore, in any life cycle stress analysis, the anticipated stresses experienced by the hardware in each mode should be evaluated and the most significant of these incorporated into the test program. The profile shows only areas of environmental concern and does not attempt to show operational use patterns. The relative frequency and duration of storage, shipping, and mission events must be considered in determining life cycle environmental test parameters. It should also be remembered that even one-shot devices (rockets, shells, etc.) must endure combinations and repetitions of all these events before they are ultimately fired.

In addition to the differences already discussed, the military has a wider variety of maintenance concepts than do most other commercial organizations. In recent years, many logistics initiatives have been initiated to reduce logistics requirements and streamline logistics support. These have resulted in many concepts recently being revised or replaced with new concepts. Two-level maintenance, outsourcing, agile logistics, and other initiatives add new, sometime stringent, sustainability requirements on new products.

Training is another area that differentiates the military customer from the commercial customer. The military experiences a relatively high turnover of maintenance personnel because many soldiers, sailors, and airman leave the service after only one or two enlistments. In times of a strong economy, in which the job market is tight, the turnover problem becomes even more acute. The results are a continuous need for training and a maintenance force of limited experience and maturity. Under such circumstances, sustainability becomes extremely important. Ideally, the military services want products that can be maintained with a minimum skill level, using simple tools and procedures, where fault detection and isolation is performed automatically. Moving one step beyond minimizing required skill levels, the military is looking to advanced technologies and

Shipping and storage	Environmental stress	Environmental stress
	Natural	Induced
	Temperatures: High/Low Humidity: High/Low Rain/Snow/Hail/Ice Sand/Dust	Road conditions: Bumps/ Holes vibration Shock: Drop/Overturn
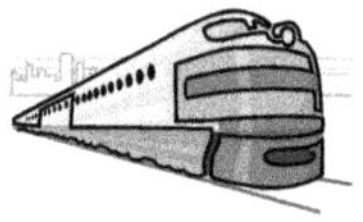	Temperatures: High/Low Humidity: High/Low Rain/Snow/Hail/Ice Sand/Dust	Vibration Shock: Drop/Overturn
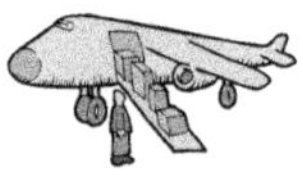	Pressurization Thermal shock	Inflight vibration Landing shock Handling shock
	Temperatures: High/Low Humidity: High/Low Rain/Snow/Hail/Ice Salt	Wave-induced vibration Wave shock Handling shock
	Temperatures: High/Low Humidity: High/Low Rain/Snow/Hail/Ice Salt Mold	Handling shock

Figure 1.3   (a) Generalized O&M life cycle histories for shipping and storage hardware and (b) generalized O&M life cycle histories for shipping and storage hardware.

Deployment and delivery	Environmental stress	Environmental stress
	Natural	Induced
	Temperatures: High/Low Humidity: High/Low Rain/Snow/Hail/Ice Sand/Dust Chemicals	Noise Vibration Shock: Drop/Overturn
	Temperatures: High/Low Humidity: High/Low Rain/Snow/Hail/Ice Sand/Dust	Vibration Shock: Drop/Overturn
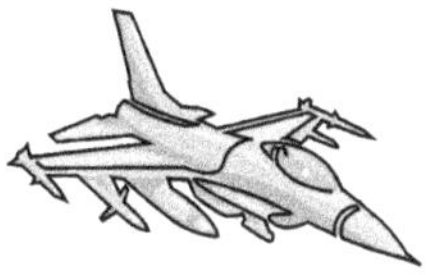	Pressurization Thermal shock	Inflight vibration Landing shock Handling shock Atmospheric effects
	Temperatures: High/Low Humidity: High/Low Rain/Snow/Hail/Ice Salt	Wave-induced vibration Wave shock Handling shock Pressurization (submarine)
	Temperatures: High/Low Humidity: High/Low Rain/Snow/Hail/Ice Salt	Handling shock Vibration Turbulence

Figure 1.3 *(Continued)*

automation in new systems, such as the CVN-21 "Supercarrier," to reduce the number of crew members (operators and maintenance personnel) required.

Finally, over the past several years, actions and discussions surrounding the acquisition of new products by the military have centered on the abolishment of many military specifications and standards as contractual requirements (except by waiver). One of the results of the 1994 Perry Memorandum* has been to focus acquisition attention on nondevelopmental item (NDI) systems and equipment. However, it is critical that manufacturers and service providers realize that the military market represents a segment of customers with a justifiably unique set of expectations and product performance objectives.

A subset of NDI is commercial-off-the-shelf (COTS) items. Essentially, COTS means buying a commercial product from a catalog, store, or distributor and then using the product as is. Table 1.10 lists some of the key advantages and disadvantages of using COTS.

Finally, many of the products and services procured by the military, especially those that are "field deployed," provide operational capabilities that allow U.S. forces to maintain technical and logistic superiority (i.e., a competitive advantage) over a potential adversary. Degradation or loss of these operational capabilities typically can result in immediate loss of human life and equipment and short- or long-term loss of competitive advantage.

1.3 A Methodology for Sustainability Engineering

Sustainability engineering is a comprehensive, iterative, problem-solving process. That process is used to transform validated customer needs and requirements into a life cycle-balanced solution set of product and process designs, to generate information for decision makers, and to provide information on the next phase in the life cycle of the entity. Sustainability engineering emphasizes the concept of concurrency, in which the requirements and approach for test, production, and logistics support are integrated with those for development, so that the solution is best suited for the entity's entire life cycle. Fundamental to this approach is the integration of all technical disciplines in a coordinated effort to provide a balanced product or service solution. Sustainability engineering addresses the entire entity and defines the requirements and design approaches for all of its elements, including hardware, software, people, data, and facilities. The approach addresses interactions between the entity and its operating environment, and it ensures that the technical and management processes address compatibility of all external and internal interfaces.

Sustainability engineering can be iteratively applied to all phases of the entity in its life cycle, regardless of how the customer's specific acquisition life cycle is defined. In each of these stages, sustainability engineering oversees development of the solution through a requirements-driven process. Sustainability engineering is not limited to any one program type. It should be performed on completely new systems, subsystems or components, system modifications, or reuse intensive development employing off-the-shelf components. Sustainability engineering is also not limited by program organization. It should be an integral part of functional organizations, integrated product team (IPT) organizations, colocated organizations, and distributed organizations.

---

* Responding to increasing criticism, Secretary of Defense William Perry issued a memorandum in 1994 that effectively eliminated the use of most defense standards. This has become known as the "Perry memo." Subsequently, many defense standards were cancelled. In their place, the Department of Defense encouraged the use of commercial standards, such as the ISO 9000 series for quality assurance.

Table 1.10 Advantages and Disadvantages of COTS

Area of Comparison	Advantages	Disadvantages
Technical, schedule, and financial risk	Decreased technical, financial, and schedule risks due to less new design of components and subsystems. Ideally no research and development costs are incurred.	When used as the components and subsystems of a product, integration of those items into the product can be difficult, expensive, and time consuming.
Performance	There can be increased confidence due to established product performance and the use of proven components and subsystems.	Sustainability trade-offs may be needed to realize advantages. Integration may be difficult.
Environmental suitability	In similar applications, proven ability to operate under environmental conditions.	In new applications, may require modifications external to the equipment to operate.
Leverage	Ability to capitalize on economies of scale, state-of-the-art technology, and products with established quality.	There may not be a perfect match between sustainability requirements and available products.
Responsiveness	Quick response to an operational need is possible because new development is eliminated or minimized.	Integration problems may reduce the time saved.
Manufacturing	If already in production, processes are probably established and proven.	Configuration or process may be changed with no advance notice.
Resupply	There is no need for (large) inventory of spares because they can be ordered from supplier.	The long-term availability of the item(s) may be questionable.
Logistics support	No organic support may be required (probably not possible). Repair procedures and rates are established.	Supplier support or innovative integrated logistics support strategies may be needed to support the product.

Standards such as MIL-STD-499 (systems engineering management) and others have historically delineated major phases, activities, documentation, and reviews to create order and to control the engineering process. However, even with these standards, the developing sustainability engineering project needs to define its specific process architecture and the next level of detail of the sustainability engineering activities, methods, and tools to complete this operation. Figures 1.4 and 1.5 illustrate the elements and concepts of a project's methodology for sustainability engineering.

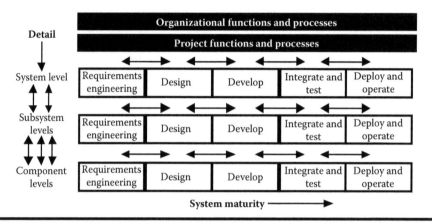

Figure 1.4 Elements of sustainability engineering methodology.

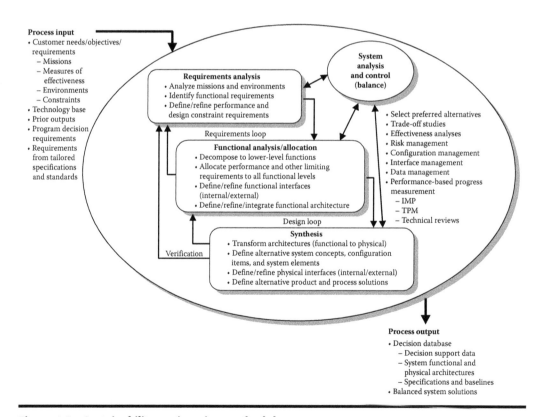

Figure 1.5 Sustainability engineering methodology.

A key role of the methodology is to oversee the entire technical effort of the project. As a management process, sustainability engineering works to ensure integration of the technical efforts of the entire team to meet program cost, schedule, risk, and performance objectives. This integrative function brings together all technical plans and activities across the entire program. Sustainability engineering has oversight for technical planning activities, which include tasking, scheduling, and skills profiles. Planning activities provide the project with a set of unified processes, tailored from this common approach, that yield a disciplined, structured approach to product or service development.

Sustainability engineering provides monitoring and controlling functions that provide feedback on the integrity of the program processes and the system under development. It defines and executes corrective actions in response to any unacceptable deficiencies. The sustainability engineering management process provides a disciplined approach to decision making. It employs risk management assessments to balance decisions with respect to cost, schedule, technical performance, and risk. It controls evolving requirements and design baselines so that changes are made only after consideration of the impact of the change and approval of the change by the responsible leaders. It provides development and maintenance of a database that documents the basis for decisions that have been made and provides this data as needed.

As stated in the ISO 9000 family of quality management standards, any activity that receives inputs and converts them into outputs can be considered a process. In this section, sustainability engineering is presented as a generic set of processes that work on inputs to produce the desired outputs. This general discussion is intended to provide process definition and a transition to the building block approach to sustainability engineering.

Sustainability engineering can be characterized as being composed of a number of processes that work together on a set of inputs to achieve the desired output. In this case, the desired output is a product entity that meets the user's needs and requirements in a near optimal manner. Sustainability engineering processes can be conducted once, or repeated in an iterative manner, to ensure that the final product reflects additional information derived as a result of actually performing the sustainability engineering processes. An overview of the methodology is shown in Figure 1.6. At this very high level, processes can be categorized as management and technical processes. Management processes consist of technical coordination, technical control, and technical management. The technical processes include analysis and concept definition, functional analysis/allocation, design synthesis, and compliance verification. These processes interact with each other and with the phase-dependent inputs (e.g., user defined requirements) to produce the phase-dependent outputs (e.g., verified and validated requirements).

1.3.1 Process Inputs

Inputs to the sustainability engineering methodology (SEM) are phase dependent, meaning that the exact inputs depend on the process being performed. These inputs can come from clients, previous work, or other drivers, as shown in Figure 1.7.

The client inputs to sustainability engineering processes establish the requirements of the product as seen from the perspective of the people that have the need for the product, either directly as users of the product or as the user's agents. The client defines the mission to be performed, the concept of operation for the product, goals and objectives, and the operational environment and constraints. The client also defines the measures of effectiveness by which the success of the product will be judged. Other drivers for the product include decision requirements on performing the program, specifications and standards that must be addressed, and the technology base of the

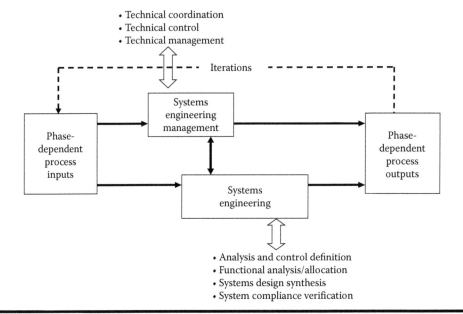

Figure 1.6 Sustainability engineering methodology overview.

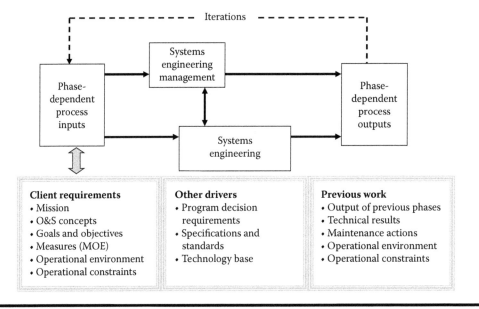

Figure 1.7 Phase-dependent inputs.

client and developing organizations. Previous work output generated from other projects or from prior phases within the overall process can also influence the sustainability engineering processing. This type of input also can include technical results, management actions, and hard and soft products generated within or outside the project. Typically, the client input is dominant at the start of the project but should continue through product acceptance and operations.

1.3.2 Process Outputs

Outputs from the SEM are also phase dependent. The outputs from sustainability engineering processes include technical outputs, management outputs, and products, as shown in Figure 1.8.

The technical outputs from sustainability engineering processes include analyses, design concepts, requirements and decisions databases, standards and practices, and requirements and product verification. The management output from sustainability engineering includes plans, directions, work and product approvals, progress measurements, budgets, and schedules. Products from systems engineering include the systems architecture, system products and designs, test systems and test results, and documentation of the process conclusions and results. Typically, management outputs are most evident in the very early phases of the project, while technical and product output start slightly later and continue at a high level through conclusion of the sustainability engineering activities.

1.3.3 Management Processes

The sustainability engineering management processes provide the technical coordination, technical control, and technical management of the sustainability engineering technical processes, as shown in Figure 1.9. The plan for conducting the sustainability engineering management process for a project is documented in a sustainability engineering management plan (SEMP). Sustainability engineering management processes provide direction to the sustainability engineering technical activities on a continuing basis during the entity's life cycle.

Technical coordination works to ensure that the individual sustainability engineering processes work together as a consistent and concurrent total process. This includes planning and coordinating the engineering specialties that are applied and the technical documents that are produced. Technical coordination is also accomplished by the creation and maintenance of the sustainability engineering database, which contains requirements allocations and related design and test information.

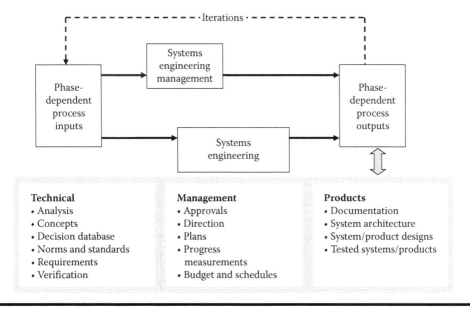

Figure 1.8 Phase-dependent outputs.

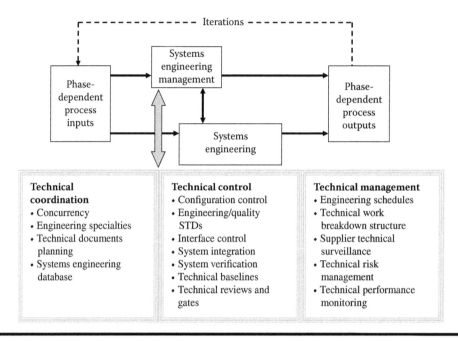

Figure 1.9   Sustainability engineering management processes.

Technical control is responsible for ensuring that the design is consistent and meets expectations and standards. This is accomplished through the adoption of appropriate engineering and quality standards, overseeing performance of technical reviews at defined technical gates, and maintaining technical baselines. Sustainability engineering also oversees configuration control, system integration, and system verification activities. Technical management processes assists in the planning and monitoring of the engineering and technical activities for the project. Sustainability engineering management is involved in setting the engineering schedules and technical work breakdown structure (WBS). Sustainability engineering management also assists project management in technical risk management, technical performance monitoring, and supplier technical surveillance.

1.3.4 Technical Processes

Sustainability engineering technical processes include the performance of the system analysis and concept definition tasks necessary to determine user needs and expectations and to convert them into system requirements. Sustainability engineering is then responsible for conducting the analysis and allocations necessary to establish the functional architecture. The functional architecture is used as input to the system synthesis, which produces the design to be developed. During and following development activities, sustainability engineering has oversight and responsibilities for verification of the system compliance to the design and requirements. Figure 1.10 illustrates the components of the sustainability engineering technical processes.

The sustainability engineering analysis and concept definition processes are aimed at analyzing user requirements and defining the user's concept of operations for the system. The results of these activities are baseline requirements and definition of requirements to be fulfilled by suppliers. The functional analyses and allocation activities include functional and data decomposition of the requirements and concept of operation. This leads to a functional architecture with requirements

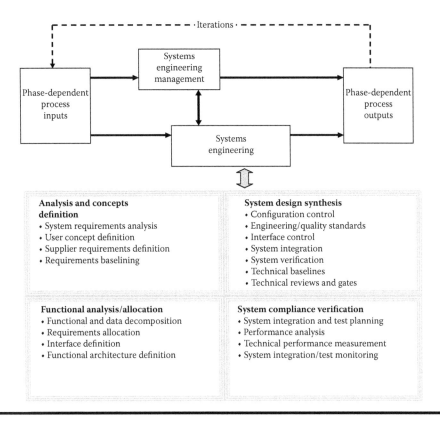

Figure 1.10 Sustainability engineering technical processes.

allocations to the architectural components (i.e., functional, hardware, software, and manual) and definition of the system interfaces (i.e., internal and external). Design synthesis activities convert the requirements, concept of operations, and functional architecture into a physical architecture definition through trade-offs and design optimization. Configuration items are identified with associated requirements allocated to each. As the project progresses, technology changes are monitored and assessed for potential improvement to the design. The system compliance activities involve planning for and monitoring system integration and testing. These activities also include performance analysis and technical performance measurement.

The sustainability engineering technical processes form building blocks. These building blocks can be performed sequentially or iteratively to create various life cycles. For example, the building blocks might be performed once, as in a typical waterfall life cycle, or several times (e.g., once per increment) for the spiral or evolutionary life cycle models. Figure 1.11 illustrates this concept.

1.3.5 Phase Dependency

Sustainability engineering processes span the entire project development life cycle from concept initiation to deployment and operation. However, the level of activity associated with specific sustainability engineering processes as well as the requisite skills and products vary as the project progresses through its phases, either once or successively in recursive life cycles. Table 1.11 shows

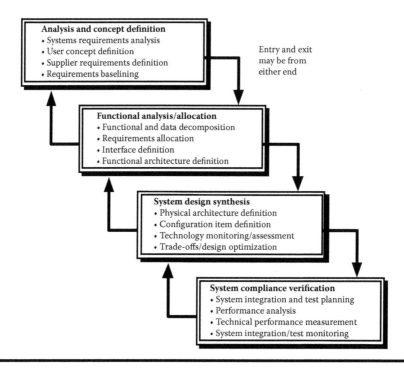

Figure 1.11 **Interaction among sustainability engineering technical processes performed per increment and/or per system.**

Table 1.11 **Typical Sustainability Engineering Processes, Skills, and Products**

Element Process	Primary Sustainability Engineering	Typical Skills Most Required	Product Examples
Planning and requirements engineering	■ Technical management ■ Analysis and concept definition	■ Sustainability engineering planning ■ Requirements engineering	■ SEMP ■ WBS ■ Requirements specification ■ Concept of operations
Design	■ Technical coordination ■ Technical management ■ Functional analysis/allocation ■ System design synthesis (architecture)	■ Sustainability engineering management ■ Systems architecture ■ Engineering specialties (e.g., communications, hardware, security) ■ Requirements engineering	■ Requirements baseline ■ Trade-offs ■ Configuration items ■ Physical architecture ■ Sustainability engineering database ■ Risk items

Table 1.11 Typical Sustainability Engineering Processes, Skills, and Products (*Continued*)

Element Process	Primary Sustainability Engineering	Typical Skills Most Required	Product Examples
Development	■ Technical control ■ Technical coordination ■ System compliance verification ■ System design synthesis	■ Sustainability engineering management ■ Configuration management ■ Test planning ■ System architecture ■ Performance measurement ■ Issues analysis and resolution	■ Configuration database ■ Test plans ■ Performance measurements ■ Technical performance monitoring ■ Risk management
Integrate and test	■ Technical control ■ Technical coordination ■ System compliance verification	■ Sustainability engineering management ■ Configuration management ■ Testing ■ Performance analysis	■ Configuration database ■ Test results analysis

typical sustainability engineering processes, personnel skills, and products for the generic elements of the SEM that were depicted in Figure 1.5.

In many cases, products are delivered to the client for review at end of each phase and approval to proceed to the next phase obtained before proceeding. The degree of formality and detail of the products delivered as well as the associated review and approval process will vary with the customer requirements and life cycle chosen for the project.

References

Anthony, M., and A. Dirik. 1995. "Simplified Quality Function Deployment for High-Technology Product Development." *Visions* April: 9–12.

Burgelman, R. A., and Maidique, M. A. 1988. *Strategic Management of Technology and Innovation*. New York: McGraw-Hill Irwin.

Dean, E. B. 1993. "Quality Function Deployment for Large Systems." In *Transactions of the Fifth Symposium on Quality Function Deployment*, Novi, MI, June 21–22, 1993, pp. 165–74.

Gillespie, L. K. et al. 1990. *Quality Function Deployment as a Mechanism for Process Characterization and Control*. DE90-014755, KCP-613-4276. Kansas City, MO: Allied-Signal Aerospace Co.

Guinta, L. R., and N. C. Praizler. 1993. *The QFD Book*. New York: American Management Association.

Reed, B. M., D. A. Jacobs, and E. B. Dean. 1994. "Quality Function Deployment: Implementation Considerations for the Engineering Manager." In *Proceedings of the IEEE International Engineering Management Conference*, Dayton, OH, October 17–19, 1994, pp. 2–6.

Reliability Analysis Center (RAC). 1995. *Benchmarking Commercial Reliability Practices*. Rome, NY: Handle. http://handle.dtic.mil/100.2/ADA310563. Accessed 10 December 2011.

Schubert, M. A. 1989. "Quality Function Deployment—A Comprehensive Tool for Planning and Development." In *Proceedings of the IEEE 1989 National Aerospace and Electronics Conference*, NAECON, Dayton, OH, May 22–26, pp. 1498–1503.

Wang, C. W. 1990. "Concept of Durability Index in Product Assurance Planning." In *Proceedings Annual R&M Symposium*, Los Angeles, CA, January 1990, pp. 221–27.

Chapter 2

Structuring a Sustainability Engineering Program

2.1 Customer Requirements

Understanding the customer's requirements has two objectives:

1. To establish contractual product-level sustainability requirements that, if met, will ensure that the operational sustainability of the product will meet the user's mission needs and be consistent with operating and support cost constraints.
2. To allocate the product-level requirements down to the level needed to be meaningful to the design and manufacturing process engineers. This level may be subsystem, component, or even lower.

The user must define the operational, product-level requirements prior to the beginning of the procurement or acquisition cycle. In the initial phases of a totally new acquisition, the requirements may be stated as "goals" with firm requirements not imposed until later. Based on the results of the early phases, the initial requirements may have to be adjusted, but a valid starting point is essential. The user may translate the operational requirements to contractual requirements or require the contractors to do so as part of their proposals. Allocation of requirements down to a given level must be done after a product sustainability model has been developed and before design efforts, at that level, begin.

Users, commercial and military, measure the performance of products in their own ways to suit their own needs. One customer may be most concerned with low operating costs and few visits to the repair shop. Another customer may be most concerned, after safety, with rapid turnaround. These measures may or may not include factors within the control of the contractor. So, the way in which a customer measures the sustainability of a product in use may not be meaningful in a specification, and a translation from the user's measures to measures more appropriate to the specification may be needed. Table 2.1 shows how operational (the user's) sustainability and contractual sustainability differ.

Table 2.1 Operational and Contractual Sustainability Contrasted

Contractual Sustainability	Operational Sustainability
Used to define, measure, and evaluate contractor's program	Used to describe performance when operated in planned environment
Derived from operational needs	Not used for contract requirements
Selected such that achieving them allows projected satisfaction of operational reliability and management (R&M)	Used to describe needed level of R&M performance in actual use
Expressed in inherent values	Expressed in operational values
Account only for events subject to contractor control	All events must be accounted for, regardless of the cause
Include only effects of design and manufacturing	Includes combined effects of item design, quality, installation environment, maintenance policy, repair, delays, and so on.
Typical terms • Mean time to repair (MTTR) • A_i	Typical terms • Mean downtime (MDT) • A_o

A two-step conversion might be needed to translate an operational need to a contractual parameter.* Consider the following example in which we know what the operational availability† must be, as defined by the equation given below, and we want to specify sustainability.

$$A_O = \frac{\text{MTBM}}{\text{MTBM} + \text{MDT}} \times 100\%$$

where MTBM is the mean time between maintenance and MDT is the mean downtime. Assume we have a maximum allowable or desirable MDT, which includes the actual repair time plus logistics delay time. Solve for MTBM. MTBM includes maintenance to repair failures, preventive maintenance, such as lubrication, and inspections. We have now "translated" A_o into MTBM and MDT. MTBM and MDT are operational measures that take into account factors beyond the control of development contractors. So MTBM and MDT must be translated into contractual terms (mean time between failures [MTBFs] and mean time to repair [MTTR], for example) for which contractors can be held accountable.

The process cannot end with the translation to a contractual value. The translated requirements must be evaluated for realism. The following questions have to be answered: Are the requirements compatible with the available technology and do the requirements unnecessarily drive the design (conflict with product constraints such as weight and power)? Answering these questions usually

---

* Note that the translation of operational to contractual terms is not always necessary. What is required is that the parameters used to describe operational requirements be meaningful to the user and tracked for the product being replaced (or tracking must be planned if no prior product exists).

† Operational availability (A_o) is a measure of the degree to which an entity is in an operable state at the start of its mission when the mission is called for at a random point in time. A_o can also be defined as the probability that the entity will be ready to perform its specified function in its specified and intended operational environment when called for at a random point in time.

involves a review of previous studies and data for similar or comparative products (if any exist). The requirements may need to be adjusted to account for improvement of technology, different operating environments, and different duty cycles.

The customer normally specifies the operational product sustainability requirements at the "product level." Now, recall the definition of sustainability:

> Sustainability is the relative ease and economics of time and resources with which an entity (enterprise or product) can be retained in, or at least restored to, a specified and satisfactory condition by personnel having the necessary skill levels and ability, using prescribed procedures and resources.

Understanding the definition will help us develop better requirements. To better understand the definition, we need to examine how maintenance is categorized. Figure 2.1 depicts the various categories of maintenance, preventive and corrective, and the terms used to describe the categories. Maintenance is required because a failure has been indicated or because some preventive action is mandated by policy, servicing needs, an impending failure (indicated by performance trending monitoring), or other nonfailure reasons. A failure indication will always result in some maintenance activity. Sometimes that activity will stop after retesting the supposedly failed item because the failure cannot be confirmed. In fact, unconfirmed failures (called Retest OK [RTOK] or Can Not Duplicate [CND], depending on where and when the attempt to confirm occurs) account for the majority of maintenance actions on some complex subsystems and components.

In addition to the time actively spent making a repair and servicing an item (called active maintenance time), some time is spent waiting for parts, crew shift changes, and administrative reasons. This time is called inactive maintenance time or delay time. It is important to recognize

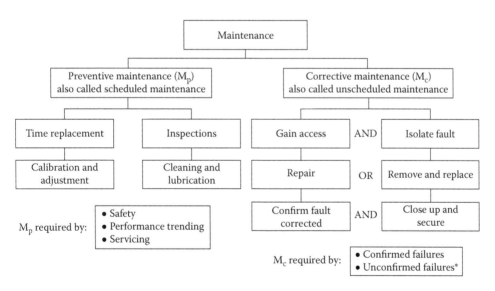

*Unconfirmed failures result from false alarms in the built-in-test, intermittent failures, or test equipment failures. Unconfirmed failures will trigger some unscheduled maintenance actions, ranging from confirming no fault exists (attributed to false alarm or Cannot Duplicate) to removing and replacing the item only to later find (at another level of maintenance) that the item is good (Retest OK).

Figure 2.1 Major categories of maintenance.

these distinctions between categories of time, just as it is to recognize the distinctions between categories of maintenance, because the various sustainability measures are derived from these distinctions. Typical operational measures are maintenance DT and mean time between scheduled maintenance.

As is the case for reliability, the operational sustainability requirements must usually be translated into contractual terms. It is possible, however, that no translation will be needed. For example, if the product being procured is to be operated and maintained by the contractor, the differences shown in Table 2.1 "disappear." In any event, requirements actually put in the specification are referred to as contractual requirements, whereas those stated in user's terms are referred to as operational requirements (or user's needs).

Operational requirements can be expressed for sustainability, or they may combine maintainability with reliability in the form of availability. Some typical contractual sustainability requirements are

- Mean downtime
- Maximum corrective maintenance time
- Mean corrective maintenance time
- Maximum preventive maintenance time
- Mean preventive maintenance time
- Mean active maintenance time
- Maximum active corrective maintenance time
- Mean time to repair

Product-level requirements are not sufficient to support the design effort. For example, a requirement that a truck have an MTTR of 2.1 hours does not help the designers of the transmission, engine, and other components. How sustainable must these components be? Consequently, the requirement process involves allocating the sustainability requirements to lower levels. In some cases, the process is iterative, requiring several attempts to satisfy all requirements. In other cases, the requirements cannot be satisfied (to meet the product-level requirement, components are needed with higher-than-possible levels of sustainability) and dialogue and trade-offs with the user are required.

A sustainability model can be used to support sustainability allocation and calculations. Such a model should be developed as soon as hardware definition permits, ideally before specifications for subsystem or components are developed and before predictions of sustainability are made. The model may range in complexity from a simple functional flow or block diagram (a reliability block diagram, for example) to a mathematical equation relating system parameters and system performance. Sustainability models are verified through analysis.

The need for a sustainability model is largely based on the complexity of the product. The cost of creating a model for a very simple, small product may not be justified by the value to be gained. For large, complex products, a model is almost always mandatory. When a model is deemed necessary, it should be developed to the lowest level at which maintenance and repair will be performed. If it has been previously determined, for example, that repairs will not be made below a certain level, then the model is not required below that level.

Figure 2.2 is an example of a very simple functional block diagram for a product.

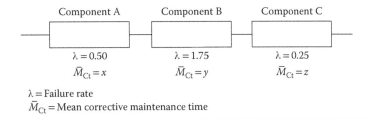

λ = Failure rate
$\bar{M}_{Ct}$ = Mean corrective maintenance time

Figure 2.2   Simple functional block diagram.

Table 2.2   Data for Figure 2.2

Component	Quantity (Q)	FR	Contribution of Total Failures $(C_f) = Q \times FR$	Component $\bar{M}_{Ct}$	Contribution of $\bar{M}_{Ct}(C_i) = \bar{M}_{Ct} \times C_f$
A	1	0.50	0.50	x	$0.50x$
B	1	1.75	1.75	y	$1.75y$
C	1	0.25	0.25	z	$0.25z$
			$C_f = 2.50$		

To allocate a maintainability/sustainability factor to a product, one of three approaches could be used.

1. If the inherent availability (A_i) and reliability are specified for the product and the product reliability has been allocated, sustainability can be allocated on the basis of availability.
2. If the relative failure rate (FR) of each component is known, the sustainability requirement can be allocated on that basis (higher FR components would be allocated higher sustainability—lower repair time—requirements).
3. If the relative mission criticality of each component is known, the sustainability requirements can be allocated on that basis (higher criticality components are allocated higher sustainability—lower repair time—requirements).

Using the second approach, we can allocate a mean corrective maintenance time ($\bar{M}_{Ct}$) for the product represented in Figure 2.2 as follows. Assume that the product-level requirement is 0.675 hours. Using the data from Figure 2.2, we can create Table 2.2 using the following equation.

$$\bar{M}_{Ct} = \frac{\sum \text{Component } \bar{M}_{Ct} \times C_f}{\sum C_f}$$

where C_f is the contribution of total failures.

$$0.675 = \frac{0.50x + 1.75y + 0.25z}{2.50}$$

$$\text{or}: \ 0.50x + 1.75y + 0.25z = 1.6875$$

In the above equations, x, y, and z are the mean corrective maintenance times for components A, B, and C, respectively, as shown in Figure 2.2.

Component B is the least reliable component so allocate the lowest $\bar{M}_{Ct}$ to it. Component A will be assigned an $\bar{M}_{Ct}$ that is 3.5 times higher than that of B (1.75/0.50), and Component C is assigned a value 7 times higher (1.75/0.25). Thus, $x = 3.5y$ and $z = 7y$. Therefore, $5.25y = 1.6875$ so the mean corrective maintenance times are $y = 0.32$, $x = 1.12$, and $z = 2.24$.

Sustainability requirements can be verified through analysis and sometimes testing. Comparative product analysis, life cycle cost modeling, mission modeling, and simulations are some of the techniques used to determine the operational level of sustainability needed. In a completely new acquisition, some testing of prototypes and breadboard designs may be done. When such testing is performed, the results can be used to "refine" initial sustainability requirements.

2.2 Participants in the Process

This section contains high-level descriptions of a project organization and roles required to support the sustainability engineering process. The focus of the project organization is the efficient development of quality products by application of both project management processes and sustainability engineering technical processes. Although the thrust of this book is keyed to the technical aspects of the development of systems or the execution of system tasks by the project, the same general principles apply to subcontracted projects or tasks as well.

A notional sustainability-engineering-related organization is identified in Figure 2.3. Depending on the size of the effort and where in the project life cycle a process element takes place, individuals may have more than one responsibility. Descriptions of the functions the elements of the organization perform are provided in the paragraphs that follow. Representatives of these functions are their "agents." References to the applicable "agents" are provided in each of the processes discussed in Chapter 4.

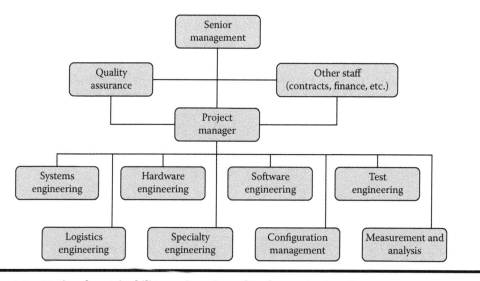

Figure 2.3 Notional sustainability-engineering-related organizational structure.

2.2.1 Senior Management

Project leadership is responsible for ensuring that the process, resources, trained staff, and management support are in place to support the development of systems that will satisfy the desired mission capabilities. Senior management is to review and approve all development support commitments and changes made to external groups, and senior management should review the activities for the development tasks both independently and with the project leadership on a regular basis. These activities include the following:

■ Integrated project management
■ Risk management
■ Quality assurance (QA)
■ Support requirements management
■ System support status
■ Subcontractor management (as appropriate)
■ Configuration management (CM)
■ Task-specific training program
■ System process development and improvement
■ Project team coordination

2.2.2 Project Leadership

The project leadership has the critical role of establishing and implementing a sustainability engineering approach that includes all stakeholders and leads all participants to translate operational needs and capabilities into technically feasible, affordable, and operationally effective and suitable increments of a system. The sustainability engineering approach should permeate concept definition, technology/manufacturing maturation, competitive prototyping, design, production, test and evaluation, and life cycle system support. Project leaders must exercise leadership, decision making, and oversight throughout the system life cycle. Implementing a sustainability engineering approach adds discipline to the process and provides the program manager with the information necessary to make valid trade-off decisions to balance cost, schedule, and performance throughout a program's life cycle.

The project leader is the individual who has overall responsibility for a project or task, including planning and management through the development of appropriate plans. Once the plans have been implemented, the project leader is responsible for monitoring and controlling cost, schedule, risks, and resources for the project. The project management plan (PMP) provides high-level project information such as costs, schedules, project organization, resources, risk, and a WBS. The project leader should use an appropriate tool to develop the PMP. The project leader has to perform the following activities:

■ Identify relevant stakeholders.
■ Establish and maintain the project's defined processes.
■ Establish and maintain estimates of project planning parameters.
■ Ensure that the stakeholders' needs and expectations are translated into customer requirements.
■ Ensure that all requirements are managed.
■ Generate a comprehensive PMP.
■ Obtain commitment to the plan.

- Monitor actual performance and progress of the project against the PMP, including cost and schedule.
- Manage corrective actions to ensure that all actions are completed as necessary.
- Ensure that the product is properly prepared for integration.
- Ensure that the PMP documents the procedures for risk management or that they are addressed in a risk management plan.
- Ensure that a SEMP is completed.
- Ensure that a base electronic sustainability engineering plan (BESEP) is completed, if appropriate.
- Ensure that an installation design plan (IDP) is completed, if appropriate.
- Ensure that a system operational verification test (SOVT) plan is completed, if appropriate.
- Ensure that a cut-over plan is completed, if appropriate.
- Monitor actual performance and progress of the project against the PMP.
- Mitigate project risks.
- Participate in Configuration Control Board (CCB) meetings.
- Manage corrective actions to ensure that all actions are completed as necessary.
- Establish measurement objectives and practices.
- Report measurement results.
- Ensure that noncompliance issues are objectively tracked, communicated, and resolved.
- Ensure that supplier agreements are established when necessary.
- Ensure that agreements with the suppliers are satisfied by all parties.
- Assign specific responsibilities to project team members.
- Ensure that project team members have received appropriate training.
- Ensure all appropriate work products are placed under CM.
- Involve stakeholders in reviews and decision making.
- Ensure that process and system work products are evaluated objectively.
- Review the status of all relevant activities with Senior Management on a regular basis, depending upon the complexity of the project or task.

2.2.3 Sustainability Engineer

Each project should have a lead or chief sustainability engineer to implement the engineering process. Personnel from nonsustainability engineering functions or from outside the project management structure is often required to support activities related to sustainability engineering. Most program personnel should see themselves as participants in the sustainability engineering processes. Sustainability engineering activities include defining architectures and capabilities and conducting functional analyses. Warfighters, sponsors, maintainers, and planners also actively participate throughout the systems acquisition life cycle. The lead sustainability engineer insures coordination and integration of all these various functional area representatives.

Sustainability engineering is responsible for helping the project leader with the development of the SEMP. Throughout the sustainability engineering process, the lead sustainability engineer supports the project leader by

- Generating a SEMP
- Generating a base electronic sustainability engineering plan (BESEP), if appropriate
- Generating an IDP, if appropriate

- Generating a cut-over plan, if appropriate
- Preparing a sustainability engineering master schedule and a sustainability engineering detailed schedule
- Preparing the integrated data package
- Monitoring and reporting technical progress of the project
- Preparing for, and conducting product reviews
- Establishing and maintaining an integrated database
- Supporting the continuous improvement of products and processes

Sustainability engineering is responsible for identifying potential product solutions and then selecting from the alternatives. After selecting a solution, sustainability engineering, with support from hardware engineering and software engineering, is responsible for identifying the products for the project or task. Sustainability engineering normally develops all requirements for the project, and a member of the sustainability engineering team serves as the requirements manager.

Sustainability engineering is responsible for reviewing the list of deliverables and contractually required formats and incorporating them into an SEMP. With assistance from test engineering, the same considerations are incorporated into the system and software test plans. As a joint effort, the systems, software, hardware, and test engineering disciplines are also responsible for establishing integrated schedules in support of these plans. Sustainability engineering, hardware engineering, and software engineering provide inputs to the appropriate sections of the WBS.

2.2.4 Logistics Engineer

Logistics engineering is responsible for three major activity areas: (1) influencing the design for supportability, (2) performance of logistics support analysis, and (3) obtaining the system enabling resources for the following logistics elements:

- Spares and repair parts
- Support equipment
- Technical manuals and other publications
- Training materials
- Computer resources support
- Facilities
- Packaging, handling, storage, and transportation

2.2.5 Software Engineer

Software engineering takes the solution selected by sustainability engineering and develops the software design for the product and implements it. As part of the implementation, software engineering is responsible for identifying the software-related products required for the project or task, including reviewing the list of deliverables and contractually required formats and incorporating them into the software development plan (SDP) and software test plan (STP). The SDP is normally developed by software engineering, whereas the STP is developed by test engineering. Software engineering is also responsible for establishing integrated schedules in support of these plans. Software engineering provides inputs to the appropriate sections of the WBS. Software engineering establishes the quality, productivity, and performance metrics, and develops software inputs for system integration.

2.2.6 Hardware Engineer

Hardware engineering is responsible for identifying the hardware-related products required for the project or task. Hardware engineering must integrate and coordinate implementation, schedules, and testing with sustainability engineering, software engineering, logistics, and other specialty engineering participants.

2.2.7 Test Engineer

Test engineering is responsible for performing the independent testing, as part of the overall set of verification activities, of the system to determine whether all the allocated requirements have been met. All test plans, including the SOVT plan, are developed by test engineering.

2.2.8 Measurement Analyst

The measurement analyst is responsible for identifying and analyzing measures to support the measurement objectives of the project. The issues of interest for the analyst include

- Resources and cost performance
- Schedule and progress performance
- Technical performance
- Product size and requirements stability
- Product quality
- Process effectiveness and efficiency performance

The measurement analyst establishes and maintains the measurement and analysis plan and associated standard operating procedures. The measurement analyst is also responsible for documenting the measures in support of objectives, the procedures for collecting and storing the measurement data, as well as the procedures for analyzing and reporting the measurement data. Additionally, the measurement analyst supports the project leader by collecting and reporting the results of measurement and analysis activities to all relevant stakeholders. The measurement analyst also collects and reports improvement information derived from planning and performing the process to support the future use and improvement of the processes and process assets.

2.2.9 Configuration Manager

CM is responsible for developing a project configuration management plan (CMP) and for providing a disciplined approach to controlling baseline work products including hardware, software, and documents. Within CM, the baseline is identified and documented; changes are implemented in a rigorously defined manner. Every change to documentation, hardware, software, and firmware must be identified, documented, reviewed, and approved by the appropriate authority. Configuration status accounting reports are used to record and report on the configuration of the product throughout the change; through an audit, the completed change can be verified as functionally and physically correct. Among other things, a CMP is required for the project to control changes to the system work products.

2.2.10 QA Engineer

QA engineer is responsible for objectively ensuring that the development process has been followed, that the products meet the established technical standards, and the requirements they were produced to satisfy. QA accomplishes these activities through a planned and systematic pattern of measurements, reviews, inspections, and audits. The QA engineer has a direct reporting chain to senior management to ensure the independence of the QA.

2.2.11 Specialty Engineer

Specialty engineering includes such functional areas as the following:

- Reliability, availability, and maintainability
- Human sustainability engineering
- Information assurance
- Security
- Safety
- Producibility and manufacturing engineering

2.2.12 Integrated Product Teams

Integrated Product Teams (IPTs) are multidisciplinary teams made up of functional area subject matter experts, some of which were described in the capabilities listed above. IPTs are cross-functional teams that are formed for the specific purpose of delivering a product for an external or internal customer. IPT members should have complementary skills and be committed to a common purpose, performance objectives, and approach for which they hold themselves mutually accountable. Members of an IPT represent technical, manufacturing, business, and support functions and organizations that are critical to developing, procuring, and supporting the product. Having these functions represented concurrently permits teams to consider more and broader alternatives quickly, and in a broader context, enables faster and better decisions. Once on a team, the focus of an IPT member changes from a given discipline to a product and its associated processes. All team members should offer their own unique expertise to the team and understand and respect the expertise available from each other. Team members work together to achieve the team's objectives.

The project leader forms the sustainability engineering integrated product team (SE-IPT) with members or representatives of the overall project team as needed for the specific process and/or life cycle phase of the project. An SE-IPT is often formed with a core set of functional area members and supplemented with other members as needed. The SE-IPT ensures that the processes described in this book are implemented to translate user-defined needs into defined technical requirements, and operational system specifications consistent with cost, schedule, and performance constraints. The methodology to guide the SE-IPT should include a structured sustainability engineering approach beginning with a thorough understanding of the requirements; a structured and documented decision support process; measurable objectives; and a shared digital data environment.

2.3 Influences and Factors

The process of structuring a sustainability engineering program can be influenced by internal and/ or external factors. The purpose of this section is to highlight some of these influences. Figure 2.4 illustrates some of the legislative, doctrinal, and environmental factors and influences that have the potential to affect, or be affected by, the process.

2.3.1 Enterprise Environment

The enterprise is the high-level organizational context in the process. It is the source for project starts, completions, and cancellations. It is also the source of resource support. The enterprise level manages multiple projects to effectively apply its scarce resources and infrastructure. The enterprise also establishes mechanisms to enable the use of technologies and support systems in product lines. It also enables project performance. It is within this context that the enterprise prepares policies, processes, procedures, tools, and equipment to create projects, review projects, or cancel projects. Thus, one cannot forget that it presents a tremendous influence over projects.

2.3.2 Project Environment

The enterprise shall identify and make available to the project: the tools, equipment, metrics, management reviews, and reporting requirements. The project environment is composed of project support and process groups, as illustrated in Figure 2.4.

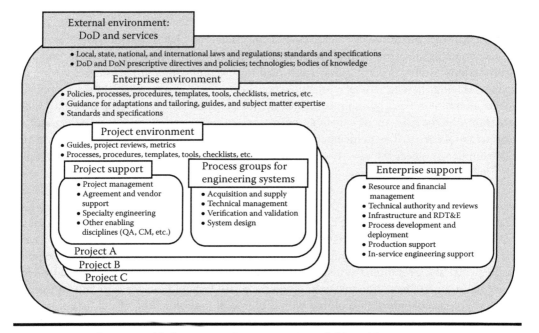

Figure 2.4   Environmental influences and usage context of this guide.

2.3.2.1 Project Support

Projects shall be in alignment with the business strategy of the enterprise and its component commands. The following support is to be expected from the enterprise and its components:

1. Investment decision support, including business needs assessments, selection of new projects, determination of project continuance, technical authority, review of projects, and allocation of financial resources for equipment, tools, and training
2. Infrastructure support, including research and development, marketing, facilities, in-service support, development and test environments, supply and contractual support, computer services, and other services that enable the project to meet its obligations
3. Resource management support, including financial management, personnel management, training for needed knowledge, skills and abilities of project personnel, work environment standards, building maintenances and support facilities, office and computer equipment, and shipping and receiving
4. Process development, deployment, and management support, including establishment of standard enterprise supply and procurement processes and methods, guidelines for tailoring adopted processes from this guide; selection and acquisition of tools and implementation mechanisms; assessments of process efficiency and effectiveness with respect to deployment, implementation, and improvement
5. Production support, including fabrication, construction, and staffing; equipment and tools; QA, and testing and evaluation in the areas of verification, product integration, and validation
6. In-service support, including installation, customer support, product upgrades, warranty service, field modifications, on-site consulting, and product certification

The availability and adequacy of these support functions affect the viability of a sustainability project, schedule of project tasks, capability to satisfy established agreements, and the availability of personnel who have the skills and knowledge to complete the project responsibilities.

2.3.2.2 Technical Process Groups

Projects, by way of a scope of work or tasking statements, provide the context in which a system is procured, engineered, fielded, and supported, or a service is provided. Prescriptive policies, directives, and procedures are prepared at the enterprise level and their component commands to direct both the project management functions and the technical efforts applicable to the specific project. Sustainability projects should follow the processes outlined in Chapter 4, or adaptations to them, as per allowable limitations (tailoring or deviations) to satisfy necessary agreements and the scope of work. As such, technical and functional support to meet the project requirements is needed. This support includes the following:

1. Acquisition, supply, and agreement support, including preparing appropriate tasking agreements between projects, or within the project, to implement the planned technical effort and providing proposal preparation support, as applicable
2. Project management, including project integration, scope management, time management, cost and schedule management, quality and CM, human resource management, communication management, risk management, and procurement management

2.3.3 External Environment

Some external environmental factors that can affect the application of processes for engineering a system for sustainability, or providing a service, include local, state, national, and international laws and regulations; Department of Defense (DoD) prescriptive directives and policies; national or international standards and specifications; available technologies; and subject matter expertise. An example is the DoD Environmental Directives and Instructions directive* that provides policy and procedures to enable DoD officials to take account of environmental considerations when authorizing or approving certain major federal actions that have the potential to do significant harm to the environment beyond the geographic borders of the United States.

System engineering activities can also be affected by external agreements with interfacing systems of systems (SoS) or family of systems (FoS) development or service efforts. The interaction and interfaces (physical or functional) between the system products and their external operational environment can also affect the implementation of the processes used for sustainability engineering. Changes in the operational environment can strongly affect system effectiveness and functionality as well as fielding and support it receives. System performance and adequacy also can be affected by the system's ability to respond both the operational environment, changes in the environment, as well as the life cycle support it receives.

2.3.4 Other Enterprise Projects and Systems

The enterprise often has multiple product and service projects going on at the same time. Two or more such programs or projects can sometimes benefit from the exchange of products: for example, parts, subassemblies, data, and/or logistics support. Agreements between such projects, resource providers, and their vendors are to be established, as appropriate.

2.4 Integrate Sustainability into the Project Life Cycle

Sustainability engineering is a top-down iterative process involving requirements definition, functional analysis and allocation, synthesis and design, and test and evaluation. By integrating the sustainability activities into this process, sustainability requirements will be addressed concurrently with other performance requirements. In this way, sustainability activities will be integrated with all engineering and design activities, thereby avoiding duplicative effort and making the best use of activity outputs. An integrated, systems approach to sustainability is essential because sustainability is related to other product characteristics.

Sustainability is a true design characteristic achieved through sound design practice, proper application of parts, and good manufacturing processes. Various analytical techniques are available for ensuring that sustainability is an integral consideration in design decisions and approaches. These methods are discussed in Chapter 5. However, other analyses performed as part of the overall design effort provide valuable information for sustainability purposes even though the analyses are not specifically considered "sustainability analyses." These include the following:

■ Safety analyses
■ Reliability analyses

---

* Retrieved from http://environ.spawar.navy.mil/Links/law3.html on February 11, 2011.

- Logistics analyses
- Life cycle cost analysis

Given that sustainability analyses can benefit from and provide information for many other analyses, it naturally follows that an integrated approach is essential.

An integrated, systemic approach to sustainability is implemented through an integrated product and process development (IPPD). The concept of IPPD was defined in a National Center for Advanced Technologies (NCAT 1993) white paper. The paper described IPPD as a management methodology that incorporates a systematic approach to early integration and concurrent application of all the disciplines that play a part throughout an entity's life cycle. The practice of IPPD requires the simultaneous use of several techniques integrated in a decision support process. The paper described the methodology as the integration of four key elements:

1. A top-down design decision support process
2. Measurement and analysis methods
3. An integrated information environment
4. Systems engineering methods

This methodology can be applied for a system, a component of the system, part-level trades, or an integrated combination. Illustrated in Figure 2.5 are the interactions of these four elements.

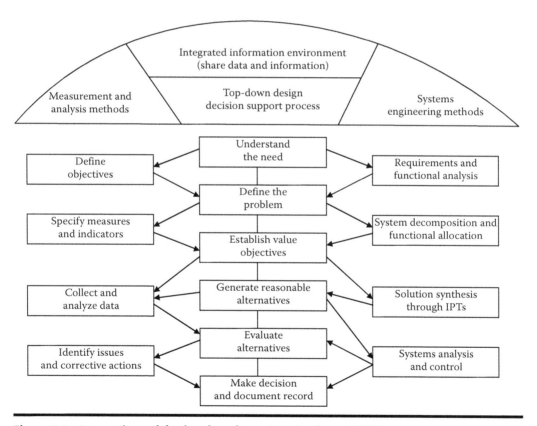

Figure 2.5   Interactions of the four key elements to implement IPPD.

The heart of the implementation is the top-down design decision support process. Decision support is an essential element that can support a trade-off process and can be used to focus efforts on design or other work-product goals. It supplies a logical, rational means for including factors that must be considered when making a decision. With the developed integrated design system, the structural concept selection can be based upon evaluation of both the product and process metrics that characterize the concepts.

The systems engineering methods on the right side of Figure 2.5 are decomposition oriented and product design driven. Systems analysis is typically based on product design metrics, but may also include process metrics. The measurement engineering methods illustrated on the left side of the figure are "re-composition" oriented, statistically based, and process design driven. A process-based assessment of the design in terms of its production costs at the major component level is used in the evaluation of the feasible alternatives. For top-level IPPD studies, several systems analysis techniques can be very useful to help define the design problem once the need has been established, and track and deploy the decision-making process.

The top of Figure 2.5 illustrates how the three previously described elements function within an integrated information sharing environment. This environment allows the interactive involvement of the three elements, indicated by the arrows between elements. The IPPD trade studies associated with the evaluation of alternative structural concepts required the integration of several design tools into a functional design system.

IPPD is a process integrated with sound business practices and common sense decision making. Organizations may undergo profound changes in culture and processes to successfully implement IPPD. IPPD is a fundamental shift from sequential development in which separate groups operating independently design the product, then the manufacturing processes, and then the support system. It has its roots in integrated design and production practices, concurrent engineering, and total quality management. In the early 1980s, U.S. industry used the concept of integrated design as a way to improve global competitiveness. Industry's implementation of IPPD expanded concurrent engineering concepts to include all disciplines, not just technical, associated with the design, development, manufacture, distribution, support, and management of products and services. Diverse segments of U.S. industry have successfully implemented this concept to become recognized leaders in IPPD practices, most notably in the auto and electronics industries. Many corporations have institutionalized the IPPD process and associated training programs. Some firms have given their own name to the IPPD approach and added or modified the approach to better fit their needs. Although the name and details may vary, the objective is consistent: to optimize the processes used to design, manufacture, market, and support a product. Under IPPD and similar approaches, all essential activities related to the product are simultaneously integrated through the use of multidisciplinary IPTs to optimize the design, manufacturing, business, and supportability processes. The following are criteria for a successful IPPD approach:

■ The need for a shift from a product development focus to IPPD is understood and encouraged.
■ Members of the IPT are empowered to make decisions and have a clear understanding on the extent of their authority.
■ Technical program reviews with senior management routinely address affordability issues.

IPPD activities focus on the customer and meeting the customer's need. Accurately understanding the various levels of users' needs and establishing realistic requirements early in the development of a new product is now more important than ever. Trade-off analyses are made among design, performance, production, support, cost, and operational needs to optimize the system (product or service) over its life cycle. To afford sufficient numbers of technologically up-to-date systems, cost must be considered a critical component of DoD system optimization. For commercial firms, product price is a function of competition, market demand, and the costs of developing and manufacturing the product. Cost should not simply be an outcome, as was often the case in DoD programs. Instead, cost should be viewed as an independent rather than dependent variable in meeting the user's needs.

Although there are common factors in all known successful IPPD implementations, IPPD has no single solution or implementation strategy. Its implementation is product and process dependent. A generic IPPD iterative process is shown in Figure 2.6.

Resources applied include people, processes, money, tools, and facilities. The IPPD process reorders decision making, brings downstream and global issues to bear earlier and in concert with conceptual and detailed planning, and relies on applying functional expertise in a team-oriented manner on a global-optimization basis. An early understanding of the processes needed to develop, produce, operate, and support the product is essential. Equally important are these processes' impacts on product design and development. The basic elements of the iterative process follow.

1. *Requirements*: This is the first step in the iterative process. Knowing what the customer's true requirements are is paramount. As discussed in Chapter 1, a QFD matrix can be used to facilitate communications between a customer and his supplier to determine "what" the customer wants (customer expectations) and "how" the supplier can meet those expectations.

2. *Disciplined Approach*: This element includes five general activities: understanding the requirements, outlining the approach, planning the effort, allocating resources, and executing and tracking the plan. Decisions made using this approach should be reevaluated as a system matures and circumstances (budgetary, marketplace, technology) change. A disciplined approach provides a framework for using tools, teams, and processes in a structured manner that is responsive to systematic improvement efforts. Tools in this IPPD process include documents, information systems, methods, and technologies that can be fit into a generic-shared framework that focuses on planning, executing, and tracking. Tools help define the

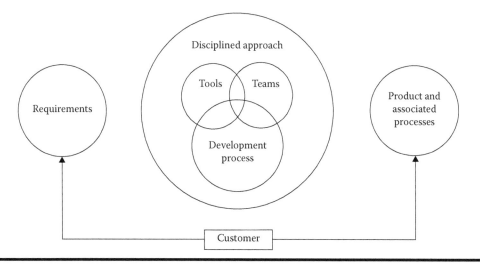

Figure 2.6   Generic IPPD iterative process.

product(s) being developed, delivered, or acted upon, and relate the elements of work to be accomplished to each other and to the end product. Examples of tools used include integrated master plans, three-dimensional (3D) design tools and their associated databases, cost models linked to process simulations, activity-based costing, development process control methods, and earned value management. Teams are central to the IPPD process. Teams are made up of everyone who has a stake in the outcome or product of the team (the "stakeholders"), including the customer and suppliers. Collectively, team members should represent the know-how needed and have the ability to control the resources necessary for getting the job done. Teams are organized and behave so as to seek the best value solution to a product acquisition.

3. *Development Processes*: This element includes those activities that lead to both the end product and its associated processes. To ensure efficient use of resources, it is necessary to understand what activities are necessary and how they affect the product and each other. Examples include requirements analysis, CM, and detailed design drawings.

4. *Product and Associated Processes*: Included in this element is what is produced and provided to the customer. Customer satisfaction with the product, in terms of functional effectiveness, as well as operating and support aspects and costs, is the ultimate measure of the team's success.

5. *Customer*: The customer is the user of the product and should be considered a team member and the ultimate authority regarding the acceptability of the product.

The generic IPPD iterative process just described is a sustainability engineering approach. It differs from the long-held view that sustainability engineering is essentially a partitioning, trade-off, control process that brings the "abilities" (including maintainability) and test functions together. The IPPD process controls the evolution of an integrated and optimally balanced system to satisfy customer needs and to provide data and products required to support management decisions that, themselves, are part of the IPPD/IPT process. The approach also transforms the stated needs into a balanced set of product and process descriptions. Within DoD, these descriptions are incrementally matured during each acquisition phase and used by the department and its contractors to plan and implement a solution to the user needs. This process balances cost, system capability, manufacturing processes, test processes, and support processes.

One way to ensure that sustainability is integrated into the systems engineering process is for the organization or individual responsible for sustainability to explicitly address how the included activities will be integrated into the product and manufacturing design processes. This organization or individual should also show how the results of the included activities will be used to support other activities, such as logistics planning and safety analyses.

Chapter 3

Foundational Concepts in Sustainability Engineering

3.1 System Concept

A system is defined as an aggregation of "end products" that enable a system to achieve a given purpose. The system to which processes are applied consists of both end products to be used by a customer for an intended purpose and the set of "enabling products" designed and developed to make the end products successful and supportable. These are designed in parallel with the end products. Enabling products are used to perform associated-process functions of the system. They include actions such as developing, producing, testing, deploying, and supporting the end products; training the operators and maintenance staff using the end products; and retiring or disposing of end products that are no longer viable for use. Both end products and enabling products are either developed or reused as appropriate. The relationships between system elements are shown in Figure 3.1.

The highly simplistic structure shown in this figure takes into account the people who perform the functions of developing, producing, testing, operating, supporting, and retiring the system. It also takes into account the people who train others involved in these functions as well as the human factors issues and concerns associated with these personnel. These issues are typically included in the processes used for development or acquisition of the product and its enabling support systems.

3.1.1 System Blocks

A system is formed from the integration of its architected and constituent parts and activities, and it forms the basis for a larger structure, as shown in Figure 3.2. This larger structure is called the "system block," and it contains the framework for sustainability processes.

Conceptually, a system block is made up of a system (striped elements in Figure 3.2), one or more end products (black elements), two or more subsystems for each end product (gray elements), and the ensemble of enabling products (white elements). Each end product and each enabling product includes one or more of the following: Hardware, software, firmware, personnel, facilities,

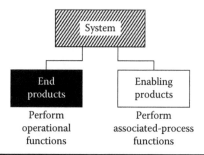

Figure 3.1   System concept.

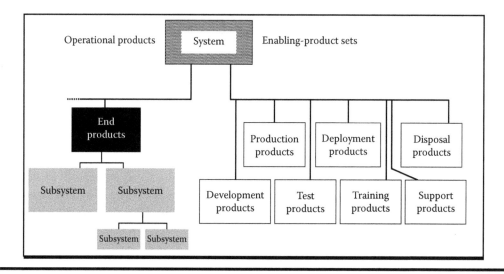

Figure 3.2   System block.

data, materials, services, and processes. As a system is developed into end products, enabling products, and systems and subsystems, the complete range of technical processes discussed in Chapter 4 is applied to each of them. These technical processes are also recursive in that they are applied to each system block and elements of the system.

Although the list is not exhaustive, the following information is typically associated with each element within a system block:

■ Configuration identification
■ Costs to be collected
■ Identification of interfacing elements inside and outside the "system block"
■ Specifications relevant to the element
■ Definition of work to be done
■ Other relevant agreement information

End products, subsystems, enabling-product sets, and the like are examples of building blocks or elements that make up a system block.

3.1.1.1 System Elements

The system element in the system block of Figure 3.2 is the object that defines stakeholder requirements. For example, the activities in the International Organization for Standardization (ISO) standard for Stakeholder Requirements Definition process could be used to define the requirements. Such ISO standards are documented in Institute of Electrical and Electronics Engineers (IEEE) ISO Standard 15288 (IEEE 2008): "This International Standard establishes a common process framework for describing the life cycle of man-made systems. It defines a set of processes and associated terminology for the full life cycle, including conception, development, production, utilization, support and retirement. This standard also supports the definition, control, assessment, and improvement of these processes. These processes can be applied concurrently, iteratively, and recursively to a system and its elements throughout the life cycle of a system" Page iii (Abstract) in IEEE (2008).*

3.1.1.2 End-Product Elements

End products of a system perform the operational functions of the system. An end product can be either a legacy product that is being reengineered or a product made by an enterprise that has the expertise to make it and has similar products already in the marketplace. These developments are identified as precedent, derivative, or next generation. When the specified end product is not known a priori or when the organization has limited experience in the development of a new system, the development is identified as precedent or a new concept. An end product is defined by customer needs. It performs the operational functions required by the customer, and it has "abilities" designed into it (e.g., producibility, testability, reliability, supportability).

These products should be developed using the solution definition subprocesses of the ISO standards for Logical Architecture Design process, Physical Architecture Design process, and Document the Design process. Products should be verified against specified requirements using the ISO Product Verification process and validated against customer requirements using the ISO Product Validation process. An end product can also be self-contained in terms of its use and operations. It can be an item that has no use outside a larger end product but is developed as an end product of a subsystem (lower-layer system building block) using the ISO Solution Design process. Examples of self-contained end products are aircrafts, automobiles, communications satellites, nuclear reactors, telecommunication switching modules, or an integrated system that is delivered to an operator. An end product can also be one of many products that make up a self-contained end product. Examples of such end products are an engine or a radio on an aircraft, a power train or a brake for an automobile, a solar panel, a transmitter for a satellite, a control panel or a control valve for a nuclear reactor, a switch or a transducer for a telecommunication switching module, or a life support package or a hatch door for a space vehicle. These end products can be found at the assembly, subassembly, line replaceable unit, component, or part levels of a system. The end-product element shown in black in Figure 3.2 represents those elements of the system block that are physically integrated with end products of upper-layer and lower-layer building blocks to form a composite end product and eventually a self-contained end product. Also, there can be more than one end product in a system block. In such cases, the system consists of an "aggregation" of end products, plus their enabling products.

---

* The ISO Standard 15288-2008 "Systems and Software Engineering System Life Cycle Processes." Retrieved from the web at http://standards.ieee.org/findstds/standard/15288-2008.html on February 22, 2011.

3.1.1.3 Subsystem Elements

If end products cannot be manufactured or are not off-the-shelf products that can be reused and purchased from another supplier, then subsystems of an end product should be developed using sustainability processes. An end product that is developed may consist of two or more subsystems (gray elements in Figure 3.2). When a subsystem is developed, another lower-layer system block is established, as depicted in Figure 3.3. The hierarchy of these system blocks is called the "system structure."

3.1.1.4 Enabling-Product Elements

Enabling-product elements perform associated processes or nonoperational functions of a system. They are designed to support and make end products successful and supportable, as depicted in Figure 3.1. Each end product has its own enabling products. Enabling products should be verified against their requirements and validated to ensure they perform their intended functions when required to support their related end product or aggregation of end products. When each set of enabling products is developed using the sustainability processes discussed in Chapter 4, another building block is formed, as depicted in Figure 3.4. Development of an enabling-product building block is normally initiated after related end products are fully defined and requirements for enabling products are identified. Further, some enabling products can be designed and produced or acquired concurrently with their end products. Examples of enabling products developed concurrently with a system are listed in Table 3.1. Both end products and enabling products exist at various layers in the system structure.

3.1.2 System Block Roles

The system block concept is used for performing the following six actions:

1. Identifying and assigning specifications for the system, end products, and subsystem elements
2. Managing interfaces
3. Enabling multidisciplinary teamwork
4. Assessing risk
5. Structuring technical reviews
6. Cost collection and reporting

Data and document management is facilitated by the system block, since each element of the system shows the source of such data and documents. Data and documents are generated as work products or deliverables as a result of technical efforts for developing each system element. Similarly, each system element is assigned a work package to direct the team performing the planned technical effort.

3.1.2.1 WBS

A WBS is a hierarchical, product-oriented tree structure that displays and defines products and services to be developed, procured, or produced. It relates the elements of work to be accomplished to each other and to end products. There is typically some use of a WBS structure for subsystems

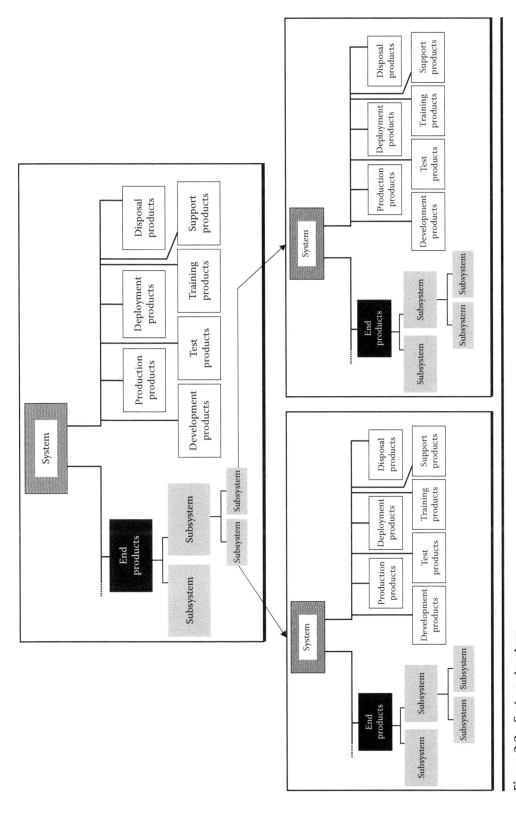

Figure 3.3 System structure.

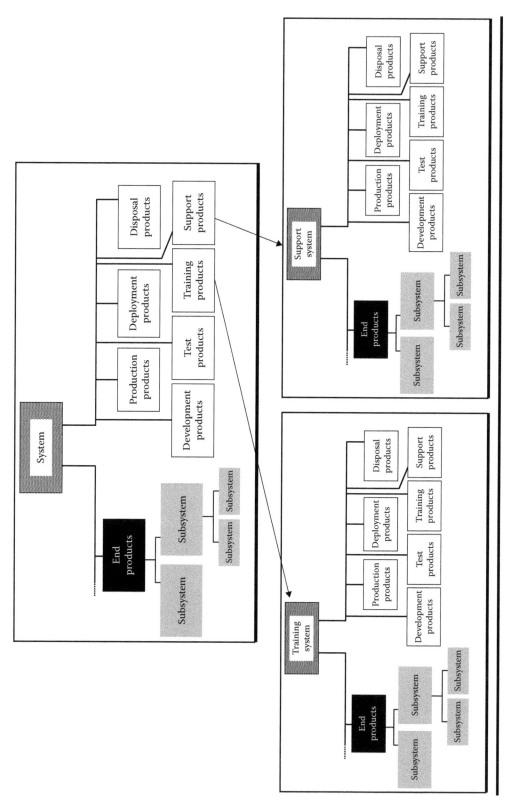

Figure 3.4 System structure for enabling products.

Table 3.1   Examples of Enabling Products for Each Associated Product Set

Associated Enabling-Product Set	Examples of Enabling Products
Development	Development plans and schedules, engineering policies and procedures, integration plans and procedures, information database, automated tools, analytical models, physical models, engineering management personnel and technology, and connecting cables and other interface structures not being developed as separate end products
Production	Production plans and schedules, manufacturing policies and procedures, manufacturing facilities, jigs, special tools and equipment, production processes and materials, production and assembly manuals, measuring devices, and manufacturing and procurement personnel
Test	Test plans (including test environment interactions) and schedules, test policies and procedures, test models, mass/volume mock-ups, special tools and test equipment, test stands, special test facilities and sites, measuring devices, simulation or analytical models, demonstration and scale test models, inspection procedures, and test personnel
Deployment	Deployment plans and schedules, deployment policies and procedures, mass/volume mockups, packaging materials, special storage facilities and sites, special handling equipment, special transportation equipment and facilities, installation procedures, installation brackets and cables, special transportation equipment, deployment instructions, ship alteration drawings, site layout drawings, and installation personnel
Training	Training plans and schedules, training policies and procedures, simulators, training models, training courses and materials, special training facilities, and trainers
Support	Support plans and schedules, support policies and procedures, special tools and repair equipment, maintenance assistance modules, special services (e.g., telephone hotline and customer access lines), special support facilities and handling equipment, maintenance manuals, maintenance records system, special diagnostic equipment (not an integral part of the end product), and repair personnel
Disposal	Disposal plans and schedules, disposal policies and procedures, refurbishment facilities and equipment, special disposal facilities and sites, special equipment for disposal of end products, and disposal personnel

and enabling-product development. The WBS is used to understand work elements, interfaces, and dependencies. It is incrementally developed through the use of engineering and technical management processes. It is the common framework that consistently links program and technical planning, resource allocation, risk management, performance measurement, technical assessment, and status reporting. The WBS is also an essential input for creating program or project schedule. The current standard for creating and using a WBS is U.S. military handbook MIL-HDBK-881A (Department of Defense 2005).

3.1.2.2 Specifications of End Products

Specifications describe the required characteristics of end products (black elements in Figure 3.2) or a group of products (gray elements in Figure 3.2). They are output from various subprocesses of the ISO Solution Design process mentioned in Section 3.1.1.2. Characteristics of the end products include the following:

- Functional and performance requirements
- Interface requirements
- Environments in which the product is required to perform its functions
- Physical characteristics and attributes
- Basis for evaluating test articles
- Methods for verifying compliance
- Intended uses
- Enabling-product requirements

The system block relationships of black- and gray-element specifications and white-element requirements, as well as appropriate interface specifications, are shown in Figure 3.5.

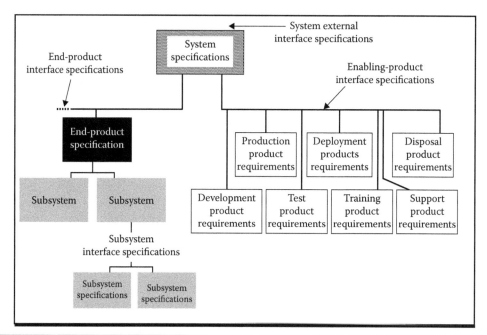

Figure 3.5 System block: Specifications.

3.1.2.2.1 Stages of Specification Maturity

The specifications for the system, end product, and subsystem elements evolve through three stages: (1) Conceptual, (2) initial, and (3) established. Conceptual specifications are used to show feasibility of a high-level initial specification (e.g., end product) and to record the characteristics of notional products. Conceptual specifications evolve into initial specifications by the application of system design processes. Initial specifications are used to direct lower-layer building block developments of subsystems. The initial specifications evolve into established specifications by the application of solution design processes. Established specifications

- Enable making valid estimates of work and resources needed for the next lower-layer building block development
- Provide a basis of communication with and among the development team, suppliers, and customers
- Provide guidance to testers for completing system verification and end products validation processes
- Provide basis for negotiation of engineering changes
- Guide preparation of detailed drawing or software development file design definitions
- Enable development of lower-layer building block specifications and solution definitions, for example, drawings, parts lists, and code lists
- Enable configuration management (control and maintenance) of solution definitions that satisfy technical requirements
- Enable definition of logistics support for spares, replacement parts, training manuals, maintenance operations, diagnostic tools, and support equipment

3.1.2.2.2 Performance Specifications

Performance specifications are used when it is appropriate to state requirements in terms of the following:

- Required results without stating the method for achieving the required results
- Function (what is to be accomplished) and performance (how well each function is to be performed)
- The environment in which products must perform the functions
- Interface and interchangeability characteristics
- The means for verifying compliance

3.1.2.2.3 Detailed Specifications

Detailed specifications are used when it is appropriate to state design requirements in terms of the following:

- Material to be used
- How a requirement is to be achieved
- How a product is to be fabricated or constructed

Examples include detailed design drawings, pseudocode, and a software dictionary in a software development file.

3.1.2.3 Defining Interfaces

Interface specifications are essential in most system development activities and are used to clarify interdependencies between system elements within the system block (internal) and other systems above, below, and at the layer of development (external). Interface specifications may be defined in an interface control document (ICD) to define and specify three elements:

1. Physical and functional relationships between system elements, including operators
2. Functional requirements resulting from these relationships
3. Constraints

3.1.2.4 Multidisciplinary Teamwork

Another role for the system block is to enable multidisciplinary teamwork. A reference structure for team assignment is shown in Figure 3.6. Groups of people assigned to a task do not ensure teamwork. How the different teams are integrated; whether they have the requisite knowledge, skills, and abilities; and how they synergize together are important. Multidisciplinary teamwork ensures the accuracy and completeness of evolving technical data packages from which test articles, preproduction prototypes, and production products are to be manufactured or created. A core system engineering team usually comprises a project technical manager along with members to be assigned to team lead positions on end-product and associated-process teams. An end-product team can comprise leaders from their respective subsystem teams. An enabling-product team can be individuals representing their respective functional disciplines. These functional specialists are also appropriately assigned to subsystem teams. A subsystem team normally comprises appropriate domain experts as well as functional specialists and other required engineering specialists (safety, security, etc.). A subsystem team becomes the core team for the next lower-layer building block development of subsystem and end products.

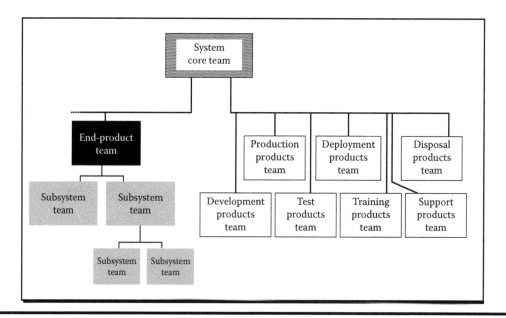

Figure 3.6   System block role: Team work.

An emerging concept is an IPT, which is discussed in Chapter 2. This team is responsible for planning their work, making cost estimates related to their work and the work product, doing the work, and identifying and managing their risks.

3.1.2.5 Recursion and Iteration

Recursion and iteration are two commonly-used, powerful methods of solving complex problems. Both methods rely on breaking up the complex problems into smaller, simpler steps that can be solved easily. Iteration is a looping process in which a problem is converted into steps/tasks that are finished one at a time, one after another. Recursion is also a looping process, but it calls upon (repeats) itself until a condition is met.

3.1.2.5.1 Use of Recursion

Recursion is the repeated application of technical sustainability processes to successively lower system elements or blocks within a system structure or to end products at successively higher levels in the system structure, as illustrated in Figure 3.7.

3.1.2.5.2 Use of Iteration

Iteration is the reapplication of processes already applied to a system block or model based on feedback indicating that identified problems need resolution. As technical difficulties, such as unacceptable risk, infeasible technology, new technologies applied in new environments, or

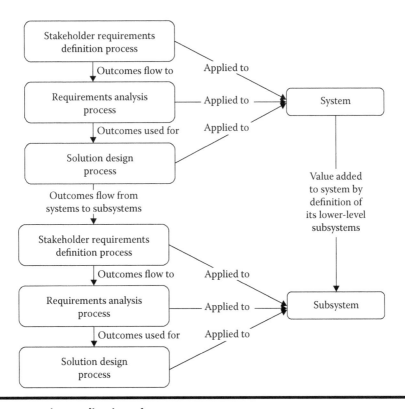

Figure 3.7   Recursive application of processes.

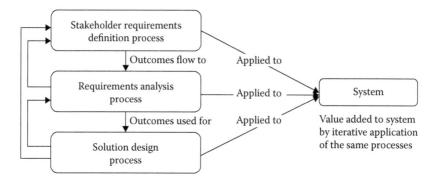

Figure 3.8 Iterative application of processes.

requirements conflicts, are discovered, these difficulties must be resolved through the reapplication of the appropriate technical process. Further, they must be completed, or the issue must be resolved, before finalizing the design solution of the end product and the subsequent flow of design specifications down to subsystems at the lower layers, as illustrated in Figure 3.8.

3.1.2.6 Assessing Risk

Another role for the system block model is to provide a structure for assessing and managing risk. Risk portrays relationships between subsystems and end products. Based on the degree of risk and the relationship among the other block elements, risk aversion plans are created and tracked.

3.1.2.6.1 Definition of Risk

Risk is the potential problem to the project or system under development. Significance of a risk is often portrayed as an interaction between the probability of occurrence of the risk and its consequences. Risk is assessed for the project, product, and process aspects of a system. This includes adverse consequences of process and subprocess variability. The sources of risk include the following four sources:

1. Technical risk (e.g., feasibility, operability, producibility, testability, and system and interface effectiveness)
2. Cost risk (e.g., estimates and goals)
3. Schedule risk (e.g., technology or material availability, technical achievements, and milestones)
4. Program risk (e.g., resources)

3.1.2.6.2 Risk Management

Risk management requires discipline. Moreover, risk management is a continuous process. Risk management is useful only to the degree that it highlights the need for action and that action leads to the potential problem being addressed quickly and thoroughly. Things that can go wrong may go well until the last phase of the project. The risk associated with arriving at the solution definition of each end product is a function of the risk assigned to each subsystem of the end product. Similarly, the totality of risk associated with the development of a system is a function of end-product risks, associated enabling-product risks, its interface risks, and risks unique to the entire integrated system.

3.1.2.7 Technical Reviews

Another advantage of the system block model is its use in efficient structuring of various reviews. At the systems level, technical reviews are scheduled and conducted during each engineering life cycle phase as appropriate. Their purpose is to review progress against the plan, against the established agreement, and against the applicable enterprise-based life cycle phase exit criteria. They are an important oversight tool, and they are conducted to determine whether to continue investing in future engineering or enterprise-based life cycle phases based on the following five factors:

1. Risks and costs associated with lower-layer developments
2. Maturity of the development to date
3. Whether requirements and technical plans being tracked are on schedule and are achievable within existing project constraints
4. Resources required for lower-layer projects
5. Readiness to proceed to external supplier availability and agreement preparations, if applicable

There are two types of review, incremental and system, which are discussed here. A typical order of reviews is shown in Figure 3.9. Planning for technical reviews, which includes the conduct, reviewing body, and presenters of specific technical reviews, is accomplished in a technical review plan during the planning process. The team associated with a specific review is assigned the task of creating and presenting the technical review.

A few examples of technical reviews, which are documented in the sustainability engineering plan, are as follows:

- Initial technical review (ITR)
- Alternate system review (ASR)
- System requirements review (SRR)
- System functional review (SFR)
- Preliminary design review (PDR) and critical design review (CDR)
- System verification review (SVR)
- Functional configuration audit (FCA) and physical configuration audit (PCA)

3.1.2.7.1 Incremental Reviews

Incremental reviews (e.g., subsystem reviews) are conducted on subsystems, associated processes for related sets of enabling products, and end products. Their purpose is to provide the status of product development and the product's readiness to initiate subsequent activities. Enabling-product readiness reviews, for example, demonstrate that an enabling product's requirements are defined and ready for initiating development. End-product reviews are held to confirm that a product has been adequately defined to approve the initiation of the development of enabling products and subsystem blocks.

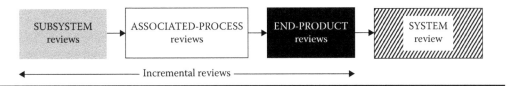

Figure 3.9   System block role: Technical reviews.

3.1.2.7.2 System Reviews

On completion of incremental reviews, a system review is conducted. System reviews are the top element of a system block (e.g., an engine, tank, or satellite). They are typically associated with the top-level system or an identified critical end product that makes up the system-level end product (e.g., a weapon or computer). For subsystem reviews, the parent end-product team is the reviewing body. End-product team members and team leads selected from other associated-process teams make up the reviewing body for associated-process reviews. These reviews can be held as a joint review. The core team is the reviewing body for end-product reviews. Reviewing bodies can be supplemented by other specialists from outside the project, as appropriate, to meet technical review objectives. The reviewing body for a system review can be designated in the agreement, project plan, or engineering plan. The system review can be held along with a project review, when intended, to meet the exit criteria for an enterprise-based life cycle phase.

The purpose of incremental and system reviews are listed in Table 3.2.

3.1.2.8 Cost Collection and Reporting

Another use of the system block structure is to facilitate the collection and reporting of costs related to engineering life cycle activities. Costs are incurred in each system element as development activities are conducted in accordance with assigned work packages. These work packages are generated during planning. The costs incurred include direct labor costs that are associated with applying engineering process tasks for requirements definition, design definition, design verification, trade-off and effectiveness analyses, fabrication, software bulk copying, technical reviews, data and document generation, integration, and testing. Technical agreement, planning, and control costs are also collected and reported as part of development of associated-process enabling products. Used in conjunction with the WBS, costs associated with system development can be easily summarized by rolling up the costs of subsystems, end products, and associated processes. When project performance is tracked by a customer, or for internal control using a cost performance measurement system, cost and performance measurements can be combined using an earned-value approach.

Table 3.2 Purpose of Selected Reviews

Review	Purpose
Subsystem	To assess progress in defining and satisfying subsystem requirements
Associated enabling-product set	1. To assess progress and identify issues associated with requirements for one associated process or group of associated processes 2. To ensure the suitability and availability of the services of enabling products when they are needed
End product	To address issues and demonstrate required building block development progress and maturity

3.2 System Structure

Although system block was used as a concept in Section 3.1, a single block model is very simple, and it rarely defines the complete solution to a customer or other stakeholders. Thus, a system structure concept is needed. A system structure provides the basis for top-down system design and bottom-up system realization. Under a system structure, enabling products are defined, developed, or acquired so that they are available when needed. An example system structure is depicted in Figure 3.10.

As shown in the figure in the form of a "V" model (explained in Section 3.2.1), if a subsystem requires further development it is done as a subordinate block development. Lower-layer block developments are initiated as soon as definite contents of a building block are determined. Specified requirements of a subsystem become assigned requirements at the next lower layer of development. Each block can have stakeholder requirements that are not related to requirements that are either assigned from above or directed by users or customers. This layered approach in the decomposition of blocks continues until end products of a block can be implemented, requirements for an end product can be satisfied by an existing product, or end products can be acquired from a supplier. The specific block structure varies with each system based on the number of end products, number of subsystems in an end product, and applicable enabling products of associated processes.

The contents of subordinate blocks are represented as end-product specifications, initial subsystem specifications, interface specifications, and requirements identified for applicable enabling products of associated processes. Conversely, subordinate blocks are connected to form the system structure, or a system block hierarchy. The relationship among building blocks in a hierarchy is shown in expanded view in Figure 3.11.

3.2.1 Top-Down Development

Figures 3.10 and 3.11 show a layered development approach for a project or program. Typically, the project receives customer requirements in a formal agreement and delivers products in accordance with that agreement. Each project can have several lower-layer subsystem block developments. An agreement is used for each lower layer of development using requirements assigned from the parent upper-layer building block. Only one engineering plan is normally required for the multiple layers of subordinate blocks or subsystems within a single project. If an external supplier is used for a lower-layer block development, a separate formal agreement is required because implementation and management of requirements and activities pertaining to the new agreement is outside project management control of the project.

Figure 3.12 depicts an example system structure again showing layered development. The top building block contains the end product that must satisfy the primary customer's requirements. This top building block represents what is often called the "prime contractor's project." Two other projects are shown in Figure 3.12: (1) Project A and (2) Project B. The top building block in each of these projects represents the top layer of development for the respective project but the second layer for the prime contractor's project. Project A spawns two layers of development, whereas Project B spawns multiple lower-layer building block development activities. Lines connecting the layers reflect the specified requirements assigned by a parent block to its subordinate block.

A project applies solution design processes to each block in the project boundary to develop the appropriate system, end product, and subsystem development specifications that are defined to satisfy assigned and other stakeholder requirements related to a single building block. The products, therefore, do not require further development. Project B's second layer of development in Figure 3.12 has one building block that requires a third layer of development, whereas the specifications of

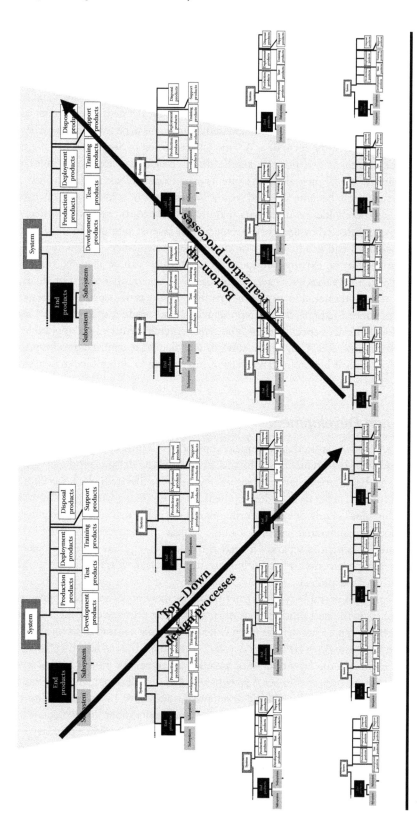

Figure 3.10 Forming a system structure.

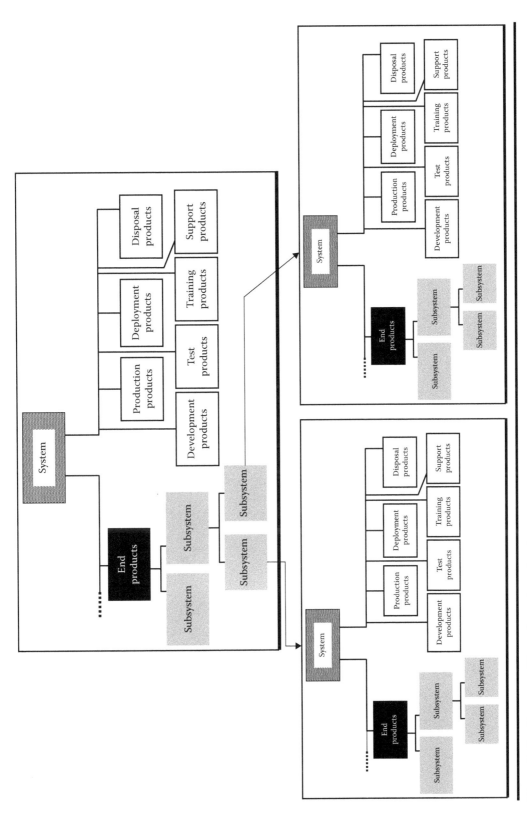

Figure 3.11 Decomposition of a system structure.

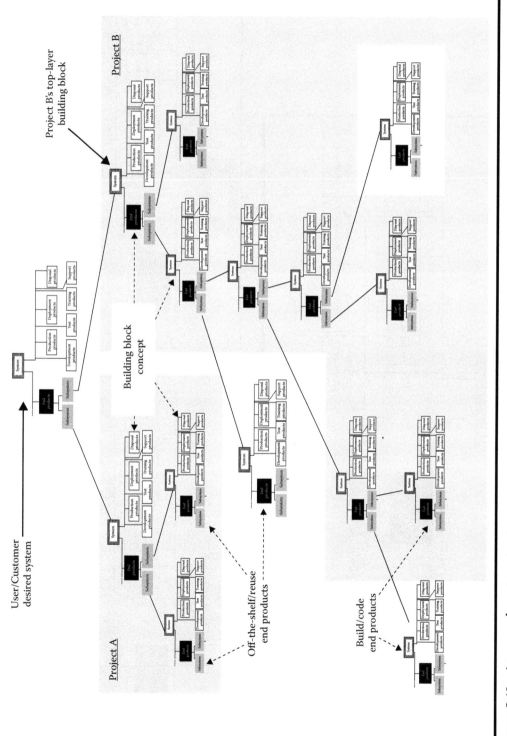

Figure 3.12   An example system structure.

the other building block's end product are satisfied by either an off-the-shelf product or a reused product. Project B requires five layers of development to complete the downward definition of end products sufficiently so that they can be built, coded, or procured off the shelf or reused. Project B relies on external suppliers for three end products: One at layer three and two at layer five.

3.2.2 Bottom-Up Realization

Section 3.2.1 explained how end products that make up the system structure are developed from the top down. Once specific end products are defined sufficiently by specifications so that a product can be used or the end products can be built or coded, product realization processes can be initiated. (Product Realization processes are described in Chapter 5 on sustainability engineering processes.) As shown in Figure 3.12, this can occur at any layer in the system structure. However, assembly, integration, verification, and validation of a product occur from the bottom up.

The bottom-up realization of end products is represented in the model depicted in Figure 3.13. The end products that are procured (or built, coded, used off-the-shelf, reused, or delivered by an external supplier) are verified using the Product Verification process (described in Chapter 5). Once verified, end products are integrated using the Product Integration process and delivered along with verification data in accordance with the established agreement. The end product is validated against its assigned requirements to ensure that it works in its intended environment or in the end-product environment. Validation is completed using the Product Validation process. Since this is an iterative realization process, integration and validation activities may occur many times before final assembly, verification, and validation occurs, and the systems block is delivered in accordance with the agreement.

The bottom-up realization approach mitigates risk. It allows the discovery of test and evaluation problems and design anomalies at the lowest level of development possible. It prevents

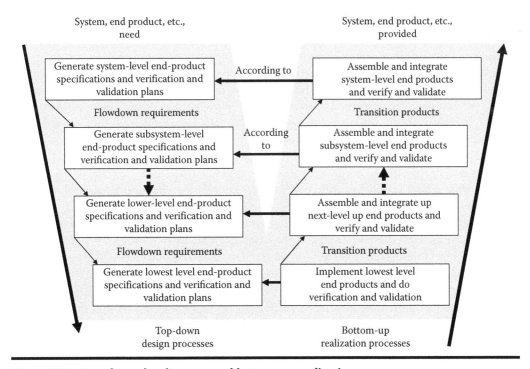

Figure 3.13   Top-down development and bottom-up realization.

lower-layer end-product defects from being buried or overlooked and then showing up during top-layer end-product verification and end-product validation (or validation of aggregation of end products). Solution design processes are applied to the affected block activities to correct anomalies uncovered by product verification or product validation processes. End products that do not meet specified requirements must be reworked (e.g., remanufactured, recoded, or reprocured) to correct the defect or nonconformance, so that a corrected test or evaluation item can be verified.

There are actually three approaches to engineering a system for sustainability: (1) Top-down, (2) bottom-up, and (3) middle-out approaches. The top-down approach is intended to flowdown requirements to the next lower layer to ensure satisfaction of project customer requirements in the upper-layer block. It is also intended to take advantage of reuse and off-the-shelf items that satisfy assigned requirements in order to lessen development costs and shorten development-cycle time. Figure 3.12 is an example. This chapter is based on the top-down approach. A bottom-up approach to development is not normally used unless it is ascertained that the requirements of the top-layer system block are not impacted. A middle-out approach is considered if the aforementioned hierarchy of building blocks start anywhere in the system structure of a project.

3.3 Life Cycle Models and Strategies

In Chapter 1, Figure 1.4 showed the interaction among organizational, project, and technical components of the system engineering life cycle. That is, organizational functions interact with project functions, which also interact with technical activities. In addition, as the development of the system progresses through its life cycle, there is interaction and feedback among the constituent steps and between levels of system detail. This produces many possible approaches to performing sustainability engineering and development. As a consequence, several prominent life cycles and models have evolved, which specify details of steps to be performed, the sequencing of the steps, and degree of feedback and interaction among the steps. This section presents the more prominent of these life cycle models and strategies.

A life cycle model is a representation of the processes, and the relationship between them, needed to evolve a system from the first identification of customer need through development, testing, production, deployment and operations, and continuing through various upgrades or evolutions until the product is disposed at the end of its life. Life cycle models are further subdivided into development life cycle models, production life cycle models, and so on, in order to provide a finer level of detail on the processes that comprise them. It is the responsibility of sustainability engineers to define the details of the life cycle that is appropriate for their program. The particular life cycle chosen for a specific program may be a variation of a generic model, such as waterfall, spiral, or evolutionary, or a combination or hybrid of these models.

A program's life cycle definition is a strategy for coordinating resources and activities in order to accomplish the end objective. It reflects the program's strategy for dealing with various issues that may not be stated in the contract but affect the course and outcome of the program. For example, the customer may need an early demonstration in order to secure support for continued funding. The ability of a subcontractor to complete his or her effort successfully may be in question. The customer may be located 5000 miles from the contractor facility and have a limited travel budget. The company may have a business development goal to market a variation of the system to another customer by a certain calendar date.

The sustainability engineer defines the project's life cycle as a strategy that addresses not only the contractual requirements but also the derived or unstated key drivers of program success. The selected life cycle drives the type, sequence, and frequency of the specific processes. Integration of sustainability engineering monitoring and control processes into the overall process is highly dependent on the model selected, especially when its selection is driven by the need for risk containment. Fast prototyping is an approach for mitigating risk and can be used with all life cycle models.

The purpose of a life cycle model is to define phases that are applicable to projects, organize process applications and activities, assess the adequacy of project activities, and monitor progress. A life cycle model forms a foundation for project processes. Requirements of a life cycle model include the following:

- Define the models and strategies for the enterprise and for individual projects.
  a. Describe the different life cycle models and strategies to be used in situations applicable to organizational units and projects.
  b. Describe the selection of life cycle models and strategies based on the needs of organizational units and projects.

3.3.1 Functions and Characteristics of Life Cycle Models

This section describes the use of life cycle models and life cycle strategies as they apply to a framework for application of processes. The primary function of a life cycle model is to form a life cycle framework for process application. Some processes are primarily applicable to specific life cycle phases. The primary function of a life cycle strategy is to specify the iterative nature of the application of processes. Some processes are repeated as some projects loop back through life cycle phases. The sustainability engineering processes, discussed in Chapter 5, require an enterprise to adhere to the following three foundational principles:

1. A life cycle model shall be established.
2. A life cycle model should comprise one or more stages as needed.
3. A life cycle model shall be assembled as a sequence of stages that may overlap and/or iterate. The stages provide a framework for the sustainability processes.

A Guide to the Project Management Body of Knowledge (*PMBOK® Guide*; Project Management Institute 2008) describes work as being accomplished by processes. The *PMBOK® Guide* provides the following eight characteristics for a project life cycle:*

1. Defines the phases that connect the beginning of a project to its end.
2. In the DoD acquisition model, the life cycle begins with an analysis of a problem before the project officially begins.
3. Phases are generally sequential; but some processes are implemented recursively or iteratively.
4. Defines what technical work (processes) to do in each phase.
5. Defines when deliverables (from processes) are to be generated and reviewed in each phase.
6. Defines who is involved (in implementing processes) in each phase.
7. Defines how work (technical process) is approved and controlled in each phase.
8. There is no "one-size-fits-all" life cycle model. The process of adjusting the life cycle to fit a project is called "tailoring."

---

* The PMBOK is consistent with other management standards, such as ISO 9000 and the Software Engineering Institute's Capability Maturity Model Integration (CMMI).

3.3.2 Standard Life Cycle Models

All projects should have a well-defined life cycle model. The model should be one of the standard life cycle models or an approved tailored version of these models, which complies with the aforementioned functions and characteristics. A block diagram is normally used to show the relationship between all activities. In support of this effort, program evaluation and review technique diagrams are extremely useful. It is good practice to identify any overarching approaches, such as design-to-cost, which will be used on the program, since these may impact the sequencing and interrelationship of the activities.

The model shown in Figure 3.14 is the DoD acquisition model, and it is provided in this chapter as a reference for comparison to the stages in the models discussed in Section 3.3.3.

The model shown in Figure 3.15 is considered the enterprise-standard life cycle model for the sustainability processes presented in Chapter 5. The model was created for the specific purpose of providing a framework for the process architecture of subprocesses. This model was created by integrating phases used in a number of models, including the DoD acquisition life cycle model, the International Council on Systems Engineering (INCOSE) life cycle model, and the ISO 15288 example model. The integration of these models provides the most complete model for the processes presented in this book, and it may also be considered suitable for some large development- and sustainment-type projects.

An intermediate-size life cycle model for medium-size projects with seven stages is provided in Figure 3.16. A simple life cycle model using some of the project management functions described in *PMBOK® Guide* that may be useful for small projects is shown in Figure 3.17.

The three life cycle models discussed in this section (Figures 3.15 through 3.17) are the standard approved models for an enterprise seeking sustainability. Sustainability projects should select one of these models, unless there is a reasonable rationale for using a different tailored model. Note that

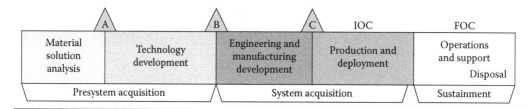

A		B		C	IOC	FOC
Material solution analysis	Technology development	Engineering and manufacturing development	Production and deployment		Operations and support	
						Disposal
Presystem acquisition		System acquisition			Sustainment	

Figure 3.14   The DoD acquisition model.

Preproject activities	Analysis and customer requirements	Project initiation and planning	Logical design	Physical design	Integration testing and verification	Production, deployment, and transition	Project/phase closeout	Operations and sustainment (O&S)	Retirement and disposal

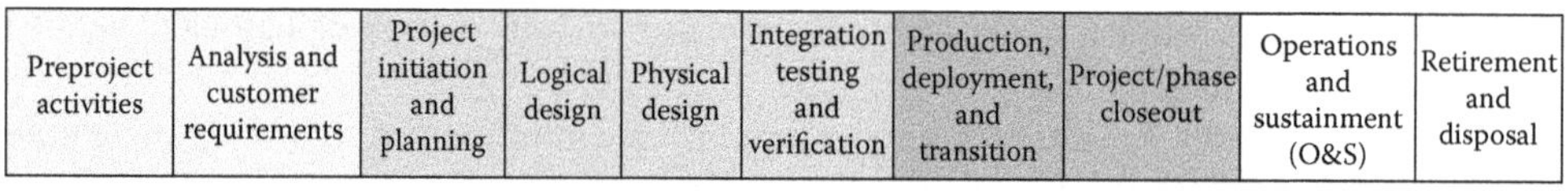

Figure 3.15   Life cycle model for large/complex projects in an enterprise.

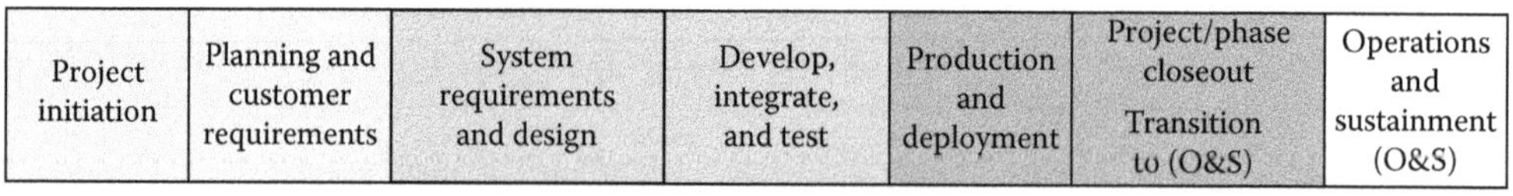

Project initiation	Planning and customer requirements	System requirements and design	Develop, integrate, and test	Production and deployment	Project/phase closeout Transition to (O&S)	Operations and sustainment (O&S)

Figure 3.16   Life cycle model for medium-size projects.

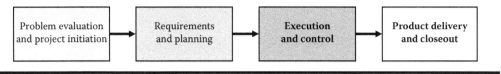

Figure 3.17 Life cycle model for very small projects.

project phase closeout is not shown consistently in these models. Closeouts are not shown because some projects close out once at the end of the project, whereas others close out at the end of selected phases.

3.3.3 Life Cycle Strategies

Life cycle strategy refers to the iterative philosophy that must be used in the planning and execution of a life cycle model. At least five life cycle strategies are in common use: (1) Once-through (waterfall), (2) V, (3) incremental, (4) spiral, and (5) evolutionary models.

3.3.3.1 Once-Through (Waterfall) Strategy

The once-through strategy is another name for the "grand design" or waterfall model. It is a program strategy associated with DoD standard STD-2167A terminology for software projects. The once-through model is a sequential development model usually applied to the creation of software in which development is seen as flowing steadily downward, similar to a waterfall, through the phases shown in Figure 3.18. The phrase waterfall has come to refer to any approach to work product creation that is seen as inflexible and noniterative. This strategy was conceived during the early 1970s as a remedy to the undisciplined "code and fix" method of software development. It is one of the original and probably the best-known life cycle models. It is a once-through, do-each-step-once strategy. In this approach, each process is performed in sequence, and each process is completed before proceeding to the next process in the sequence. However, this does not in any way preclude the iterative nature of the process. For example, design does not begin until project plans are prepared, reviewed, and completed. Similarly, implementation does not begin until the requirements and design phases are complete. The once-through model may be considered useful for some current nonsoftware projects, such as study projects or selected integration or services projects.

Risk containment in the waterfall model utilizes design reviews as "go" or "no go" decision points. At these major milestones, a decision is made whether the development to date is sufficiently mature and the risk is acceptable for work to begin on the next major building block of the life cycle. As a risk reduction step, particularly during requirements definition, fast prototyping may be incorporated into the model. This may be applied to the entire system or to portions of the system. This imparts iterations of design-build-test-design until the risk is mitigated. At this time, the process intercepts the normal waterfall model life cycle and progresses onward. The use of fast prototyping with other life cycle models is common.

3.3.3.2 V Strategy

The V strategy is a software development model, which is an extension of the waterfall model. The V concept has also been used in hardware development. Instead of moving down in a linear manner, process steps are bent upward after the coding phase to form the typical V shape shown in Figure 3.19. Progress is made by traversing the V from the upper-left user need to the upper-right

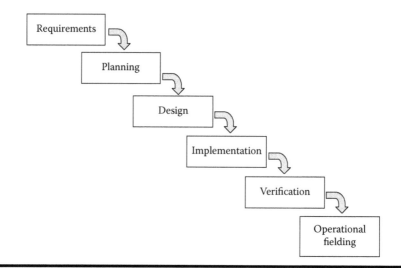

Figure 3.18 Once-through life cycle strategy.

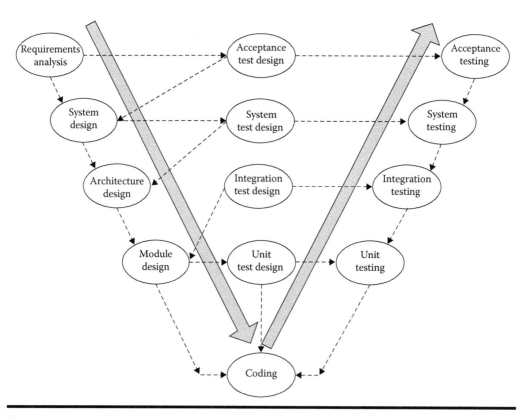

Figure 3.19 Generic V strategy.

user-validated system. On the left, at the upper side of the V, requirements decomposition and definition activities are used to create and refine system architecture, producing increased levels of requirements and design details. Development forms the base of the V, followed by integration and verification to the right of the V producing higher levels of system coherence and completeness. Thus, system detail increases when moving from top to bottom of the V, and system maturity increases when moving from left to right. The V strategy demonstrates the relationships between each phase of the development life cycle and its associated phase of testing.

Figure 3.20 illustrates the application of the V framework to the sustainability processes discussed in Chapter 5.

3.3.3.3 Incremental Strategy

The incremental strategy is a cyclical process developed in response to weaknesses of the waterfall strategy. Incremental development is a scheduling and staging strategy, as shown in Figure 3.21, in which the various parts of the system are developed at different times or rates and integrated as they are completed. This strategy is unlike the iterative development or waterfall development strategies, which historically resulted in considerable rework. The alternative to incremental development is to develop the entire system with "big bang" integration.

3.3.3.4 Spiral Strategy

The spiral life cycle strategy, shown in Figure 3.22, is a systems development method (SDM) used in information technology (IT). This strategy of development combines the features of prototyping models and the waterfall model. The spiral model is intended for large, expensive, and complicated projects. In this strategy, the effort is divided into several increments or spirals. Each spiral contains four sets of activity. The spiral strategy can be generalized by the following nine steps:

1. The new system requirements are defined in as much detail as possible. This usually involves interviewing a number of users representing all external or internal users and considering other aspects of the existing system.
2. A preliminary design is created for the new system.

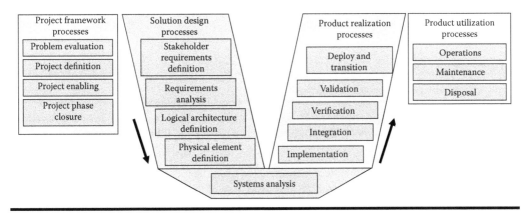

Figure 3.20   The V framework applied to sustainability processes.

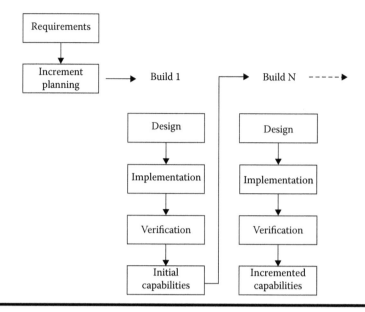

Figure 3.21   Incremental strategy.

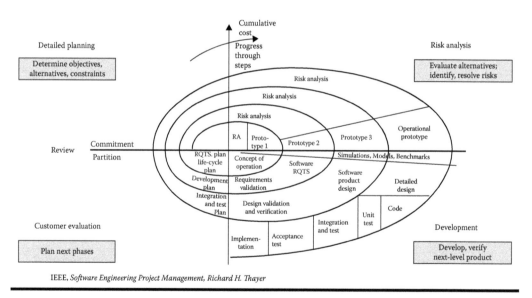

IEEE, *Software Engineering Project Management, Richard H. Thayer*

Figure 3.22   Spiral strategy. (Richard H. Thayer, Barry W. Boehm (1986). *Tutorial: software engineering project management.* **Computer Society Press of the IEEE. p.130.)**

3. A first prototype of the new system is constructed from the preliminary design. This is usually a scaled-down system and represents an approximation of the characteristics of the final product.
4. A second prototype is evolved by a fourfold procedure: (1) Evaluating the first prototype in terms of its strengths, weaknesses, and risks; (2) defining the requirements of the second prototype; (3) planning and designing the second prototype; and (4) constructing and testing the second prototype.

5. At the customer's request, the entire project can be aborted if the risk is deemed too great. Risk factors might involve development-cost overruns, operating-cost miscalculations, or any other factor that could, in the customer's judgment, result in a less-than-satisfactory final product.
6. The existing prototype is evaluated in the same manner as the previous prototype, and if necessary another prototype is developed from it according to the fourfold procedure outlined in step 4.
7. The preceding steps are iterated until the customer is satisfied that the refined prototype represents the desired final product.
8. The final system is constructed based on the refined prototype.
9. The final system is thoroughly evaluated and tested. Routine maintenance is carried out on a continuing basis to prevent large-scale failures and minimize DT.

Risks are identified at the beginning of each spiral and activities based on those risks are planned and executed for that spiral. These are evaluated to determine whether to proceed with the associated development and to identify risks and plans for the next spiral. In spiral development, sustainability engineering activities, such as requirements definition or a specific design review, are repeated multiple times at varying levels of detail. The back end of a spiral development effort looks much like that for the waterfall model, since the effort progresses to the verification and validation building block only after completion of the last spiral.

Spiral development was originally conceived as an approach to risk reduction and fast prototyping, although it is no longer limited to that use. It has the advantage of permitting the program to use fast-prototyping techniques during the early spirals as a way of rapidly getting something that can be tested by the team or the user. In most cases, the integration building block is part of each spiral and the user or team performs tests that help define risks for the next spiral during that step. Occasionally, risks are perceived at the component or unit test level, and all spirals are completed prior to entering the integration building block.

3.3.3.5 Evolutionary Strategy

An evolutionary strategy shown in Figure 3.23 is the preferred DoD strategy for rapid acquisition of mature technology. An evolutionary approach delivers capability in increments, recognizing the need for future capability improvements. This strategy is similar to spiral development in that the system is developed incrementally. It differs from spiral development in that as part of each spiral or increment, a deliverable product undergoes testing and verification and is then delivered to the customer. The system is fielded, and feedback from the customer provides input to the next increment. Each succeeding increment adds functionality to the product already fielded, which is then replaced with or updated to the new version once delivered. It is essentially a series of little waterfalls and, similar to spiral development, each increment executes the portion of the life cycle that is needed to address the next evolution of the system.

Most major acquisition category (ACAT) programs use this strategy/model. The objective is to balance needs and available capability with resources and to put capability into the hands of the user quickly. Success of the strategy depends on consistent and continuous definition of requirements and the maturation of technologies that lead to disciplined development and production of systems that provide increasing capability toward a material concept. This strategy usually calls for repeated implementation of numerous systems and sustainability engineering processes as each system increment or block of changes goes through its own life cycle. The Operation of the Defense Acquisition System, DoD Instruction 5000.02 (DoD 2008), provides top-level guidance for the evolutionary strategy.

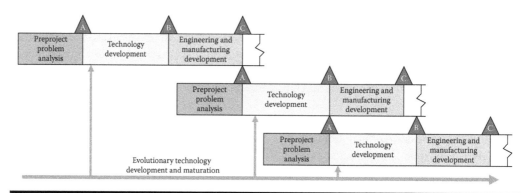

Figure 3.23 Evolutionary strategy model.

The goal of evolutionary development is managing risk. If, for example, the full set of requirements is not understood well, then developers may choose to define and implement only the highest-priority and best-understood system features. Once these are proved to provide the desired benefits, through customer use, additional features may be added. This is a particularly valuable approach when the need driving customer's requirements is changing or evolving. This may be due to new threats, as in the case of weapon systems, or may be due to new customer demands, as in the case of company information management systems. This is also a potent approach when there is a desire to insert new technology that may be maturing at a rate difficult to predict but for which there is certainty that it will be desired before the system reaches the end of its development or life. As with spiral or waterfall development, fast prototyping can be incorporated into the model.

3.3.4 Selection and Tailoring of Life Cycle Models and Strategies

Sustainability programs and projects should select an approved standard model or create a tailored life cycle model that satisfies their needs and requirements. Use of a tailored nonstandard model may be necessary to fit the needs of specific projects. The project life cycle model should be described in project technical planning documents, including the sustainability engineering plan (SEP) or project technical plans or both. The idea is to describe the rationale for the model selected or created. Approval of the technical plan by an appropriate technical authority can be considered approval of the model and strategy. The life cycle model shall form a framework for applicable project processes and identify any decision gates applicable to the project. The model may be dictated by the project sponsor. Program and project developers should also select an applicable life cycle strategy that facilitates the iterative and recursive application of some processes over the life cycle of the project. The life cycle strategy may be dictated by the sponsor.

A sustainability project can also describe the planned use of multiple life cycle models and strategies to satisfy project objectives. For example, a project may use a once-through (waterfall) strategy for early life cycle phases, switch to an incremental or evolutionary strategy for product design phases, and then switch back to a once-through strategy for product utilization phases. Note that project or phase closure may appear as the last process in some models and as an intermediary process in other models. This is because some models focus on a once-through application, whereas others are more iterative. Some projects have contract and administrative closure activities between major phases of the life cycle. This need for variation on the planned use of processes is part of the need and justification for the tailoring of life cycle models.

Table 3.3   Key Features of Life Cycle Strategies

Program Strategy	*Define All Requirements First?*	*Are There Multiple Development Cycles?*	*Is There a Field Interim Solution?*
Once-through (waterfall)	Yes	No	No
V	Yes	No	No
Incremental	Yes	Yes	Maybe
Spiral	No	Yes	No
Evolutionary	No	Yes	Yes

The Institute of Electrical and Electronics Engineers (IEEE) standard IEEE/EIA 12207 (IEEE 1996) provides an excellent discussion of life cycle strategies. The strategies when discussed in the context of software are universally applicable and serve as an excellent means of addressing the fundamental concepts of sustainability of any product. Key features assisting the selection of a life cycle strategy are summarized in Table 3.3.

References

Institute of Electrical and Electronics Engineers (IEEE). 1996. "Standard for Information Technology—Software Life Cycle Processes." IEEE/EIA (Electronic Industries Alliance) Standard 12207.0.

Institute of Electrical and Electronic Engineers (IEEE). 2008. "Systems and Software Engineering—System Life Cycle Processes." IEEE/IEC Standard 15288-2008. 2nd ed. 2008-0201.

Project Management Institute. 2008. *Project Management Body Of Knowledge—PMBOK*. Paperback. ISBN 1933890517. 459 pages. Project Management Institute Inc. Newtown Square, PA.

Richard H. Thayer, Barry W. Boehm (1986). *Tutorial: software engineering project management*. Computer Society Press of the IEEE. p.130.

U.S. Department of Defense. 2005 (July). "Work Breakdown Structures for Defense Material Items." MIL-HDBK-881a. Superseding MIL-HDBK-881 on January 02, 1998.

U.S. Department of Defense. 2008 (December). "Operation of the Defense Acquisition System." DoD Instruction 5000.02.

Chapter 4

The Sustainability Engineering Methodology

4.1 Sustainability Engineering Methodology

The SEM defines the interdisciplinary tasks that are required throughout the life cycle of the enterprise or entity to transform customer needs, requirements, and constraints into an enduring product. These tasks are then translated into a set of processes for sustainability, which are presented in the next chapter. SEM is a problem-solving methodology. It contains a series of tasks that flow from requirements analysis to systems (product) analysis. Each task contains a number of integrated subtasks that have clearly defined interfaces with known information inputs and outputs. Figure 1.5 provided a graphic illustration of this methodology. The following is an outline of the major and supporting tasks in that illustration. The subsections provided below in this chapter follow this outline.

4.1.1   Input
4.1.2   Requirements Analysis
4.1.3   Requirements Validation
4.1.4   Functional Analysis
4.1.5   Functional Verification
4.1.6   Synthesis
4.1.7   Design Verification
4.1.8   Systems Analysis and Control
4.1.8.1 Trade-Off Studies
4.1.8.2 System/Cost-Effectiveness Analysis
4.1.8.3 Risk Management
4.1.8.4 Configuration Management
4.1.8.5 Interface Management
4.1.8.6 Data Management
4.1.8.7 Integrated Master Plan (IMP)
4.1.8.8 Technical Performance Measurement (TPM)
4.1.8.9 Technical Reviews

4.1.9   Output
4.1.9.1 Specifications and Baselines
4.1.9.2 Life Cycle Support Data

A general description of the SEM tasks is contained in Table 4.1 and illustrated in Figure 4.1. Each task consists of a number of supporting processes. Chapter 5 presents these supporting processes. The tasks should be individually tailored based on the complexity of the product requirements. The methodology ensures that all design activities are properly focused on stakeholder requirements and constraints. It uses design trade-off analyses to balance customer requirements against product development constraints. These SEM tasks were adopted by the U.S. Department of Defense. They follow Standard 1220 from the Institute of Electrical and Electronics Engineers (IEEE 2007).

Table 4.1   Summary Description of the SEM Tasks

Task	Description
Requirements analysis	This task clarifies and defines the problem statement in verifiable quantitative terms. Requirements and constraints are identified and documented in the system requirements baseline
Requirements validation	This task validates and resolves conflicting requirements and assumptions from all stakeholders
Functional analysis	This task is used to identify and develop all functional tasks required to execute the requirements baseline
Functional verification	This task validates the functional architecture to ensure that it meets the minimum requirements baseline objectives
Synthesis	This task includes all the design activities necessary to achieve specified functional architecture
Design verification	This task is used to validate the system architecture against both functional and requirements baseline documentation
Systems analysis	This problem-solving task is used throughout the SEM to make decision trade-offs. Supporting tasks include 1. Trade-off studies 2. System/cost-effectiveness analysis 3. Risk management 4. Configuration management 5. Interface management 6. Data management 7. IMP 8. TPM 9. Technical reviews
Control	This management task is used to coordinate, document, and track the sustainability engineering tasks

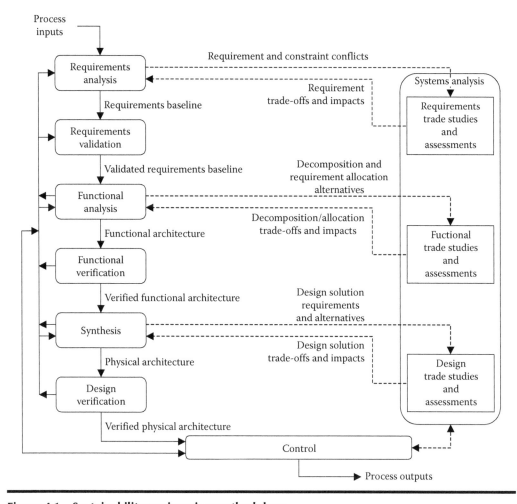

Figure 4.1   Sustainability engineering methodology.

The technical manager should ensure that all technical execution and management efforts are integrated in conformance with the SEM. Technical tasks, including task requirements from other standardization documents, are integrated to yield a single and complete task that focuses all activities on the common objective. Task planning and execution must implement and demonstrate multidisciplinary teamwork whereby all appropriate technical disciplines are applied to satisfy identified needs. The integrated technical effort is documented in a sustainability engineering tasking description (SETD), an IMP, and an integrated master schedule (IMS).

The project leader should be responsible for implementing the SEM, and other sustainability engineering activities, as documented in the SETD. The SETD represents the implementation of the methodology. It describes the methodology, as applied to the program, and the plan to execute that methodology. The project leader should also be responsible for maintaining and updating the SETD. The chief engineer should develop the technical elements of the IMP for the tasking and integrates these tasks with the IMP management elements. The technical leader develops an IMS and maintains it as detailed, time-dependent tasks evolve. Normally, an initial version of the IMS accompanies the project leader proposal. Then, the IMS is maintained during task execution to track technical progress and associated risks. The project leader extends the program work breakdown structure (PWBS) developed by the technical leader to the level necessary to complete task requirements. The project leader has the flexibility to extend the task PWBS below the reporting requirement to reflect how the work will be accomplished consistent with managing program risk.

4.1.1 Input

As was illustrated in Figure 1.5, the input includes customer needs, customer objectives, customer requirements, information supporting continued technical efforts, information supporting new or updated customer needs, technology base data, outputs from a previous phase, and program constraints. The project leader notifies the chief engineer that technical input information is needed, why it is needed, and when it is needed. The chief engineer will, in turn, inform the project leader what information can and cannot be provided. If the information will not be provided, then it must be generated using documented research, analysis, and assumptions.

When sufficient data to establish requirements are not available, technical objectives are used to provide a basis for defining and trading off relationships among need, urgency, costs, risks, and value. For some programs, technical objectives should be derived from key performance parameters in program requirements documentation. Technical objectives are identified to assist in converging on a system solution, focus on factors critical to success, and offer substantial capability payoffs for resources expended. The project leader identifies the needed technical objectives with a rationale, develops metrics and success criteria to ensure that increases in system capabilities are cost-effective when technical objectives are established for capabilities beyond requirements, and then uses critical technical objectives in TPM.

4.1.2 Requirements Analysis

The requirements analysis task is conducted to collect and translate all product requirements and constraints. This task is critical to defining the requirements early in product development. Clearly defined product requirements should prevent unnecessary analysis and design changes that could be expensive to implement once product development has progressed. Requirements analysis analyzes customer needs, objectives, and requirements in the context of customer missions, environments, and identified system characteristics to determine functional and performance requirements for each primary system function. Prior analyses are reviewed and updated to refine mission and environment definitions to support system definition. Requirements analysis is conducted iteratively with functional analysis to develop requirements that depend on additional product definition (e.g., other system items, performance requirements for identified functions) and verify that people, product, and task solutions (from synthesis) can satisfy customer requirements. Requirements analysis

1. Assists in refining customer objectives and requirements.
2. Defines initial performance objectives and refines them into requirements.
3. Identifies and defines constraints that limit solutions (e.g., missions and environments or adverse impacts on natural and human environments).
4. Defines functional and performance requirements based on customer provided measures of effectiveness (MOEs). When MOEs are not provided at the level of detail needed, the project leader develops and uses a set of MOEs relating to customer missions; environments; needs, requirements, and objectives; and design constraints.

Functional requirements that are identified in the requirements analysis, and as task inputs, are used as the top-level functions for functional analysis. Performance requirements are developed interactively across all identified functions based on system life cycle factors and characterized in terms of the degree of certainty of the estimate, the degree of criticality to product success, and relationship to other requirements. The requirements analysis task description is summarized in Table 4.2 and illustrated in Figure 4.2.

4.1.3 Requirements Validation

The requirements validation task is conducted to ensure that the requirements baseline is correct. The baseline is examined and compared to provide a "second check" before proceeding with the functional analysis task. The requirements validation task is summarized in Table 4.3 and illustrated in Figure 4.3. This task often will include the final resolution of conflicting requirements that are resolved through the systems analysis task.

4.1.4 Functional Analysis Task

The functional analysis task is conducted to translate requirements into system functions. The functional analysis task is summarized in Table 4.4 and illustrated in Figure 4.4. System functions are continuously decomposed until enough details have been developed for the specified system level (i.e., system, subsystem, cell, and equipment).

The project leader defines and integrates a functional architecture to the depth needed to support synthesis of solutions for people, products, and tasks and risk management. Functional analysis is conducted iteratively:

1. To define successively lower-level functions required to satisfy higher-level functional requirements and to define alternative sets of functional requirements.
2. With requirements analysis to define mission- and environment-driven performance parameters and to determine that higher-level requirements are satisfied.

Table 4.2   Requirements Analysis Task Description

Description	Tasks

• This task translates stakeholders, marketing, functional, performance, regulatory, and enterprise internal/external requirements and constraints into a requirements baseline
• This task establishes the requirements baseline
• The requirements baseline defines the system problems to be solved for the following areas:
  • Operational
  • Functional
  • Design
	1. Define customer expectations
	2. Define project and enterprise constraints
	3. Define life cycle task concepts
	4. Define human factors, manpower, personnel, training, human engineering, and safety requirements
	5. Define functional requirements
	6. Define performance requirements
	7. Define modes of operation
	8. Define TPMs
	9. Define design characteristics
	10. Establish requirements baseline

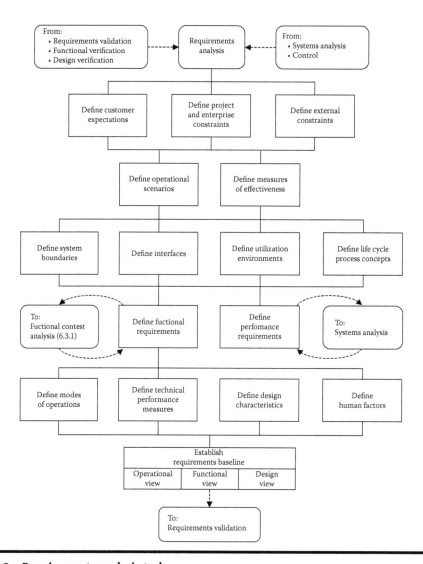

Figure 4.2   Requirements analysis task.

Table 4.3   Requirements Validation Task Description

Description	Tasks
• This task validates the requirements baseline to ensure that the baseline properly addresses the system requirements and constraints • The requirements baseline is an input to the functional analysis task	1. Compare requirements to customer exceptions 2. Compare requirements to enterprise and project constraints 3. Compare requirements to external constraints 4. Identify variances and conflicts 5. Establish validated requirements baseline

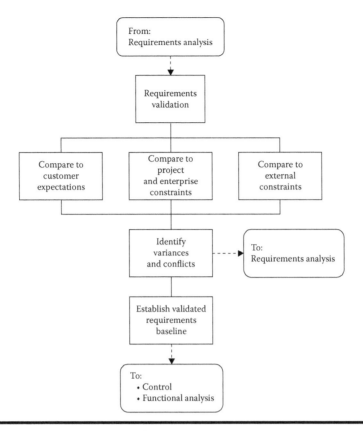

Figure 4.3 Requirements validation task.

Table 4.4 Functional Analysis Task Description

Description	Tasks
• This task takes the requirements baseline and decomposes the system functions into lower-level functions that must be performed by elements of the system design solutions • This task translates the requirements baseline into a functional architecture • The functional architecture defines the allocation of performance requirements to be solved during the synthesis task	1. Functional context analysis 2. Analyze functional behaviors 3. Define functional interfaces 4. Allocate performance requirements 5. Define external constraints 6. Define operational scenarios 7. Define MOEs 8. Define system boundaries 9. Define interfaces 10. Define utilization environments 11. Functional decomposition 12. Define subfunctions 13. Define subfunction states and modes

(Continued)

Table 4.4   Functional Analysis Task Description (*Continued*)

Description	Tasks
	14. Define functional time line
	15. Define data and control flows
	16. Define functional failure modes and effects
	17. Define safety-monitoring functions
	18. Establish functional architecture

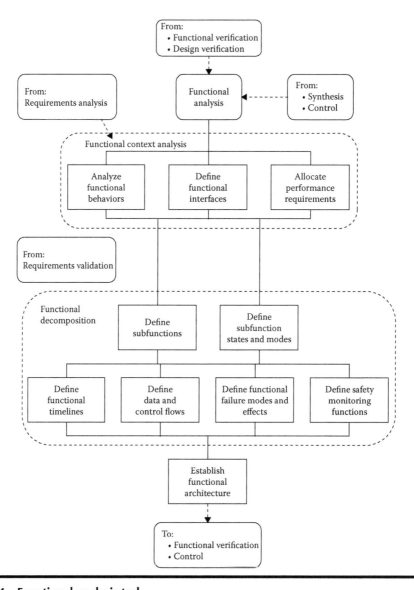

Figure 4.4   Functional analysis task.

3. To flow down performance requirements and design constraints.
4. With synthesis to define and refine feasible alternatives that meet requirements and to place derived requirements into the functional architecture.

Identified functional requirements are analyzed to determine the lower-level functions required to accomplish the parent requirement. All specified usage modes are included in the analysis. Functional requirements are arranged so that lower-level functional requirements are recognized as part of higher-level requirements. When time is critical to performance or sequencing of functions, a time line analysis is conducted. Functional requirements need to be logically sequenced, have input, have output, and have functional interface (internal and external) requirements defined. It should be traceable from beginning to end conditions and across interfaces.

The project leader successively (highest to lowest level) establishes performance requirements for each functional requirement and interface. Time requirements that are prerequisite for a function or set of functions are determined and allocated. The resulting set of requirements needs to be defined in measurable terms, applicable go/no-go criteria, and in sufficient detail for use as design criteria. Performance requirements need to be traceable throughout the functional architecture, through the analysis by which they were allocated, to the higher-level requirement they are intended to fulfill.

4.1.5 Functional Verification Task

The functional verification task is conducted to verify the functional baseline. The baseline is examined and compared to defined requirements to verify decomposition and traceability. The functional verification task is summarized in Table 4.5 and illustrated in Figure 4.5. This task resolves conflicting issues through the systems analysis task. Unresolved issues are sent to the functional analysis task for reassessment and correction. If the functional analysis task is unable to correct or resolve the issue, then the problem is sent back into the requirements analysis task. The problem is then reprocessed from the beginning of the SEM. The requirements baseline and functional architecture will be updated to capture any new changes. This method of processing unresolved problems illustrates the balancing and configuration control properties of the SEM. Often, new solutions are identified during the system analysis task, and these new solutions will require the same reprocessing cycle.

Table 4.5 Functional Verification Description

Description	Tasks
• This task verifies that the functional architecture satisfies the requirements baseline • This task ensures that all requirements and constraints can be traced to the functional architecture • The verified functional architecture is an input to the synthesis task	1. Define verification procedures 2. Conduct verification evaluation 3. Verify architecture completeness 4. Verify functional and performance measures 5. Verify satisfactory resolution of constraints 6. Identify variances and conflicts 7. Establish verified functional architecture

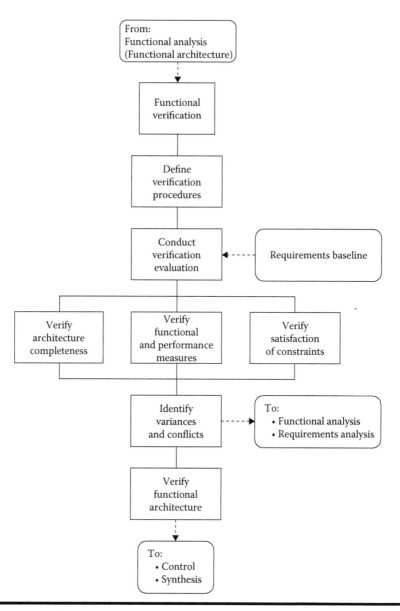

Figure 4.5   Functional verification task.

4.1.6 Synthesis Task

The synthesis task is summarized in Table 4.6 and illustrated in Figure 4.6. The synthesis task is conducted to develop design solutions from documented functional requirements. The task requires the design to be assessed from multiple points of view (e.g., safety issues, environmental concerns).

The project leader defines and designs solutions for each logical set of functional and performance requirements in the functional architecture and integrates them as a physical architecture. The project leader conducts synthesis iteratively with functional analysis to define a complete set

Table 4.6 Synthesis Description

Description	Tasks
• This task translates the functional architecture into a design architecture that provides an arrangement of system elements, the decomposition, and interfaces • The design architecture is developed for system breakdown structure (SBS) level (i.e., system, subsystem, component, or cells) • System analysis is used selectively to evaluate and manage risk, schedule, and performance impacts of alternative design options	1. Group and allocate functions 2. Identify design solution alternatives 3. Assess safety and environmental hazards 4. Assess technology requirements 5. Assess life cycle quality factors 6. Define design and performance characteristics 7. Define physical interfaces 8. Identify standardization opportunities 9. Identify make or buy alternatives 10. Develop models and prototypes 11. Assess failure modes, effects, and criticality 12. Assess testability needs 13. Assess design capability to evolve 14. Finalize design 15. Initiate evolutionary development 16. Produce integrated data package 17. Establish design architecture

of functional and performance requirements necessary for the level of design output required. Requirements analysis is used to verify that solution outputs can satisfy customer input requirements. In first defining the solution, the project leader

1. Determines the completeness of functional and performance requirements for the design and identifies derived requirements needed for completeness in terms of function and performance
2. Defines internal and external physical interfaces including required function and performance and ensures that requirements are integrated and verifiable across interfaces
3. Identifies critical parameters, then analyzes parameter variability and solution sensitivity to the variability
4. Defines people, product, and task alternatives interactively (including the concepts, techniques, and procedural data applicable to each of the primary system functions) as well as required allowances for tolerances and variability for those alternatives
5. Defines system and system element solutions to a level of detail that enables verification that required accomplishments have been met
6. Translates the architecture into a work breakdown structure, specification tree, specifications, and configuration baselines

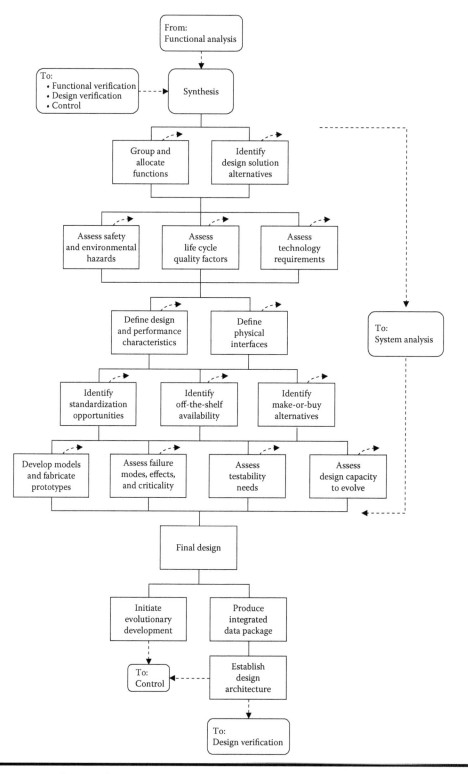

Figure 4.6   Synthesis task.

4.1.7 Design Verification

The outputs from synthesis need to describe the complete system, including the interfaces and relationships between internal and external items. For system design, the project leader:

1. Develops the information for establishing and updating applicable functional, allocated, and product baselines; develops system, task, and material specifications including commercial item descriptions; creates drawings and lists; generates interface control documentation; develops technical plans; defines life cycle resource requirements; creates procedural handbooks and instructional materials; and documents personnel task loading
2. Applies design simplicity concepts by evaluating alternatives with respect to factors such as ease of access, ready disassembly, common and noncomplex tools, decreased parts counts, modularity, producibility (e.g., assembly ready), standardization, and less demanding cognitive skills
3. Demonstrates design consistency with results from risk reduction efforts
4. Establishes and controls correlation among interdependent and functionally related elements

The project leader progressively verifies that product and process designs satisfy the requirements (including interfaces), from the lowest level of the current physical architecture up to the total system, and can be implemented. The design verification task is summarized in Table 4.7 and illustrated in Figure 4.7. The design verification task is conducted to ensure that the design architecture elements match the defined functional architecture elements. Design verification is a

Table 4.7 Design Verification Description

Description	Tasks
• This task verifies that the design architecture satisfies the functional architecture • This task ensures that all design architecture elements derived can be traced to the verified functional architecture • This task also ensures that the design architecture satisfies the requirements baseline	1. Select verification approach 2. Define inspection, analysis, demonstration, or test requirements 3. Define verification procedures 4. Establish verification environment 5. Conduct verification evaluation 6. Verify architecture completeness 7. Verify functional and performance measures 8. Verify satisfactory resolution of constraints 9. Identify variances and conflicts 10. Verify design architecture 11. Verify design architectures of the life cycle task 12. Verify system architecture 13. Establish specifications and configuration baselines 14. Develop SBS

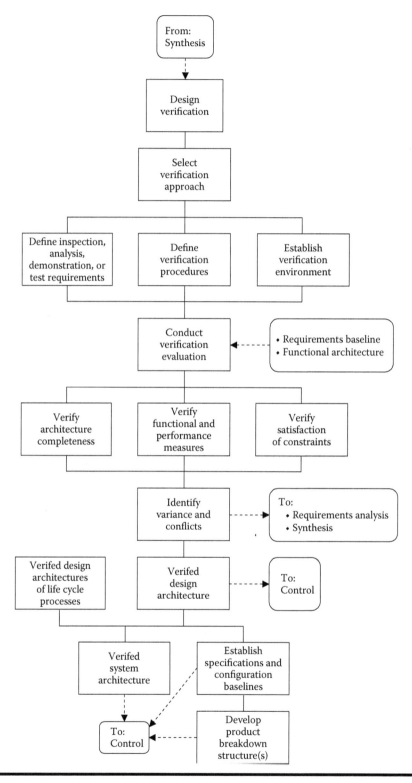

Figure 4.7   Design verification task.

critical task because an official system baseline is established when design verification is complete. Design baselines are major gating points in the program management of the SEM. Incomplete data or analyses could result in costly overruns and redesign activities in the next iteration of the SEM for lower-level product design.

4.1.8 Systems Analysis and Control

The SEM has two common tasks that are repeatedly used throughout the top-down flow of the overall SEM. The system analysis and control tasks are used to integrate and coordinate the overall SEM. These tasks are used to select and execute decisions derived from each of the previous tasks.

The system analysis task is used repetitively throughout the SEM to balance requirements and solution trade-offs. The system analysis task is described in Table 4.8 and is illustrated in Figure 4.8. The sustainability engineering tasks provide the necessary data inputs needed to provide a comprehensive evaluation of the available options.

The control task provides a very structured technical management approach to assess, manage, and document the SEM. The control task is described in Table 4.9 and is illustrated in Figure 4.9. The control task is the primary interface between the enterprise and the SEM. The control task is used to identify and allocate the resources needed to execute all SEM tasks. The SEM becomes a framework for the concurrent development and design of products and systems tasks. The SEM framework synchronizes the design of customer products and the required system tasks to produce the product, thereby creating a truly integrated product development environment.

Table 4.8   System Analysis Description

Description	Tasks
• This task is used to resolve conflicts among requirements analysis, functional analysis, and design synthesis • This task provides a rigorous quantitative basis to select a balanced set of requirements and design trade-offs and assessments • This task is part of a feedback network loop for the requirements analysis, functional analysis, and synthesis tasks	1. Assess requirement conflicts 2. Assess functional alternatives 3. Assess design alternatives 4. Identify risk factors 5. Define trade-off analysis scope 6. Select methodology and success criteria 7. Identify alternatives 8. Establish trade study environment 9. Conduct trade-off analysis 10. Analyze life cycle costs 11. Analyze system and cost-effectiveness 12. Analyze safety and environmental impacts 13. Quantify risk factors 14. Select risk handling options 15. Select alternative recommendations 16. Consider trade-offs and impacts 17. Design effectiveness assessment

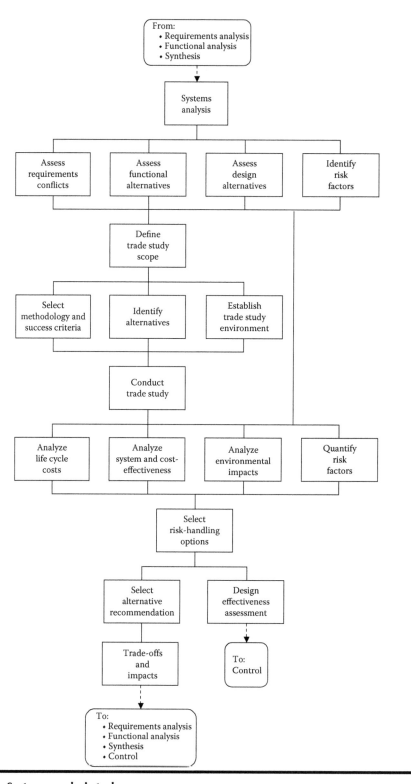

Figure 4.8 Systems analysis task.

Table 4.9   System Control Task Description

Description	Tasks
• This task identifies tasks required to manage and document the SEM • This task is the management interface to the enterprise • Resource allocation and planning is coordinated through this task	1. Technical management 2. Data management 3. Configuration management 4. Interface management 5. Risk management 6. Performance-based progress measurement 7. Track system analysis and test data 8. Track requirement and design changes 9. Track progress against project plans 10. Track product and task metrics 11. Update specifications and configuration baselines 12. Update requirements views and architectures 13. Update engineering plans 14. Update technical plans 15. Maintain technical databases

The project leader measures progress, evaluates alternatives, selects preferred alternatives, and documents data and decisions used and generated. Systems analyses are conducted including trade-off studies, effectiveness analyses and assessments, and design analyses to determine progress in satisfying technical requirements and program objectives and to provide a rigorous quantitative basis for performance, functional, and design requirements. Control mechanisms include risk management, configuration management, data management, and performance-based progress measurement including the IMP, TPM, and technical reviews. The project leader implements systems analysis and control to ensure that the following areas are covered thoroughly:

1. Decisions on solution alternatives are made only after evaluating the impact on system effectiveness, life cycle resources, risk, and customer requirements. The project leader identifies those alternatives that will provide improved system effectiveness or costs when compared with those based on program requirements.
2. Technical decisions and system-unique specification requirements are based on sustainability engineering outputs and documented results of decisions.
3. Traceability from task inputs to outputs is maintained, including changes in requirements.
4. Schedules for the development and delivery of products and tasks are mutually supportive.
5. Technical disciplines and disciplinary efforts are integrated into the sustainability engineering effort.
6. Impacts of customer requirements on resulting functional and performance requirements are examined for validity, consistency, desirability, and attainability with respect

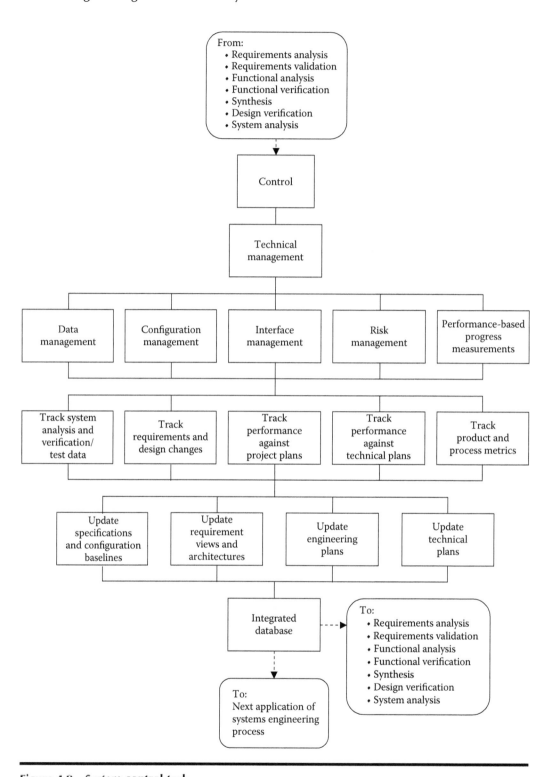

Figure 4.9 System control task.

to technology availability, physical and human resources, human performance capabilities, life cycle costs, schedule, risk, applicable statutes, designated hazardous material lists, and other identified constraints. This examination needs to either confirm existing requirements or determine that more appropriate requirements need to be defined for the system.

7. Product and process design requirements are directly traceable to the functional and performance requirements that the design requirements were designed to fulfill and vice versa.

4.1.8.1 Trade-Off Studies

Desirable and practical trade-offs among user requirements, technical objectives, design, program schedule, functional and performance requirements, and life cycle costs are identified and conducted. Trade-off studies are defined, conducted, and documented at the various levels of the functional or physical architecture in enough detail to support decision making. The level of detail of each study needs to be commensurate with cost, schedule, performance, and risk impacts.

4.1.8.1.1 Requirements Analysis Trade-Off Studies

The project leader needs to conduct requirements analysis trade-off studies to establish alternative performance and functional requirements to both resolve conflicts with and satisfy customer requirements.

4.1.8.1.2 Functional Analysis/Allocation Trade-Off Studies

The project leader needs to conduct trade-off studies within and across functions to

1. Support functional analyses and allocation of performance requirements
2. Define a preferred set of performance requirements that satisfy identified functional interfaces
3. Determine performance requirements for lower-level functions when higher-level performance and functional requirements cannot be readily resolved to a lower level
4. Evaluate alternative functional architectures

4.1.8.1.3 Synthesis Trade-Off Studies

The project leader needs to conduct synthesis trade-off studies to

1. Support decisions for new products and process developments versus non-developmental products and tasks.
2. Establish product configuration(s).
3. Assist in selecting system concepts, designs, and solutions (include people, parts, and materials availability).
4. Support materials selection and make-or-buy, process, rate, and location decisions
5. Examine proposed changes.
6. Examine alternative technologies to satisfy functional/design requirements including alternatives for moderate- to high-risk technologies.

7. Evaluate environmental and cost impacts of materials and tasks.
8. Evaluate alternative physical architectures to select preferred products and tasks.
9. Select standard components, techniques, services, and facilities that reduce system life cycle cost and meet system effectiveness requirements. Government and commercial databases should be used to provide historical information for evaluation decisions.

4.1.8.2 System/Cost-Effectiveness Analysis

The project leader should plan and implement a systems analysis effort as an integral part of the SEM. The project leader would develop, document, implement, control, and maintain a method to control analytic relationships and MOEs. Critical MOEs used for decision making are identified for TPM. System/cost-effectiveness assessments are used to support risk impact assessments. The project leader would analyze each primary system function to support the following areas.

1. Identification and definition of performance and functional requirements for the primary system functions to which system solutions must be responsive.
2. Selection of preferred product and process design requirements that satisfy performance and functional requirements.

4.1.8.3 Risk Management

The project leader should establish and implement a risk management program. Risks are assessed for the program, its products, its tasks (e.g., process variability), and its product/process interrelationships. Risk is also assessed for program-identified variations, uncertainties, and evolutions in system environments. The risk management program is conducted to

1. Identify potential sources of technical risk including critical parameters that can be risk drivers.
2. Quantify risks, including risk levels, and the impacts on cost (including life cycle costs), schedule, and performance. Include design, cost, and schedule uncertainties and sensitivity to program, product, and process assumptions.
3. Determine sensitivity of interrelated risks.
4. Determine alternative approaches to handle moderate and high risks.
5. Take actions to avoid, control, or assume each risk and adjust the SETD as necessary.
6. Ensure that risk factors are evaluated as a part of decision making including the selection of specification requirements and design and solution alternatives.

4.1.8.4 Configuration Management

The project leader should manage the configuration of identified products and tasks. This effort includes configuration:

1. Identification that involves selecting the documents to compose the baseline for the system, and configuration items (CIs) involved, and the numbers and other identifiers affixed to the items and the documents
2. Control, including the systematic proposal, justification, evaluation, coordination, approval, or disapproval of all proposed changes to the CIs after the baseline(s) for the CI has been established

3. Status accounting to include recording and reporting the information needed to manage CIs
4. Audits, including verification that the CI conforms to its current approved configuration documentation

4.1.8.5 Interface Management

The project leader should manage the internal interfaces within their program task responsibility. The project leader should support activities that are established to ensure that external interfaces are managed and controlled. The project leader delineates the design compatibility of external and internal engineering interfaces as interface requirements in the specifications. Interface controls should be established, coordinated, and maintained in terms of the interface requirements, documents, and drawings. All applicable project leader, vendor, and subcontractor contract items, furnished equipment, computer programs, facilities, and data should be included. Interfaces need to be controlled to ensure accountability and timely dissemination of changes.

4.1.8.6 Data Management

The project leader should establish and maintain an integrated data management system for the decision database to

1. Capture and organize all inputs as well as current, intermediate, and final outputs
2. Provide data correlation and traceability among requirements, designs, solutions, decisions, and rationale
3. Document engineering decisions, including procedures, methods, results, and analyses
4. Be responsive to established configuration management procedures
5. Function as a reference and support tool for the sustainability engineering effort
6. Make data available and shareable as called out in the task

4.1.8.7 Integrated Master Plan

The project leader should implement the IMP for top-level process control and progress measurement to ensure completion of required accomplishments; demonstrate progressive system and development achievements and maturity; ensure that integrated, multidisciplinary information is available for decision and demonstration events; provide an event-based, accomplishment-oriented framework for measuring progress; and demonstrate control of cost, schedule, and performance risks in satisfying accomplishments, requirements, and objectives. IMP accomplishments with supporting criteria are devised and structured to ensure that

1. Critical technical inputs and decision data are available for technical and program decision points, demonstrations, reviews, and other identified events.
2. Required progress and system maturity are demonstrated prior to continuing technical efforts dependent on that progress and maturity.
3. An IMP accomplishment is complete when all the associated criteria have been demonstrated.

4.1.8.8 Technical Performance Measurement

The project leader establishes and implements TPM to evaluate the adequacy of evolving solutions and to identify deficiencies that impact the ability of the system to satisfy a performance requirement. Actions taken to redress deficiencies depend on whether the technical parameter is a requirement or an objective. The TPM level of detail and documentation needs to be commensurate with the impact on cost, schedule, performance, and risk.

4.1.8.8.1 Implementation of TPM

The project leader determines the achievement-to-date for each technical parameter. Technical progress is assessed in terms of both allowed variation and the trend in achievement-to-date compared with the planned value profile. When progress in the technical effort supports revising the current estimate, a new profile and current estimate is developed. Risk assessments and analyses are updated to reflect changes in planned value profiles and current estimates and the impacts on related parameters.

4.1.8.8.2 TPM on Requirements

For identified deficiencies, analyses are performed to determine the cause(s) and to assess the impacts on higher-level parameters, interfaces, and system cost-effectiveness. Alternative recovery plans are developed with cost, schedule, performance, and risk impacts fully explored. For performance in excess of requirements, the marginal cost benefits and opportunities for reallocation of requirements and resources are assessed and an appropriate course of action is defined.

4.1.8.8.3 TPM on Objectives or Decision Criteria

The project leader should perform TPM on the objectives and decision criteria as defined in the SETD.

4.1.8.9 Technical Reviews

The project leader should plan and conduct the technical reviews necessary to demonstrate that required accomplishments have been successfully completed before proceeding beyond critical events and key program milestones. Technical reviews are conducted for the system- and program-identified CIs. Technical reviews occur at key events identified in the IMP when the project leader is ready to demonstrate completion of all the IMP accomplishments associated with the event as measured by the associated criteria.

4.1.8.9.1 Technical Review Content

System and CI technical reviews need to be integrated reviews that include all disciplines, all primary system functions, and all products and tasks of the item being reviewed. Reviews are structured within the total system context to assess the following areas.

1. Confirm that the effects of technical risk on cost, schedule, and performance have been addressed, as well as risk reduction measures, rationale, and assumptions made in quantifying the risks.
2. Demonstrate that the relationships, interactions, interdependencies, and interfaces between required items and externally interfacing items, system functions, subsystems, CIs, and system elements, as appropriate, have been addressed.

3. Ensure that performance, functional, design, cost, and schedule requirements and objectives and TPMs and technical plans are being tracked, are on schedule, and are achievable within existing programmatic constraints.
4. Confirm that continued development is warranted, and when it is not, that executable alternatives have been defined.

4.1.8.9.2 Response to Change

The project leader should define the total program impact of the identified changes to technical requirements with respect to cost, schedule, performance, and risk. Technical, cost, and schedule problems are diagnosed, and the impacts are determined. The impacts of collateral effects induced by solutions and solution alternatives on the technical program, including interfaces, are determined. The project leader informs the tasking activity of changes in cost, schedule, performance, and risk that impact the implementation (on time, within budget, meets requirements) of the program. The project leader processes all resulting changes to program requirements and configuration baselines in accordance with established change control procedures. The project leader ensures that supporting data are accessible to the tasking activity and documented in the decision database.

4.1.9 Output

Output from the sustainability engineering effort is acquisition phase dependent. The project leader should develop and implement a decision database that handles the following tasks:

1. Documents and organizes data used and generated by the sustainability engineering effort
2. Provides an audit trail of results and rationale from identified needs to verified solutions for traceability of requirements, designs, decisions, and solutions

4.1.9.1 Specifications and Baselines

The project leader should generate the required product and CI-unique documentation. General criteria are necessary in the following areas.

1. Documentation used to establish configuration baselines (functional, allocated, product) is developed progressively.
2. Specifications are formalized to establish configuration baselines commensurate with the program effort.
3. Configuration baselines are documented, controlled, and audited in accordance with program configuration management practices.
4. Essential requirements for tasks are included in item specifications.
5. Specification requirements need to be verifiable. Traceability to the verification criteria and methods is maintained.
6. The project leader presents the specifications for approval only when:
   a. The cost, schedule, and performance risks associated with the item and its tasks have been determined and the risk levels are acceptable.

    b. Item costs have been determined and those costs satisfy established design-to-cost targets or other prescribed affordability limits.

    c. Completeness and design attainability have been confirmed.

  7. System functional and CI development specifications need to be performance based.

4.1.9.2 Life Cycle Support Data

The project leader should identify, annotate, and track those elements in the decision database necessary for life cycle management of the system. They include

1. Product performance monitoring, analysis, problem identification, and corrective action recommendations.
2. Life cycle supportability analysis to identify operational and support resource requirements, to include any changes in requirements due to changes in the user community, missions, operational tempo, and operational strategy.
3. Identification of drivers of system readiness degraders and excessive total ownership cost (TOC) contributors. Analysis of alternative courses of actions and recommended actions to improve material readiness and/or reduce TOC.
4. Provide product support services to user organizations.

4.2 Concurrency and Integration

The SEM can be used to develop both the product and its production system concurrently. Product design data should be captured in the baseline drawing packages and architectures to allow this concurrency to happen. The product development management documentation and plans should include an IMS that contains concurrent product and technical tasks. The product development strategy should incorporate the enterprise strategy for proper program alignment and consistency in the SEM implementation.

The integration of internal and external environments with the SEM is important to a successful implementation. This integration is illustrated in Figure 4.10. Product design is executed in the project environment. The scheduling and management of all design activities should be synchronized with enterprise-wide tasks to ensure that cost-effective product and system development, production, operation, and sustentation are achieved.

The industry standard for enterprise sustainability engineering integration is ANSI/EIA Standard 632, "Processes for Engineering a System" (EIA 1999). This standard outlines the tasks required to implement and integrate a project-level SEM. The project-level SEM interfaces with the enterprise-level tasks through the control task. The enterprise standard is intended to be a high-level standard that "outlines essential technical activities and tasks deemed to be essential to the engineering of a system" (Martin 2000).

The SEM provides the basis to use an integrated product development team approach to concurrent engineering implementation. The integrated product development teams are multifunctional teams. The engineering should influence initial design concepts using techniques such as design for manufacturing and design for assembly. The engineers could design the manufacturing system while assisting in the design of the product. The SEM could be used in the integration and concurrent engineering decisions. The SEM could also be used to integrate enterprise-wide operations and implement continuous process improvements.

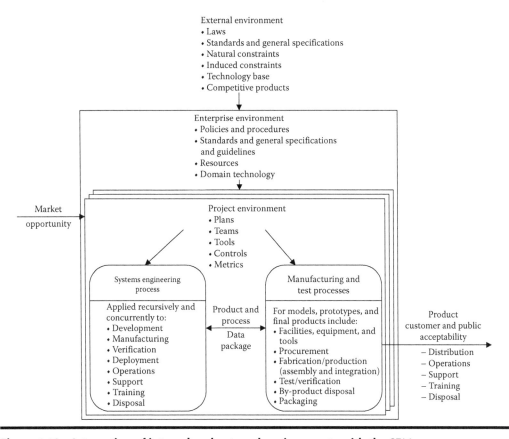

Figure 4.10 Integration of internal and external environments with the SEM.

References

Electronic Industries Alliance (EIA). 1999 (January). *Processes for Engineering a System.* Arlington, VA: Engineering Department, Government Electronics and Information Technology Association, American National Standards Institute ANSI/EIA Standard 632-1998.

Institute of Electrical and Electronics Engineers (IEEE). 2007. "Systems Engineering—Application and Management of the Systems Engineering Process." IEEE Standard 1220™-2005 (Revision of IEEE Standard 1220-1998).

Martin, J. N. 2000. "Process for Engineering a System: An Overview of the ANSI/EIA Standard and its Heritage." *Systems Engineering* 3 (1). 1–26.

Chapter 5

Sustainability Engineering Processes

Introduction

This chapter provides a set of processes and activities that could be used to help engineer a product or service for sustainability. Basically, these processes lay out the specific and detailed steps that are required to implement the sustainability engineering methodology that was presented in Chapter 4. The processes are applicable to the end products, as well as the development of the "enabling" products that provide life cycle support to the end products. They are applicable to any product or service development regardless of its place in the hierarchy of the product structure or the life cycle of the product. Five categories consisting of 36 processes are presented. Each process is designated SEP-*n* (sustainability engineering process-*n*):

1. Project framework processes
   1.1. Problem evaluation (SEP-1)
   1.2. Project definition (SEP-2)
   1.3. Supply (SEP-3)
   1.4. Acquisition (SEP-4)
   1.5. Work directives (SEP-5)
   1.6. Life cycle model management (SEP-6)
   1.7. Technical measurement (SEP-7)
   1.8. Project phase closure (SEP-8)
2. Solution design processes
   2.1. Stakeholder requirements definition (SEP-9)
   2.2. Requirements analysis (SEP-10)
   2.3. Logical architecture design (SEP-11)
   2.4. Physical architecture design (SEP-12)
   2.5. Document the design (SEP-13)

3. Technical management processes
    3.1. Technical planning (SEP-14)
    3.2. Process implementation strategy (SEP-15)
    3.3. Technical assessment (SEP-16)
    3.4. Sustainability engineering technical reviews (SEP-17)
    3.5. Requirements management (SEP-18)
    3.6. Interface management (SEP-19)
    3.7. Decision analysis and management (SEP-20)
    3.8. Data management (SEP-21)
    3.9. Configuration management (SEP-22)
    3.10. Quality management (SEP-23)
    3.11. Risk and opportunity management (SEP-24)
    3.12. Supplier performance management (SEP-25)
4. Product realization processes
    4.1. Design implementation (SEP-26)
    4.2. Product integration (SEP-27)
    4.3. Product verification (SEP-28)
    4.4. Product analysis (SEP-29)
    4.5. Testing (SEP-30)
    4.6. Product validation (SEP-31)
    4.7. Product readiness (SEP-32)
    4.8. Product deployment and transition (SEP-33)
5. System utilization processes
    5.1. Operations (SEP-34)
    5.2. Maintenance (SEP-35)
    5.3. Disposal (SEP-36)

Figure 5.1 is a graphical illustration of the 5 categories with the 36 processes.

Each process is detailed in the sections that follow using a format specifically designed to satisfy the following five objectives:

1. To utilize a standard format
2. To facilitate user efficiency in finding and using the process content
3. To capture all of the critical process components/attributes
4. To maximize the use of tables, graphics, and bullet lists so that adapting the sustainability engineering methodology, which was presented in Chapter 4, would be simplified.

The application of these 36 processes follows a product life cycle strategy. One example illustration is the "Vee" strategy, as depicted in Figure 5.2 and discussed in Chapter 3. The processes are applied recursively and iteratively to define the products (from the top down) and then to implement and transition the products (from the bottom up) to be deployed to the customer. Although the processes are presented sequentially, in practice many associated tasks are concurrent, are iterative, and have dependencies that lead to the alteration of previously established technical

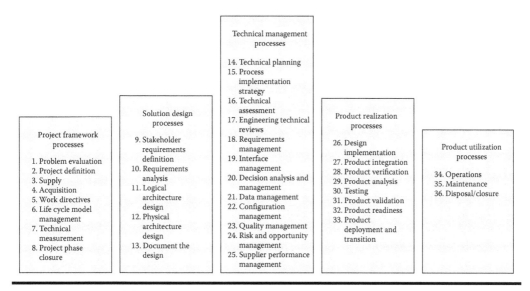

Figure 5.1   Sustainability engineering processes.

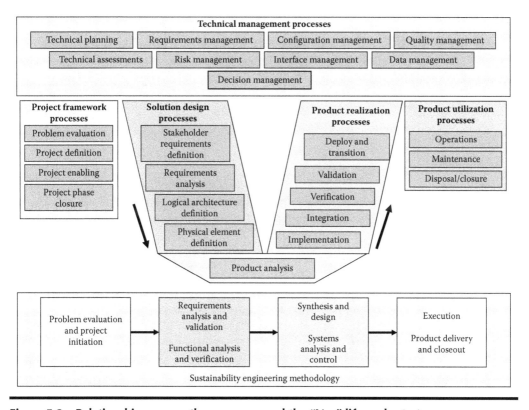

Figure 5.2   Relationships among the processes and the "Vee" life cycle strategy.

requirements. Figure 5.2 also shows the relationship of these processes to the sustainability engineering methodology.

5.1 Key Terms and Definitions

The definitions of some key terms are provided in this section.

5.1.1 Provider

The term *provider* means the provider of products, services, and other outputs of the processes in this chapter. A provider can be an individual or organizational element that enters into an agreement with the customer or the end user for the supply of a product or service. The provider delivers a product or service (end products, enabling products, or both) or a group of products or services to the customer. A supplier (external or internal to the customer's organization) can be a vendor that has a product that does not need development or a developer that must develop the desired product.

5.1.2 System

A system is a combination of interacting elements organized to achieve one or more stated purposes. A system may be considered a product or services provided by a provider. A complete system includes all of the associated equipment, facilities, material, computer programs, firmware, technical documentation, services, and personnel required for operations and support to the degree necessary for self-sufficient use in its intended environment.

5.1.3 Process

A process is a series of activities, actions, changes, or functions bringing about a result. A process uses resources to transform inputs into outputs. A process contains a set of components or characteristics, such as purpose, roles, inputs, outputs, and activity descriptions.

5.1.4 Integrity

Integrity is a concept of the consistency of actions, values, methods, measures, principles, expectations, and outcomes. Integrity represents the degree of confidence that a product meets the expectations of the customer in terms of project characteristics that define value to the stakeholder. The characteristics may include desired performance, risk, safety, security levels, reliability, and cost.

5.2 Project Framework Processes

There are eight processes associated with establishing the framework for the initiation of the project. The processes are listed in Figure 5.3.

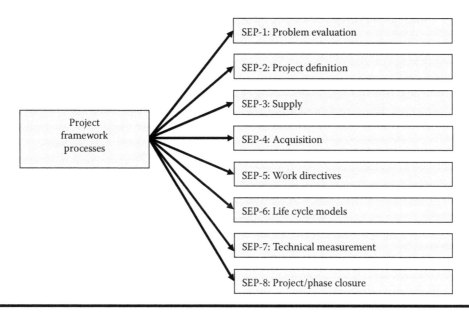

Figure 5.3 Project framework processes.

Table 5.1 Problem Evaluation Process (SEP-1)

Process purpose: The problem evaluation process defines the need or problem, performs the appropriate analysis to assess alternative solutions, and identifies one or more recommended solutions for filling the need. Additionally, the key quantitative metrics for the system, product, or service should be identified.

Roles and agents: • Project leader • Integrated product team

Inputs:	**Outputs:**
• A needs statement from a potential sponsor or stakeholder	• A description of the problem being assessed • Identification of the tasks to be completed • Identification of the capabilities required • An assessment of how well current capabilities meet the needs • As assessment of risks where gaps exist • Recommendations for nonmaterial solutions • Recommendations for material solutions

(Continued)

Table 5.1   Problem Evaluation Process (SEP-1) (*Continued*)

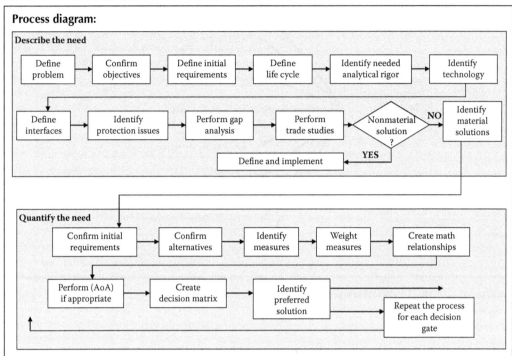

Figure 5.4   Problem evaluation process (SEP-1) diagram.

	Actions
Step	*Action*
1	**Describe the need:** Throughout the acquisition process, sustainability engineering provides the technical foundation for the acquisition program. Particularly in the early stages of an acquisition, sustainability engineering analysis is vital to the program's ability to assess appropriately the feasibility of addressing the user needs, the technology needs of potential solutions, and robust estimates of cost schedule and risk, all leading to predictable, disciplined acquisition.
1.1	**Action statement:** Define the problem to be assessed. **Notes and guidance:** List the objectives and effects to be achieved. 1. For small projects, this information is received from a potential project sponsor in discussions concerning a potential tasking. 2. For large projects, this is the initial step of a capabilities-based assessment (CBA).
1.2	**Action statement:** Define or confirm the initial objectives, desired results, and/or concept of operations (CONOPS).

Table 5.1   Problem Evaluation Process (SEP-1) (*Continued*)

Step	Action
1.3	**Action statement:** Define the initial product-level requirements and/or solution design concepts. **Notes and guidance:** 1. For small projects, this information is received from a potential project sponsor in discussions concerning initial detail on work products from a potential tasking. 2. For large projects, the CONOPS is confirmed and mission and functional threads are developed with users. Beginning with engineering analysis and product requirements definition, a strong foundation in the user CONOPS and mission threads is vital, and a working relationship with the users is essential to achieve a balance between user requirements (eventually documented in the capability development document [CDD]).
1.4	**Action statement:** Define the product/project life cycle. **Notes and guidance:** 1. For small projects, this information is received from a potential project sponsor in discussions concerning a directed life cycle model for a potential tasking. 2. For large projects, another important early step is to define the initial product life cycle to determine answers to the following questions: • What should the design time be? • What should the production time be? • How long should the product be expected to be in-service? • What quantities of products are expected to be procured? • How many product configurations are expected to be in use? • How do answers to these questions affect product support strategies?
1.5	**Action statement:** Determine the needed level of analytical rigor for this evaluation. **Notes and guidance:** 1. For small projects, this information is received from a potential project sponsor in discussions concerning the level of detail desired for a proposal concerning a potential tasking. 2. For large projects, there is a need to define the level of analytical rigor needed. Recent history indicates that these analyses suffer from too much detail and a lack of timeliness. The rigor associated with an analysis is a function of the uncertainties of the scenarios (futures) considered, the consequences of operational failure, and the complexity of the mission being assessed; for example, • Focus on the need for product replacement/recapitalization/evolution • Focus on the need for new information systems or evolution of existing products • Link the need for a comprehensive analysis to levels of new mission risk and uncertainty

(Continued)

Table 5.1   Problem Evaluation Process (SEP-1) (*Continued*)

Step	Action
1.6	**Action statement:** Identify any critical technology elements (CTEs). **Notes and guidance:** 1. For small projects, this information is received from a potential project sponsor in discussions concerning any technology issues for work products from a potential tasking. 2. For large projects, the CTEs are identified: The program team, as part of its product solutions analysis, conducts a technology maturity assessment of the hardware and software options with a focus on the CTEs.
1.7	**Action statement:** Define the product-level external interfaces, context, and boundaries. (systems-of-systems view) **Notes and guidance:** 1. For small projects, this information is received from a potential project sponsor in discussions concerning any interfaces between this project and any other projects relating to a potential tasking. 2. For large projects, external interfaces and interoperability have been determined: The team needs to understand the context in which potential products will be employed (based on CONOPS and mission/functional threads) and how this context affects the product acquisition, including programmatic and technical interfaces and interdependencies. A sustainability engineering focusing on external interfaces and interoperability facilitates an understanding of end-to-end product performance and its implication to the capability development document (CDD).
1.8	**Action statement:** Identify any critical product protection issues. **Notes and guidance:** 1. For small projects, this information is received from a potential project sponsor in discussions concerning information assurance, security, or information classification applicable to work products from a potential tasking. 2. For large projects, critical protection issues are identified: It is imperative that critical protection issues be identified in the initial stages of sustainability engineering so that their impact on possible product solutions and requirements can be addressed early and not compel a product redesign after substantial investment has been made.
1.9	**Action statement:** Perform capability gap analysis. **Notes and guidance:** 1. For small projects, this information is received from a potential project sponsor in discussions concerning existing capabilities and the new needed capabilities to assist with preparation of a proposal for a potential tasking.

Table 5.1   Problem Evaluation Process (SEP-1) (*Continued*)

Step	Action
	2. For large projects, a study must then perform the operational assessment of the current and programmed military force to provide the required capabilities, identifying capability gaps and potential military force redundancies for each scenario. Finally, the analysis assesses the potential operational risk associated with each gap.
	a. The gaps must be described in terms of the scenarios assessed and the impact on achieving the relevant military objectives. It is likely that the gaps will be inconsistent across scenarios, so it is essential to link the gaps to their operational context.
	b. The capability gaps are assessed in terms of the risk to mission (the ability to achieve the objectives of the scenario), the risk to the military force (the potential losses due to the capability gap), and other important considerations, such as resourcing risks and its effect on allies. The conditions and standards developed for the associated tasks provide the basis for the assessments.
	c. Using the programmed military force and doctrinal approaches, the capability gaps can be characterized as to whether they are due to the following:
	i. Proficiency (ability to achieve the relevant effect in particular conditions)
	ii. Sufficiency (ability to achieve the effect but inability to bring the needed military force to bear due to force shortages or other commitments)
	iii. Lack of existing capability
	iv. Need for replacement due to aging of an existing capability
	v. Policy limitations (inability to use the military force as needed due to policy constraints).
	d. Since the validation authority will ultimately decide which gaps are pervasive or important enough to develop solutions, the gaps must be directly linked to operational situations and consequences of failing to meet objectives. The analysis must explain the methodology for determining the priorities of the gaps and ensure that the linkage to strategic priorities is clear. While the analysis must present its conclusions concisely, it must also document the significant driving factors behind the recommended priorities to give the validation authority the information they need if they choose to make adjustments.
	e. The capability gap is assessed based on its impact in several areas: ability to achieve the strategic objectives; operational timelines; resources; unanticipated requirements; the military force provider resourcing; and component functions, the military force management, and institutional capacity.

(*Continued*)

Table 5.1   Problem Evaluation Process (SEP-1) (*Continued*)

Step	Action
1.10	**Action statement:** Perform initial alternative trade-off studies. **Notes and guidance:** 1. For small projects, this information is received from a potential project sponsor concerning the need for evaluating alternative solutions concerning a proposal for a potential tasking. SP-17 may be useful in performing and documenting initial trade-off studies. 2. For large projects a formal analysis of alternatives (AoA) is required. General AoA guidance and a methodology are shown in Figure 5.5. **Figure 5.5 Analysis of alternatives methodology.**
1.11	**Action statement:** Identify any nonmaterial solutions. **Notes and guidance:** 1. Not applicable for small projects. This action concerns changing the way current systems or products are utilized to solve gaps in current capability. 2. For large projects, the analysis team will determine if a nonmaterial approach can mitigate any of the gaps. Common nonmaterial approaches are as follows: a. Alternative doctrinal approaches and alternative CONOPS. Investigating alternative CONOPS is a requirement. The baseline assessment should only consider doctrinal CONOPS, but the nonmaterial approach assessment should consider doctrinal alternatives, particularly those documented in an approved joint concept. b. Policy alternatives. When considering policy alternatives, the CBA must document which policies are contributing to capability gaps and under which circumstances. A policy change that allows new applications of existing capabilities or modifies the military force posture to increase deterrence is always of interest and should be considered. Policy alternatives requiring interagency or multinational cooperation must contain support for their feasibility since the DoD cannot act unilaterally in these cases.

The figure contents (Figure 5.5):

AOA plan

Exploratory
- Define and scope problem
- Define decision criteria
- Characterize alternatives
- Screen against criteria

Effectiveness
- Model scenarios and requirements
- Evaluate alternatives

Supportability
- Reliability
- Maintainability
- Material readiness

Affordability
- Acquisition cost
- Life cycle cost
- Risk and uncertainty

Synthesis
- Cost and effectiveness

AOA report

Table 5.1 Problem Evaluation Process (SEP-1) (*Continued*)

Step	Action
	c. Organizational and personnel alternatives. A CBA cannot redesign the military force, but it can suggest ways in which certain functions can be strengthened to eliminate gaps and point out mismatches between force availability and force needs. Finally, note that operating the programmed military force under substantially different organizational or personnel assumptions will generally require the development of an alternative CONOPS to support those assumptions.
1.12	**Action statement:** Define material solution recommendations. **Notes and guidance:** 1. For small projects, this information is received from a potential project sponsor in discussions concerning material work products from a potential tasking. 2. For large projects, the final step in the analysis is to offer recommendations for materiel approaches. Materiel initiatives tend to fall into three broad types (listed in terms of fielding uncertainty from low to high): a. Development and fielding of information systems (or similar technologies with high obsolescence rates) or evolution of the capabilities of existing information systems b. Evolution of existing products with significant capability improvement (this may include replacing an existing product with a newer more capable product, or simple recapitalization) c. Breakout products that differ significantly in form, function, operation, and capabilities from existing products and offer significant improvement over current capabilities or transform how the mission is accomplished.
2	**Quantify the need:** This is an often underplayed step in early needs analysis. Without accurate and relevant quantitative criteria to measure the different alternatives, teams cannot select an alternative that will meet the often conflicting needs and requirements of the stakeholders. The key to defining the most cost-effective performance of a product may very well lie in the quantitative aspects of the analysis undertaken to define these decision criteria.
2.1	**Action statement:** Confirm the initial view of the product or project requirements. **Notes and guidance:** 1. For small projects, this step confirms information received from a potential project sponsor in discussions concerning measures of success for the project to be used in a proposal for a potential tasking. 2. For large projects, this involves identification of feasible methods for modeling the operational scenarios from activities 1.2 and 1.3 above.

(*Continued*)

Table 5.1   Problem Evaluation Process (SEP-1) (*Continued*)

Step	Action
2.2	**Action statement:** Confirm the baseline and other alternatives. **Notes and guidance:** 1. For small projects, this step confirms information received from a potential project sponsor in discussions concerning any alternative being considered for work products from a potential tasking. 2. For large projects, this step focuses on converting user needs into quantifiable and measurable technical requirements.
2.3	**Action statement:** Define the criteria, the Key performance parameters (KPPs), that distinguish one alternative from another. **Notes and guidance:** 1. For small projects, this information is received from a potential project sponsor in discussions concerning what metrics will be used in measuring degrees of success (cost, schedule, and product quality) associated with a potential tasking. 2. For large projects, KPPs are those product attributes considered most critical or essential for an effective military capability. The CDD and the capability production document (CPD) must contain sufficient KPPs to capture the minimum operational effectiveness, suitability, and sustainment attributes needed to achieve the overall desired capabilities for the product or system of systems (SoSs) during the applicable increment. The analysis identifies the attributes that contribute most significantly to the desired operational capability in threshold-objective format. Whenever possible, attributes should be stated in terms that reflect the range of military operations that the capabilities must support and the joint operational environment intended for the product (family of systems [FoSs] or SoSs). There are compatibility and interoperability attributes (e.g., databases, fuel, transportability, ammunition) that might need to be identified for a capability to ensure its effectiveness. These statements will guide the acquisition community in making trade-off decisions between the threshold and objective values of the stated attributes. Because testing and evaluation throughout a product's life cycle will assess the ability of the product(s) to meet the production threshold values as defined by the KPPs, key system attributes (KSAs), and other performance attributes, these attributes must be measurable and testable.
2.4	**Action statement:** Identify the relative importance (weight) of each parameter, as well as thresholds and targets. **Notes and guidance:** For small projects, this information is received from a potential project sponsor in discussions concerning relative importance of factors such as cost, schedule, and product quality related to work products associated with a potential tasking. Obtain a clear mutual understanding of how project success is viewed.

Table 5.1   Problem Evaluation Process (SEP-1) (*Continued*)

Step	Action
2.4	For large projects, the methodology for developing KPPs includes identification of their relative importance, as shown in the following steps: 1. List required capabilities for each mission or function, as described in the proposed CDD or CPD. This review should include all requirements that the product described in the CDD/CPD is projected to meet, including those related to other products in an FoS or SoS context. It shall also include all relevant performance metrics identified in an initial capabilities document (ICD) for which the CDD/CPD is providing a capability. 2. Prioritize these capabilities. 3. Review the list of performance attributes associated with each of the functions for applicability. Compile a list of the potential attributes as a starting point, and include any other performance attributes that are essential to the delivery of the capability. Cross walk this list with the capabilities in step 2 to assist in identifying potential performance attributes to be considered for designation as KPPs. 4. For each mission or function, build at least one measurable performance attribute using the list from step 3 as a starting point. 5. Determine the attributes that are most critical or essential to the product and designate them as KPPs. (Note: A KPP need not be created for all missions and functions for the product. In contrast, certain missions and functions may require two or more KPPs.) 6. Document how the KPPs are responsive to the capability performance attributes identified in the ICDs in support of the mission outcomes and associated desired effects.
2.5	**Action statement:** Create the math (quantitative relationships) to identify the objective function (OF). **Notes and guidance:** The OF is • A single number • Mathematical combination of KPPs • Negotiated with stakeholders to determine what makes one alternative better than others The OF is used for the following: • Evaluation of viable alternatives • A customer criterion to define success For small projects, this is the single number that sums up how the project will be measured in relation to project progress and project success relative to other similar projects.

(*Continued*)

Table 5.1   Problem Evaluation Process (SEP-1) (*Continued*)

Step	Action
	For large projects, this involves selection or creation of models for simulations and life cycle cost calculations. A KPP will normally be a mathematical rollup of a number of key system attributes (KSAs). The KPPs and KSAs will likely be traded off to deliver the overall performance required by adjusting their values in models and simulations.
2.6	**Action statement:** Perform an analysis of alternatives (AoA) if appropriate. **Notes and guidance:** 1. Not applicable for small projects. 2. For large projects, refine the values associated with the KPPs and other measures by rerunning AoA models to search for an optimum balance in identification of targets and thresholds for KPPs.
2.7	**Action statement:** Identify the preferred alternative. **Notes and guidance:** For small projects, the sponsor is likely to already have a preferred solution. If alternatives were analyzed as described in activities above, the preferred alternative should be justified by the analysis. For large projects, the preferred alternative will be justified by the structured analysis, as discussed in activities above. However, the key is in creating agreement on the part of all relevant stakeholders. Involvement of these stakeholders at each step of the process is essential to obtaining agreement on the selection of the preferred alternative. This idea is shown in Figure 5.6. 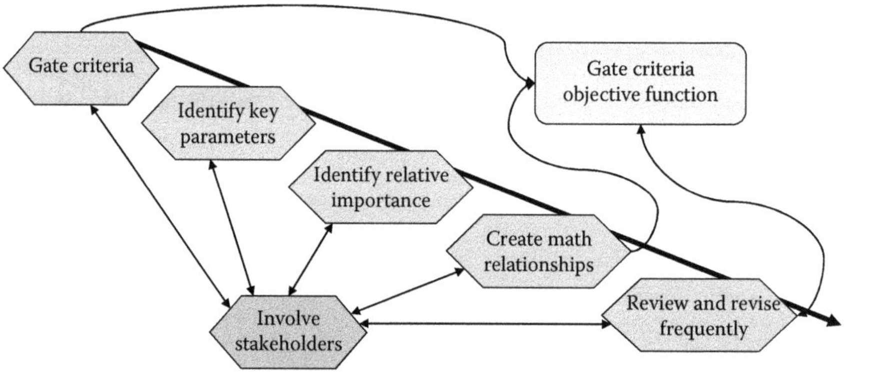 **Figure 5.6 Stakeholder agreement.**
2.8	**Action statement:** Repeat the process as necessary at each decision gate. **Notes and guidance:** For small project, a simplified or highly tailored version of this process may be applicable for presentation at life cycle gate reviews.

Table 5.1   Problem Evaluation Process (SEP-1) (*Continued*)

Step	Action
	For large projects, an updated set of KPPs and supporting analysis data to support each milestone decision is a solid requirement.
	Process Task Outcomes

This table provides an informative set of representative tasks and their expected outcomes.

- For small projects, considerable detailed information will have been gathered for preparation of a winning project proposal, approval for project startup, and development of the project technical plan.

- For large programs, the results of this early sustainability engineering analysis will provide critical technical information to the program planning effort for the technology development phase, particularly in determining the plan for critical technology elements risk reduction, prototyping, and competing preliminary designs in terms of how much scope, for what objective, and performed by whom (industry or government). This technical planning is an essential element of the technology development strategy (TDS) and is in a sense the program's initial acquisition strategy. The technical planning is the basis cost estimation, and program objective memorandum inputs are prepared. Technical planning outputs are used in developing the sustainability engineering management plan, TDS, and test and evaluation strategy and requests for proposals.

Table 5.2   Project Definition Process (SEP-2)

Process purpose: The organization shall implement the following activities and tasks in accordance with applicable organizational policies and procedures with respect to the project definition process. This process commits the investment of adequate organization funding and resources, and it sanctions the authorities needed to establish the selected projects.

Roles and agents:

- Project leader
- Sustainability engineer
- Integrated product team

Inputs:	Outputs:
• Project proposal	• Initial project plan • Approval to initiate execution of project plans

(*Continued*)

Table 5.2   Project Definition Process (SEP-2) (*Continued*)

Process diagram:

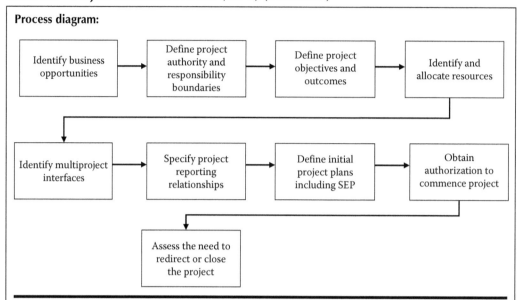

Figure 5.7   Project definition process.

Step	Action
1	**Action statement:** Identify, prioritize, select, and establish new business opportunities, ventures, or undertakings consistent with the business strategy and action plans of the organization. **Notes and guidance:** Prioritize the projects to be started and establish criteria and thresholds to determine which projects will be executed.
2	**Action statement:** Define projects, accountabilities, and authorities. **Notes and guidance:** N/A (none available)
3	**Action statement:** Identify the expected vision, goals, objectives, and outcomes of the project. **Notes and guidance:** N/A
4	**Action statement:** Identify and allocate resources for the achievement of project goals, objectives, and the work breakdown structure. **Notes and guidance:** N/A
5	**Action statement:** Identify any multiproject interfaces and dependencies that must be managed or supported by the project. **Notes and guidance:** This includes the use of enabling products used by more than one project and the use of common product elements by more than one project.
6	**Action statement:** Specify the project reporting requirements, and review milestones that will govern the execution of the project. **Notes and guidance:** N/A

Table 5.2   Project Definition Process (SEP-2) *(Continued)*

Step	Action
7	**Action statement:** Define initial project authorities, accountabilities, and plans, including technical plans. **Notes and guidance:** N/A
8	**Action statement:** Obtain authorization for the project to commence execution of approved project plans, including the technical plans. **Notes and guidance:** N/A
9	**Action statement:** Assess the need to redirect or close the project when either of three conditions exist: 1. Act to continue or redirect projects that are satisfactorily progressing or can be expected to progress satisfactorily by appropriate redirection. 2. Where agreements permit, act to cancel or suspend projects whose disadvantages or risks to the organization outweigh the benefits of continued investments. 3. After completion of the agreement for products and services, act to close the project per organizational policies and procedures and the agreement. **Notes and guidance:** For additional detail on project closure, refer to the project phase closure process.
Process Task Outcomes	

As a result of the successful implementation of the project definition management process:

1. Business venture opportunities, investments, or necessities are qualified, prioritized, and selected.
2. Resources and budgets for each project are identified and allocated.
3. Project management accountability and authorities are defined.
4. Projects meeting agreement and stakeholder requirements are sustained.
5. Projects not meeting agreement or stakeholder requirements are redirected or terminated.

Table 5.3   Supply Process (SEP-3)

Process purpose: The acquisition and supply processes are used by a developer to arrive at an agreement with another party to accomplish specific work and to deliver required products, or with another party or parties to have work done to obtain desired products. The parties can either be inside the developer's own enterprise (another project, functional organization, or project team) or be in a different enterprise. The acquisition and supply processes can be initiated as a result of a project go-ahead or approval decision or by the receipt of an acquisition request, offer, or directive. A project go-ahead can be given within an enterprise as a result of a market-needs analysis, a technology breakthrough, a perceived market opportunity, a customer requirement, an internal project directive, or a similar stimulus. Although a project or development effort can be initiated by casual means, an agreement is, nevertheless, useful to ensure that all parties involved understand the purpose, goals, and expectations of the work.

(Continued)

Table 5.3   Supply Process (SEP-3) (*Continued*)

An agreement can be between enterprises and between organizational elements within an enterprise, to include between projects, between projects and functional units, and between units within a project. The agreement within an enterprise can take the form of a work directive, work package, work authorization, or project memorandum of agreement. Agreements between enterprises can take the form of a formal contract for the delivery of a product or a memorandum of agreement that establishes the working relationship between two or more enterprises on a common project.

Regardless of the form or purpose of the agreement, certain information should be included, for example:

   a. Work to be performed

   b. Cost and schedule constraints

   c. Concept of operations

   d. Requirements to be satisfied, including known functional, performance, and interface requirements, attributes, and characteristics

   e. Product and data to be delivered

   f. Information pertaining to the cost, schedule, planning, delivery information, training and user manual, product structure, packaging and handling instructions, or installation instructions

   g. Appropriate technical plans

   h. Applicable financial structure, management, and authority provisions

   i. Exit criteria for relevant enterprise-based life cycle phases

   j. Identification of applicable engineering life cycle phases

   k. Required technical reviews

A developer can be developing a product without any contractual relationship to the user or customer (e.g., commercial product development). However, much of the information above must be available to the developing organization in order to proceed.

The customer can be either one of the following:

   a. Internal to enterprise—for example, another project, marketing organization, parent project of a product team, the project team itself, executive manager, or supervisor

   b. External to enterprise—for example, procurement agency, prime contractor, another developer, buyer, customer, end user, owner, or purchaser

The supplier can be either one of the following:

   a. Internal to enterprise—for example, another project, functional organization, or product team

   b. External to enterprise—for example, another developer, prime contractor, producer, seller, subcontractor, or vendor

---

Roles and agents:

   • Contracts

   • Business development

   • Customer

Table 5.3 Supply Process (SEP-3) (*Continued*)

- Legal
- Security
- Sustainability engineering
- Logistics
- Manufacturing
- Technical writing
- Specialty engineering

Inputs:	Outputs: (all outputs are to be archived)
• Acquisition strategy • Solicitation (request for proposal [RFP], statement of work, or statement of objectives) • Customer offer • Requests for clarification • Customer agreements, signed • Task work statements, signed	• Supplier proposal • Supplier agreement, signed • End products • Enabling products

Process diagram:

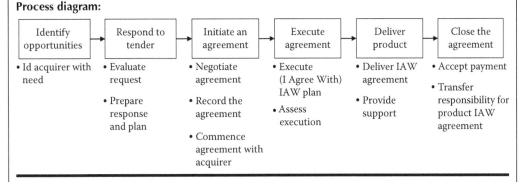

Figure 5.8 Supply process.

Step	Action
Identify opportunities	
1	**Action statement:** Identify a customer. Determine the existence and identity of a customer who has or who represents an organization having a need for a product or service. **Notes and guidance:** For a product or service developed for consumers, an agent, which is a business development or marketing function within the supplier organization, may represent the supplier.

(*Continued*)

Table 5.3   Supply Process (SEP-3) (*Continued*)

Step	Action
2	**Action statement:** Assess the acquisition request. Assess the acquisition request, offer, or directive to determine the capability to meet the acquisition documentation requirements.
	Notes and guidance: The supplier develops a business strategy and surveys the marketplace for business opportunities (Commerce Business Daily, Federal Business Opportunities [FedBizOpps] announcements, Sources Sought, etc.). The supplier obtains a RFP or quotation and allocates resources to review the RFP or quotation. For larger projects, the supplier would put together a team of personnel from various disciplines, such as engineering, financial, logistics, and management. For some efforts, a field activity may be used. In the event that another military service is used, a military interservice procurement request would be used. The team would review the RFP, determine what the requirements are, and then come up with their solution to meet all the requirements of the proposal. Some of the items that may be included in their proposal would include the following: 1. Executive overview 2. Technical approach 3. Sustainability engineering 4. Production ability 5. Cost 6. Schedule 7. Performance 8. Specifications 9. Training 10. Program management 11. Support equipment (common and peculiar) 12. Technology risks 13. Human systems integration 14. Packaging and handling 15. Technical data 16. Configuration management approach 17. Work breakdown structure (WBS) 18. Site activation 19. Industrial facilities 20. Initial spares and initial repair parts
3	**Action statement:** Prepare a response. Prepare a response that satisfies the request. **Notes and guidance:** N/A

Table 5.3 Supply Process (SEP-3) (*Continued*)

Step	Action
4	**Action statement:** Establish the agreement. Establish a satisfactory agreement within legal, regulatory, organizational, and project bounds.
	Notes and guidance: The supplier determines if the capability to meet the acquisition requirements exists, allocates resources needed to prepare the proposal/quotation, prepares proposal/quotation, submits (or presents orally) proposal/quotation, responds to proposal/quotation clarification questions from the customer, and modifies the proposal in response to the requests. The established agreement would also delineate any subcontracting that the prime contractor may enter into and any flow-down requirements.
	This agreement may range in formality from a written contract to a verbal understanding. Negotiate the differences, where applicable, between the acquisition request or tasking statement and the capability expressed in the response. The supplier confirms that the requirements, delivery milestones, and acceptance conditions are achievable, that exception handling and change control procedures and payment schedules are acceptable, and that they establish a basis for executing the agreement without unnecessary risks. In the agreement or project plans, the supplier should define or select a life cycle model appropriate to the scope, magnitude, and complexity of the project. Ideally, this is performed by using an organizationally defined life cycle model.
5	**Action statement:** Record the agreement. Record the established agreement in the form appropriate to the effort.
	Notes and guidance: N/A
6	**Action statement:** Commence the agreement with customer.
	Notes and guidance: N/A
7	**Action statement:** Execute the agreement. Execute the agreement in accordance with the supplier's established project plans and in accordance with the agreement.
	Notes and guidance: • A supplier may adopt, or agree to use, customer processes. • Communication with the customer is maintained throughout the execution of the agreement.
8	**Action statement:** Assess the execution of the agreement.
	Notes and guidance: N/A
9	**Action statement:** Deliver the products. Deliver the products and other deliverables as specified in the established agreement.
	Notes and guidance: Supplier performs work required by the contract, while customer monitors supplier's performance and compliance with requirements. Supplier develops and documents the final product design. Supplier manufactures and tests product. Supplier develops required product documentation and other technical data as delineated in the supplier signed agreement.

(Continued)

Table 5.3   Supply Process (SEP-3) (*Continued*)

Step	Action
10	**Action statement:** Provide support. Provide assistance to the customer in support of the delivered product or service in accordance with the agreement criteria. **Notes and guidance:** N/A
11	**Action statement:** Accept payment. Accept and acknowledge payment or other consideration. **Notes and guidance:** N/A
12	**Action statement:** Transfer product responsibility. Transfer the responsibility for the product or service to the customer, or other party, as directed by the agreement to obtain closure of the agreement. **Notes and guidance:** N/A

Measures:

- Effectiveness: Accomplishment of the process objective, within cost, schedule, and quality goals
- Efficiency: Time to accomplish this process, including wait time, and rework

Process Task Outcomes	
Representative Tasks	*Process Task Outcomes*
Assess acquisition request, offer, or directive	The capability of the enterprise, organization, project, or team to provide a product, or portion thereof, that meets acquisition document requirements within the stated constraints and the enterprise strategic plan and business strategy, or within the project plan and constraints, or within the team charter, as applicable, is determined. It includes, as appropriate, the following: 1. Engineering and other applicable technical and project plans that allow determination of engineering and management tasks, costs, and schedules, resource requirements, and technical capabilities and capacities (invoke applicable planning process tasks) 2. Decision whether to work with the customer to provide the desired product, or a portion thereof, based on establishment enterprise criteria or on project or team capability 3. Resolution of added or changed requirements and areas of concern 4. Preparation and submission of an appropriate technical and cost response in accordance with acquisition requirements, enterprise business strategy, and enterprise policies and procedures or with project plans, policies, and directives

Table 5.3   Supply Process (SEP-3) (*Continued*)

Representative Tasks	Process Task Outcomes
Negotiate agreement	A satisfactory agreement is established based on the bounds determined by, as applicable, the following: 1. Applicable legal, regulatory, policies, procedures, and practices that will affect negotiation strategy or conduct 2. The type of agreement to be negotiated 3. Negotiation strategy 4. Conditions identified from the plans for the procurement work effort that could affect negotiations and agreement performance 5. Constraints identified from the plans for the procurement work effort that could affect negotiations and agreement performance
Record agreement	Established agreement is captured in a form and medium appropriate to the effort.
Implement agreement	A project established and processes (including replanning, as necessary) activated to complete the requirements of the agreement.
Deliver product and other deliverables per agreement	Agreement requirements satisfied by the delivery of required products and other deliverables in accordance with agreement instructions.

Table 5.4   Acquisition Processes (SEP-4)

Process purpose: The provider performs the acquisition process to obtain a product or service from a supplier in accordance with the customer's requirements. The supplier is typically thought of as a prime contractor, but it may be a team within the military or other government activity. The acquisition may be competitive or sole source. There are different procedures that must be followed depending on whether the acquisition is competitive or sole source.

For major weapon systems, the acquisition process initiates with Milestone A within the service or field commander-in-chief's ongoing mission area need analysis effort, which may result in an initial capabilities document (ICD)—formerly mission needs statement (MNS). By certifying a mission need, the ICD may result in a concept decision to explore material solutions. The program then enters the concept refinement phase, during which product alternatives are explored. The next phase occurs after Milestone A and is known as technology development (formerly component advanced development). The preferred

(*Continued*)

Table 5.4 Acquisition Processes (SEP-4) (*Continued*)

product concept is defined by a set of product performance requirements, and the technology is demonstrated to show that any significant technical and acquisition risk areas identified have been brought under sufficient control to warrant entering the next program phase. Program initiation begins at Milestone B, which is the beginning of the system development and demonstration (SDD) Phase. The SDD phase includes the system integration and the system demonstration work efforts, which are separated by the programmatic design readiness review, and the product critical design review. The preliminary design and detailed designs are completed during the system integration work effort, and tests are performed during the system demonstration work effort.

Following Milestone C, the product enters the production and deployment phase, during which low-rate initial production and full-rate production takes place. After initial operating capability occurs, the operations and support phase is entered and modifications and product improvements are usually implemented. At the end of the product service life, it is disposed of in accordance with applicable classified and environmental laws, instructions, regulations, and directives. Disposal activities also include recycling, material recovery, salvage reuse, and disposal of by-products from development and production.

At the conclusion of the first three phases, the requirement for the program is recertified by the Milestone Decision Authority before additional resources are authorized. At each review, the decision authority may also direct a tailored program to omit or combine specific phases. These special cases are normally based on the decision authority being convinced that the technology and design maturity support such a decision.

Roles and agents:

- Project leader
- Sustainability engineer
- Integrated product team

Inputs:	**Outputs:**
• Supplier proposal	• Cost, schedule, and performance constraints
• Supplier signed agreement (contract or program directive)	• Acquisition strategy
• End products	• Solicitation (request for proposal [RFP], SOW, or SOOs with cost/ schedule requirements)
• Enabling products	
• Supplier performance management plan	• Customer offer
• Work breakdown structure	• Request for clarification
• Integrated master schedule	• Request for information
• Test and evaluation management plan	• Customer signed agreement (contract or program directive)
• Source selection plan (SSP)	
• Team work plan	• Integrated logistics support certification
• Statement of objectives (SOOs)	
• Statement of work (SOW)	
• ICD—formerly MNS	

Table 5.4 Acquisition Processes (SEP-4) (*Continued*)

Inputs:	
• Capability development document or capability production document—formerly operational requirements document • Specified requirements • Operational test readiness review certification message • Cost, schedule, and performance constraints • Acquisition strategy	

Process diagram:

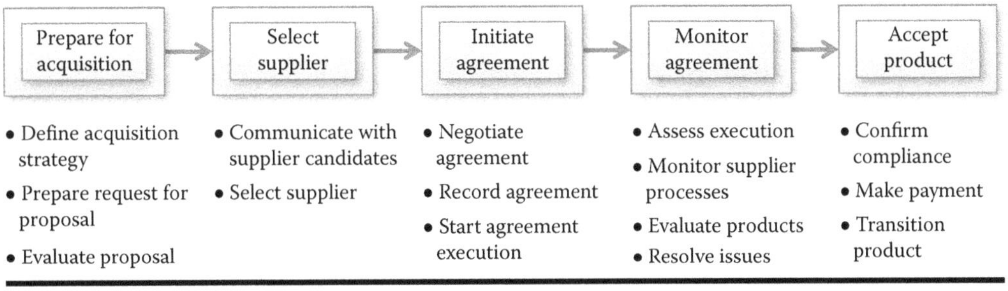

Figure 5.9 Acquisition process.

Step	Action
1	**Action statement:** Prepare for the acquisition.
1.1	**Action statement:** Develop a strategy. Establish a strategy for how the acquisition will be conducted. **Notes and guidance:** This strategy includes reference to the life cycle model, a schedule of milestones, and the selection criteria if the supplier is external to the acquiring organization.
1.2	**Action statement:** Prepare a request. Prepare a request for the supply of a product or service that includes the definition of requirements. **Notes and guidance:** Provide a definition of requirements to one or more suppliers. If a supplier is external to the organization, the request can include the business practices with which a supplier is expected to comply, and the criteria for selecting a supplier. 1. The contracting process begins with planning efforts. Planning includes the development of an RFP, specifications, a SOW or SOOs, a SSP, and the contract data requirements list (CDRL). The SOW is a statement of the work to be done. A SOOs can be utilized to obtain a SOW or an equivalent during the selection process.

(*Continued*)

Table 5.4   Acquisition Processes (SEP-4) (*Continued*)

Step	Action
1.2	2. The RFP is the solicitation for proposals. The government distributes it to potential contractors. The RFP delineates the need and what the supplier must do to be considered for the contract. It establishes the basis for the contract that will be put in place.
	3. The information required to be in the proposals responding to the solicitation is also key for the sustainability engineer. The engineering team decides the technical and technical management merits of the proposals. The directions to the suppliers must be clearly and correctly stated; otherwise, the proposal will not contain the information needed to evaluate the suppliers.
	4. The acquisition package contains the documents that will be provided to the suppliers as part of the RFP. The RFP normally includes the following:
	a. CDRL
	b. Contract schedule, specification
	c. SOW or SOOs
	d. Proposal requirements
	e. Contract security classification
	f. Supplier performance management plan (optional but recommended)
	There are other documents that are part of the acquisition package, which are kept internal to the government and must remain as part of the contract file. These documents typically include the following:
	1. Procurement request
	2. Funding authorization document
	3. Procurement planning schedule
	4. Source list
	5. Proposal evaluation plan
1.3	**Action statement:** Communicate the request. Communicate the request for the supply of a product or service to identified suppliers.
	Notes and guidance: This may include supply chain management partnering, which exchanges information with related suppliers and customers to achieve a harmonized or collective approach to common technical and commercial issues.
2	**Action statement:** Satisfy. Select the supplier.
2.1	**Action statement:** Evaluate the offer. Evaluate the supplier's response to the acquisition request, offer, or directive.
	Notes and guidance: The process begins with the development of a SSP, which relates the organizational and management structure, the evaluation factors, and the method of evaluating the suppliers' responses. The evaluation factors and their priority are transformed into information provided to the suppliers in sections L and M of the RFP. The supplier's proposals are then evaluated with the procedures delineated in the SSP. These evaluations establish which suppliers are

Table 5.4   Acquisition Processes (SEP-4) (*Continued*)

Step	Action
	conforming, guide negotiations, and are the major factor in contractor selection. The sustainability engineering area of responsibility includes support of SSP development by preparing the technical and technical management parts of evaluation factors; organizing technical evaluation teams; and developing methods to evaluate the supplier's proposals (technical and technical management).
	Source selection detemines which supplier will be the contractor, so this decision will have profound impact on program risk. The sustainability engineer should approach the source selection with great care since, unlike many planning decisions made early in product life cycles, the decisions made relative to source selection can generally not be easily changed once the process begins. Laws, regulations, directives, and instructions governing the fairness of the process require that changes be made very carefully and frequently at the expense of considerable time and effort on the part of program management and contractor personnel. In today's environment, even minor mistakes can cause distortion of proper selection. Because of the importance of this process, one organization, Naval Air (NAVAIR), has a source selection office chartered with the responsibility to ensure the source selection process is properly executed.
2.2	**Action statement:** Select one or more suppliers.
	Notes and guidance: To obtain competitive solicitations, proposals to supply are evaluated and compared against the selection criteria. Where proposals include offerings that are not covered by the criteria, then the proposals are compared with each other to determine their order of suitability and thus supplier preference. The justification for rating each proposal is declared, and suppliers may be informed why they were or were not selected.
3	**Action statement:** Initiate the agreement.
3.1	**Action statement:** Negotiate an agreement with the supplier.
	Notes and guidance: This agreement may range in formality from a written contract to a verbal understanding. Appropriate to the level of formality, the agreement establishes requirements, development and delivery milestones, verification, validation and acceptance conditions, exception-handling procedures, change control procedures, and payment schedules so that both parties of the agreement understand the basis for executing the agreement. Rights and restrictions associated with technical data and intellectual property are noted in the agreement. The negotiation is complete when the customer accepts the terms of an agreement offered by the supplier.
3.2	**Action statement:** Record the established agreement in the form appropriate to the effort.
	Notes and guidance: Upon completion of the source selection process, and after any negotiations are finished, a contract is prepared and sent to the contractor(s) for signature. After the contractor signs, the contract is returned to the procurement contracting officer for signature on behalf of the government.

(Continued)

Table 5.4 Acquisition Processes (SEP-4) (*Continued*)

Step	Action
4	**Action statement:** Monitor the agreement.
4.1	**Action statement:** Assess the execution of the agreement. **Notes and guidance:** This includes confirmation that all parties are meeting their responsibilities according to the agreement. Projected cost, performance, and schedule risks are monitored, and the impact of undesirable outcomes on the organization is evaluated regularly. Variations to the terms of the agreement are negotiated as necessary.
4.2	**Action statement:** Select, monitor, and analyze processes used by the supplier. **Notes and guidance:** In situations where there must be tight alignment between some of the processes implemented by the supplier and those of the project, monitoring these processes will help prevent interface problems. The selection must consider the impact of the supplier's processes on the project. On larger projects with significant subcontracts for development of critical components, monitoring of key processes is expected. For most vendor agreements where a product is not being developed or for smaller, less critical components, the selection process may determine that monitoring is not appropriate. Between these extremes, the overall risk should be considered in selecting the processes to be monitored. The processes selected for monitoring should include engineering, project management (including contracting), and support processes critical to successful project performance. Monitoring, if not performed with adequate care, can at one extreme be invasive and burdensome or at the other extreme be uninformative and ineffective. There should be sufficient monitoring to detect issues, as early as possible, that may affect the supplier's ability to satisfy the requirements of the supplier agreement. Analyzing selected processes involves taking the data obtained from monitoring selected supplier processes and analyzing it to determine whether there are serious issues. Perform the following subactivities: 1. Identify the supplier processes that are critical to the success of the project 2. Monitor the selected supplier's processes for compliance with requirements of the agreement 3. Analyze the results of monitoring the selected processes to detect issues as early as possible that may affect the supplier's ability to satisfy the requirements of the agreement
4.3	**Action statement:** Evaluate the work products. Select and evaluate work products from the supplier of custom-made products. **Notes and guidance:** The scope of this specific practice is limited to suppliers providing the project with custom-made products, particularly those that present some risk to the program due to complexity or criticality. The intent of this specific practice is to evaluate selected work products produced by the supplier

Table 5.4 Acquisition Processes (SEP-4) (*Continued*)

Step	Action
	to help detect issues as early as possible that may affect the supplier's ability to satisfy the requirements of the agreement. The work products selected for evaluation should include critical products, product components, and work products that provide insight into quality issues as early as possible. Perform the following subactivities: 1. Identify those work products that are critical to the success of the project and that should be evaluated to help detect issues early. Examples of work products that may be critical to the success of the project include the following: a. Requirements b. Analyses c. Architectured d. Documentation 2. Evaluate the selected work products. Work products are evaluated to ensure the following: a. Derived requirements are traceable to higher level requirements. b. The architecture is feasible and will satisfy future product growth and reuse needs. c. Documentation that will be used to operate and to support the product is adequate. d. Work products are consistent with one another. e. Products and product components (e.g., custom-made, off-the-shelf, and customer-supplied products) can be integrated. 3. Determine and document actions needed to address deficiencies identified in the evaluations.
4.4	**Action statement:** Provide data needed by the supplier and resolve issues in a timely manner. **Notes and guidance:** N/A
5	**Action statement:** Accept the product or service.
5.1	**Action statement:** Confirm that the delivered product or service complies with the agreement. **Notes and guidance:** Exceptions that arise during the conduct of the agreement or with the delivered product or service are resolved according to the procedures established in the agreement.
5.2	**Action statement:** Make payment or provide other agreed consideration to the supplier for the product or service rendered that is required for closure of the agreement. **Notes and guidance:** N/A

(*Continued*)

Table 5.4 Acquisition Processes (SEP-4) (*Continued*)

Step	Action
5.3	**Action statement:** Transition the acquired products from the supplier to the project.
	Notes and guidance: Refer to the product integration process for more information about integrating acquired products. Perform the following subactivities:
	1. Ensure that there are appropriate facilities to receive, store, use, and maintain the acquired products.
	2. Ensure that appropriate training is provided for those involved in receiving, storing, using, and maintaining the acquired products.
	3. Ensure that storing, distributing, and using the acquired products are performed according to the terms and conditions specified in the supplier agreement or license.

Process Task Outcomes
As a result of the successful implementation of the acquisition process:
a. A strategy for the acquisition is established.
b. One or more suppliers are selected.
c. Communication with the supplier is maintained.
d. An agreement to acquire a product or service according to defined acceptance criteria is established.
e. A product or service complying with the agreement is accepted.
f. Payment or other consideration is rendered.

Table 5.5 Work Directives (SEP-5)

Process purpose: The provider shall create work directives that assign and authorize the implementation of the planned technical effort.
Scope: The sustainability engineering process uses a requirements loop and a design loop in a progressive, iterative analytical approach to make operational requirements and design decisions at successively lower levels. As this process iterates, requirements are planned, documented, developed, identified, controlled, tracked, and verified within the configuration management (CM) process. CM provides the common approach necessary to minimize variation and improve information integrity.

Roles and agents:
• Project leader
• Sustainability engineer
• Integrated product team

Table 5.5   Work Directives (SEP-5) (*Continued*)

Inputs:	Outputs:
• Process implementation strategy • Life cycle phase chart • Total life cycle cost objectives • Life cycle phase exit criteria • Organizational structure • Integrated master schedule • Inputs to earned value management system • Cost, schedule, and performance constraints • System technical requirements	• Team work plan (TWP) • Statement of objectives (SOOs) • Statement of work (SOW)

Process diagram:

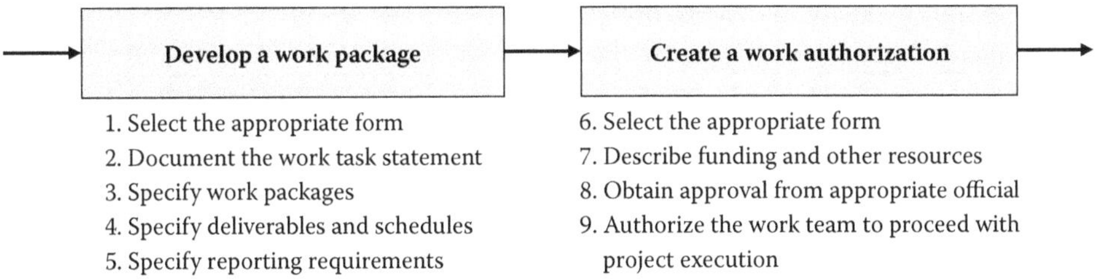

Develop a work package	**Create a work authorization**
1. Select the appropriate form 2. Document the work task statement 3. Specify work packages 4. Specify deliverables and schedules 5. Specify reporting requirements	6. Select the appropriate form 7. Describe funding and other resources 8. Obtain approval from appropriate official 9. Authorize the work team to proceed with project execution

Figure 5.10   Work directives process.

Step	Action
1	**Action statement:** Develop the work package. Develop individual project team or organization work packages that describe the work to be done, resource sources, schedules, budget, and reporting requirements. **Notes and guidance:** a. *Statement of work (SOW)*: The SOW is a portion of a contract, which establishes and defines all nonspecifications requirements for contractors' efforts either directly or with the use of specific cited documents. b. *Statement of objectives (SOOs)*: The SOOs is a portion of a contract, which establishes a broad description of the governments' required performance objectives. c. *Team work plan (TWP)*: The TWP addresses labor by category, material, travel, flight costs, expendables, range requirements, and laboratory requirements. The TWP might include a program summary, cancellations, references, and/or enclosures; technical instructions; schedule; reports and documentation to be provided; future planning information; contractual authority; source and disposition of equipment; and security classifications.

(*Continued*)

Table 5.5   Work Directives (SEP-5) (*Continued*)

Step	Action
2	**Action statement:** Generate the work request/authorization. Generate work authorizations for the team or organization that provide approval for applicable teams or organizations to complete their work package requirements and to release applicable resources.
	Notes and guidance: Work requests (WRs) and authorizations should establish the process and procedures within the enterprise for the assignment of its personnel to Work Teams. It should document the method to be used to describe the work to be done, resources, schedules, funding, and reporting requirements for competency support. The program office may use a different mechanism for setting their internal resource requirements.
	The final product is a WR that meets both the program and the competency requirements. The WR should address the following: tasks, functions, products, and/or services to be provided; funding summary; availability/duration of resources; authority/empowerment level; training requirements and agreements; collocation requirements; performance evaluation inputs required; administrative functions delegated to team leadership; and the issue resolution process to be employed.

Process institutionalization:

The project management plan(s) will describe how policy, planning, resources, responsibility assignment, stakeholder involvement, monitoring, control, status reviews, feedback, quality assurance, evaluations, and audits are to be used to ensure institutionalization of this process.

Process tailoring:

- The following process elements may *not* be tailored: process owner and process objective.
- The following process elements may only be tailored by obtaining a process waiver: process activities.
- The following process elements may be tailored: guidance and notes, process roles, process implementation assets, and process-related measures.

Process Task Outcomes	
Representative Tasks	*Process Task Outcomes*
Develop work package	The work required, input sources, schedules, budget, and reporting requirements to implement, execute, and control the work are defined and documented.
Generate work authorizations	Approval/disapproval of work packages is assigned, and work authorizations are documented.

Table 5.6 Life Cycle Model Management (SEP-6)

Process purpose: The organization and/or implementer shall perform the life cycle model (LCM) management process to define, maintain, and utilize LCMs that satisfy organizational and implementer objectives.

Roles and agents:

- Project leader
- Sustainability engineer
- Integrated product team

Inputs:	**Outputs:**
• Organizational policies and guidance related to LCM	• Standard LCM models • Defined model stages and gates • Defined exit criteria for LCM stages and gates

Process diagram:

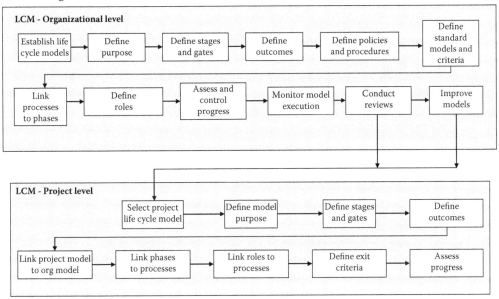

Figure 5.11 Life cycle model management process.

Step	Action
1	**Perform LCM Management: Organizational Level**
1.1	**Action statement:** Establish LCM. Establish standard LCMs applicable to the organization and projects in the organization. **Notes and guidance:** The actual range and detail of the life cycle implementation within a project will be dependent upon the complexity of the work, the methods used, and the skills and training of personnel involved in performing the work. A project tailors policies and procedures according to its requirements and needs.

(Continued)

Table 5.6 Life Cycle Model Management (SEP-6) (*Continued*)

Step	Action
1.2	**Action statement:** Define the LCM purpose. **Notes and guidance:** N/A
1.3	**Action statement:** Define stages and gates. Define the stages and decision gates applicable to the organizations' projects. **Notes and guidance:** N/A
1.4	**Action statement:** Define expected outcomes. Define the expected outcomes of each life cycle stage. **Notes and guidance:** N/A
1.5	**Action statement:** Define LCM policies and procedures. Define organizational policies and procedures related to LCMs. **Notes and guidance:** N/A
1.6	**Action statement:** Identify processes applicable to phases. Identify the standard processes that are applicable to the life cycle phases. **Notes and guidance:** N/A
1.7	**Action statement:** Define roles. Define the roles associated with the processes allocable to the identified life cycle phases and processes. **Notes and guidance:** N/A
1.8	**Action statement:** Define criteria. Define the exit criteria applicable to progression through the life cycle phases. **Notes and guidance:** Establish the decision-making criteria regarding entering and exiting each life cycle stage and for other key milestones. Express these in terms of business achievement.
1.9	**Action statement:** Establish standard LCMs for the organization that are comprised of stages and the purposes and outcomes for each stage. **Notes and guidance:** The LCM comprises one or more stage models, as needed. It is assembled as a sequence of stages that may overlap and/or iterate, as appropriate for the product-of-interest's scope, magnitude, complexity, changing needs, and opportunities. The stages are illustrated in Chapter 3 using commonly encountered examples of life cycle stages. The life cycle processes and activities are selected, tailored as appropriate, and employed in a stage to fulfill the purpose and outcomes of that stage.
1.10	**Action statement:** Assess and control progress. Define how project progress through the life cycle phases will be controlled. **Notes and guidance:** This should include feedback from the projects regarding the effectiveness and efficiency of the processes.
1.11	**Action statement:** Monitor model execution. Monitor process execution of the model. **Notes and guidance:** N/A

Table 5.6 Life Cycle Model Management (SEP-6) (*Continued*)

Step	Action
1.12	**Action statement:** Conduct reviews. Conduct periodic reviews of the standard model. **Notes and guidance:** Confirm the continuing suitability, adequacy, and effectiveness of the LCMs used by each project and make improvements as appropriate. This includes the stages, processes, and achievement criteria that control progression through the life cycle.
1.13	**Action statement:** Improve LCM models. Identify improvement opportunities based on reviews. **Notes and guidance:** N/A
2	**Perform LCM Management: Project Level**
2.1	**Action statement:** Select project LCM. Select or create the project LCM based on tailoring guidance. **Notes and guidance:** Refer to the standard LCMs and tailoring guidance in Chapter 3.
2.2	**Action statement:** Define model purpose. Define the purpose of the project LCM. **Notes and guidance:** Refer to the standard model purposes in Chapter 3
2.3	**Action statement:** Define stages. Define the stages and phase gates applicable to the project. **Notes and guidance:** Refer to the standard life cycle stages and gates required by policy and guidance in Chapter 3.
2.4	**Action statement:** Define outcomes. Define the planned project outcomes of each life cycle stage. **Notes and guidance:** N/A
2.5	**Action statement:** Link the project LCM to organizational standards. Describe how the project LCM and processes are related to organizational policies and procedures. **Notes and guidance:** N/A
2.6	**Action statement:** Link processes to phases. Identify how the project processes are related to the project life cycle phases. **Notes and guidance:** Utilize the process implementation strategy process in selecting processes applicable to the project. Describe use of these processes and how they are applied in various phases of the project life cycle.
2.7	**Action statement:** Link roles to processes and phases. Define how the roles associated with sustainability engineering processes are related to the identified life cycle processes. **Notes and guidance:** Describe roles as they apply to execution of the processes and also roles associated with the assessment of the project (technical assessment process). Use processes during phases and gates of the project life cycle.

(*Continued*)

Table 5.6 Life Cycle Model Management (SEP-6) (*Continued*)

Step	Action
2.8	**Action statement:** Define exit criteria. Define the project exit criteria applicable to progression through the life cycle phases. **Notes and guidance:** N/A
2.9	**Action statement:** Assess progress. Assess the projects progress in achieving life cycle phase gate criteria (refer to the technical assessment process). **Notes and guidance:** N/A

Process Task Outcomes
As a result of the successful implementation of the LCM management process: a. Policies and procedures for the management and deployment of LCMs and processes are provided. b. Responsibility, accountability, and authority for life cycle management are defined. c. Life cycle processes, models, and procedures for use by the organization are defined, maintained, and improved. d. Prioritized process, model, and procedure improvements are implemented.

Table 5.7 Technical Measurement Process (SEP-7)

Process purpose: The project shall perform a measurement process to collect, analyze, and report data relating to the products developed and processes implemented within the organization, to support effective management of the processes, and to objectively demonstrate the quality of the products.

The project shall implement the following activities and tasks in accordance with applicable organization policies and procedures with respect to the measurement process.

Roles and agents:

- Project leader
- Sustainability engineer
- Integrated product team

Inputs:	Outputs:
• Requests for measurement reports	• Measurement objectives
• Raw measurement base measure data	• Specifications of base and derived measures
• Requirements documents	• Data collection and storage procedures
• Product key performance parameters	
• Documented measures of effectiveness	• Data collection tools
• Critical-to-customer requirements	• Analysis specifications

Table 5.7   Technical Measurement Process (SEP-7) (*Continued*)

Inputs:	Outputs:
• Overarching plans needed to plan this process, establish schedules, assign responsibilities, establish monitoring procedures, and provide for improvement of this process • Project measurement plan • Training is completed for project personnel performing measurement and analysis	• Data analysis tools • Data integrity test results • Base and derived data sets • Analysis results and draft reports • Data storage database • Final analysis reports (delivered) • Information to aid in interpretation of results

Process diagram:

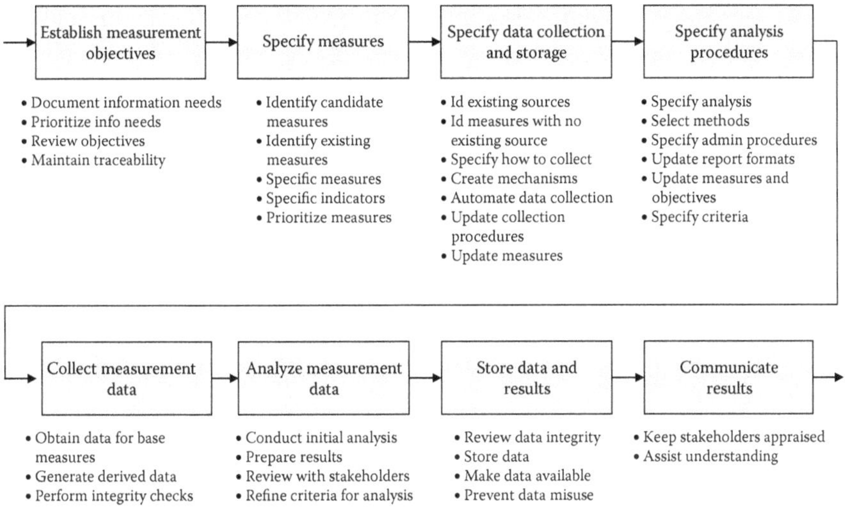

Figure 5.12   **Technical measurement process.**

Step	Action
1	**Establish Measurement Objectives**
1.1	**Action statement:** Document measurement information needs and objectives. **Notes and guidance:** Measurement objectives document the purposes for which measurement and analysis are done, and specify the kinds of actions that may be taken based on the results of data analyses. The sources for measurement objectives may be management, technical, project, product, or process implementation needs. The measurement objectives may be constrained by existing processes, available resources, or other measurement considerations. Judgments may need to be made about whether the value of the results will be commensurate with the resources devoted to doing the work. Modifications to identified information needs and objectives may, in turn, be indicated as a consequence of the process and results of measurement and analysis.

(*Continued*)

Table 5.7 Technical Measurement Process (SEP-7) (*Continued*)

Step	Action
1.2	**Action statement:** Prioritize information needs and objectives. **Notes and guidance:** It may be neither possible nor desirable to subject all initially identified information needs to measurement and analysis. Priorities may also need to be set within the limits of available resources.
1.3	**Action statement:** Document, review, and update measurement objectives. **Notes and guidance:** It is important to carefully consider the purposes and intended uses of measurement and analysis. The measurement objectives are documented, reviewed by management and other relevant stakeholders, and updated as necessary. Doing so enables traceability to subsequent measurement and analysis activities and helps ensure that the analyses will properly address identified information needs and objectives. It is important that users of measurement and analysis results be involved in setting measurement objectives and deciding on plans of action. It may also be appropriate to involve those who provide the measurement data.
1.4	**Action statement:** Provide feedback for refining and clarifying information needs and objectives as necessary. **Notes and guidance:** Identified information needs and objectives may need to be refined and clarified as a result of setting measurement objectives. Initial descriptions of information needs may be unclear or ambiguous. Conflicts may arise between existing needs and objectives. Precise targets on an already existing measure may be unrealistic.
1.5	**Action statement:** Maintain traceability of the objectives to the identified information needs. **Notes and guidance:** There must always be a good answer to the question, "Why are we measuring this?" Of course, the measurement objectives may also change to reflect evolving information needs and objectives.
2	**Specify Measures**
2.1	**Action statement:** Identify candidate measures based on documented measurement objectives. **Notes and guidance:** The measurement objectives are refined into specific measures. The identified candidate measures are categorized and specified by name and unit of measure.
2.2	**Action statement:** Identify existing measures that already address the measurement objectives. **Notes and guidance:** Specifications for measures may already exist, perhaps established for other purposes earlier or elsewhere in the organization.

Table 5.7   Technical Measurement Process (SEP-7) (*Continued*)

Step	Action
2.3	**Action statement:** Specify complete definitions of the measures. **Notes and guidance:** Operational definitions are stated in precise and unambiguous terms. They address two important criteria: • Communication: What has been measured, how was it measured, what are the units of measure, and what has been included or excluded? • Repeatability: Can the measurement be repeated, given the same definition, to get the same results?
2.4	**Action statement:** Specify complete definition of the indicators. **Notes and guidance:** N/A
2.5	**Action statement:** Prioritize, review, and update measures. **Notes and guidance:** Proposed specifications of the measures are reviewed for their appropriateness with potential end users and other relevant stakeholders. Priorities are set or changed, and specifications of the measures are updated as necessary.
3	**Specify Data Collection and Storage Procedures**
3.1	**Action statement:** Identify existing sources of data that are generated from current work products, processes, or transactions. **Notes and guidance:** Existing sources of data may already have been identified when specifying the measures. Appropriate collection mechanisms may exist whether or not pertinent data have already been collected.
3.2	**Action statement:** Identify measures for which data are needed but are not currently available. **Notes and guidance:** N/A
3.3	**Action statement:** Specify how to collect and store the data for each required measure. **Notes and guidance:** Explicit specifications are made of how, where, and when the data will be collected. Procedures for collecting valid data are specified. The data are stored in an accessible manner for analysis, and it is determined whether they will be saved for possible reanalysis or documentation purposes. Questions to be considered typically include the following: • Have the frequency of collection and the points in the process where measurements will be made been determined? • Has the timeline that is required to move measurement results from the points of collection to repositories, other databases, or end users been established? • Who is responsible for obtaining the data? • Who is responsible for data storage, retrieval, and security? • Have necessary supporting tools been developed or acquired?

(*Continued*)

Table 5.7   Technical Measurement Process (SEP-7) (*Continued*)

Step	Action
3.4	**Action statement:** Create data collection mechanisms and process guidance. **Notes and guidance:** Data collection and storage mechanisms are well integrated with other normal work processes. Data collection mechanisms may include manual or automated forms and templates. Clear, concise guidance on correct procedures is available to those responsible for doing the work. Training is provided as necessary to clarify the processes necessary for collection of complete and accurate data and to minimize the burden on those who must provide and record the data.
3.5	**Action statement:** Support automatic collection of the data where appropriate and feasible. **Notes and guidance:** Automated support can aid in collecting more complete and accurate data. Examples of such automated support include the following: • Time-stamped activity logs • Static or dynamic analyses of artifacts However, some data cannot be collected without human intervention (e.g., customer satisfaction or other human judgments), and setting up the necessary infrastructure for other automation may be costly.
3.6	**Action statement:** Prioritize, review, and update data collection procedures. **Notes and guidance:** Proposed procedures are reviewed for their appropriateness and feasibility with those who are responsible for providing, collecting, and storing the data. They also may have useful insights about how to improve existing processes or be able to suggest other useful measures or analyses.
3.7	**Action statement:** Update measures and measurement objectives as necessary. **Notes and guidance:** Priorities may need to be reset based on the following: • The importance of the measures • The amount of effort required to obtain the data Considerations include whether new forms, tools, or training would be required to obtain the data.
4	**Specify Analysis Procedures**
4.1	**Action statement:** Specify and prioritize the analysis that will be conducted and the reports that will be prepared. **Notes and guidance:** Early attention should be paid to the analyses that will be conducted and to the manner in which the results will be reported. These should meet the following criteria: • The analyses explicitly address the documented measurement objectives • Presentation of the results is clearly understandable by the audiences to whom the results are addressed • Priorities may have to be set within available resources

Table 5.7 Technical Measurement Process (SEP-7) (*Continued*)

Step	Action
4.2	**Action statement:** Select appropriate data analysis methods and tools. **Notes and guidance:** Issues to be considered typically include the following: • Choice of visual display and other presentation techniques (e.g., pie charts, bar charts, histograms, radar charts, line graphs, scatter plots, or tables) • Choice of appropriate descriptive statistics (e.g., arithmetic mean, median, or mode) • Decisions about statistical sampling criteria when it is impossible or unnecessary to examine every data element • Decisions about how to handle analysis in the presence of missing data elements • Selection of appropriate analysis tools • Descriptive statistics are typically used in data analysis to do the following: • Examine distributions on the specified measures (e.g., central tendency, extent of variation, or data points exhibiting unusual variation) • Examine the interrelationships among the specified measures (e.g., comparisons of defects by phase of the product's life cycle or by product component) • Display changes over time
4.3	**Action statement:** Specify administrative procedures for analyzing the data and communicating the results. **Notes and guidance:** Issues to be considered typically include the following: • Identifying the persons and groups responsible for analyzing the data and presenting the results • Determining the timeline to analyze the data and present the results • Determining the venues for communicating the results (e.g., progress reports, transmittal memos, written reports, or staff meetings)
4.4	**Action statement:** Review and update the proposed content and format of the specified analysis and reports. **Notes and guidance:** All of the proposed content and format are subject to review and revision, including analytic methods and tools, administrative procedures, and priorities. The relevant stakeholders consulted should include intended end users, sponsors, data analysts, and data providers.
4.5	**Action statement:** Update measures and measurement objectives as necessary. **Notes and guidance:** Just as measurement needs drive data analysis, clarification of analysis criteria can affect measurement. Specifications for some measures may be refined further based on the specifications established for data analysis procedures. Other measures may prove to be unnecessary, or a need for additional measures may be recognized. The exercise of specifying how measures will be analyzed and reported may also suggest the need for refining the measurement objectives themselves.

(*Continued*)

Table 5.7   Technical Measurement Process (SEP-7) (*Continued*)

Step	Action
4.6	**Action statement:** Specify criteria for evaluating the utility of the analysis results and for evaluating the conduct of the measurement and analysis activities.
	Notes and guidance: Criteria for evaluating the utility of the analysis might address the extent to which the following apply:
	• The results are (1) provided on a timely basis, (2) understandable, and (3) used for decision making.
	• The work does not cost more to perform than is justified by the benefits that it provides.
	Criteria for evaluating the conduct of the measurement and analysis might include the extent to which the following apply:
	• The amount of missing data or the number of flagged inconsistencies is beyond specified thresholds.
	• There is selection bias in sampling (e.g., only satisfied end users are surveyed to evaluate end user satisfaction, or only unsuccessful projects are evaluated to determine overall productivity).
	• The measurement data are repeatable (e.g., statistically reliable).
	• Statistical assumptions have been satisfied (e.g., about the distribution of data or about appropriate measurement scales).
5	**Collect Measurement Data**
5.1	**Action statement:** Obtain the data for base measures.
	Notes and guidance: Data are collected as necessary for previously used as well as for newly specified base measures. Existing data are gathered from project records or from elsewhere in the organization.
	Note that data that were collected earlier may no longer be available for reuse in existing databases, paper records, or formal repositories.
5.2	**Action statement:** Generate the data for derived measures and indicators.
	Notes and guidance: Values are newly calculated for all derived measures.
5.3	**Action statement:** Perform data integrity checks as close to the source of the data as possible.
	Notes and guidance: All measurements are subject to error in specifying or recording data. It is always better to identify such errors and to identify sources of missing data early in the measurement and analysis cycle.
	Checks can include scans for missing data, out-of-bounds data values, and unusual patterns and correlation across measures. It is particularly important to do the following:
	• Test and correct for inconsistency of classifications made by human judgment (i.e., to determine how frequently people make differing classification decisions based on the same information, otherwise known as "inter-coder reliability").

Table 5.7 Technical Measurement Process (SEP-7) (*Continued*)

Step	Action
	• Empirically examine the relationships among the measures that are used to calculate additional derived measures. Doing so can ensure that important distinctions are not overlooked and that the derived measures convey their intended meanings (otherwise known as "criterion validity").
6	**Analyze Measurement Data**
6.1	**Action statement:** Conduct initial analysis, interpret the results, and draw preliminary conclusions. **Notes and guidance:** The results of data analyses are rarely self-evident. Criteria for interpreting the results and drawing conclusions should be stated explicitly.
6.2	**Action statement:** Conduct additional measurement and analysis as necessary, and prepare results for presentation. **Notes and guidance:** The results of planned analyses may suggest (or require) additional, unanticipated analyses. In addition, they may identify needs to refine existing measures, to calculate additional derived measures, or even to collect data for additional base measures to properly complete the planned analysis. Similarly, preparing the initial results for presentation may identify the need for additional, unanticipated analyses.
6.3	**Action statement:** Review the initial results with relevant stakeholders. **Notes and guidance:** It may be appropriate to review initial interpretations of the results and the way in which they are presented before disseminating and communicating them more widely. Reviewing the initial results before their release may prevent needless misunderstandings and lead to improvements in the data analysis and presentation. Relevant stakeholders with whom reviews may be conducted include intended end users and sponsors, as well as data analysts and data providers.
7	**Store Data and Results**
7.1	**Action statement:** Review the data to ensure their completeness, integrity, accuracy, and currency. **Notes and guidance:** N/A
7.2	**Action statement:** Store the data according to the data storage procedures. **Notes and guidance:** N/A
7.3	**Action statement:** Make the stored contents available for use by only appropriate groups and personnel. **Notes and guidance:** N/A
7.4	**Action statement:** Prevent the stored information from being used inappropriately. **Notes and guidance:** N/A

(Continued)

Table 5.7   Technical Measurement Process (SEP-7) (*Continued*)

Step	Action
8	**Communicate Results (Report and Use)**
8.1	**Action statement:** Keep relevant stakeholders apprised of the measurement results on a timely basis.
	Notes and guidance: It may be appropriate to review initial interpretations of the results and the way in which they are presented before disseminating and communicating them more widely.
	Reviewing the initial results before their release may prevent needless misunderstandings and lead to improvements in the data analysis and presentation.
	Relevant stakeholders with whom reviews may be conducted include intended end users and sponsors, as well as data analysts and data providers.
8.2	**Action statement:** Assist relevant stakeholders in understanding the measurement results.
	Notes and guidance: Results are reported in a clear and concise manner appropriate to the methodological sophistication of the relevant stakeholders. They are understandable, easily interpretable, and clearly tied to identified information needs and objectives.
	The data are often not self-evident to practitioners who are not measurement experts. Measurement choices should be explicitly clear about the following:
	• How and why the base and derived measures were specified
	• How the data were obtained
	• How to interpret the results based on the data analysis methods that were used
	• How the results address information needs

Measures:

- Percent customer satisfaction with usefulness of the measurement program
- Amount of resources expended on performing the measurement program per reporting period

Process Task Outcomes

As a result of successful implementation of the measurement process:

a. The information needs of technical and management processes are identified.

b. An appropriate set of measures, driven by the information needs, are identified and/or developed.

c. Measurement activities are identified and planned.

d. The required data is collected, stored, analyzed, and the results interpreted.

e. Information products are used to support decisions and provide an objective basis for communication.

f. The measurement process and measures are evaluated.

g. Improvements are communicated to the measurement process owner.

Table 5.8   Project Phase Closure Process (SEP-8)

Process purpose: The supplier is to perform the closeout function and to provide the completed work products to the users and bring the project or phase to an orderly end.

Roles and agents:

- Project leader—responsible for ensuring compliance with this procedure
- Configuration management—responsible for assembling all life cycle documentation
- Equipment custodians—responsible for identifying all related equipment
- Resource manager—responsible for transferring custody of all related equipment and funds

Inputs:	**Outputs:**
• Completed deliverable work product(s) or provided service	• Integrated logistics or product support plan
• Project management plan to include closeout activities	• Product delivered or service provided to intended user and accepted
• Contract documentation	• Equipment returned
• Enterprise environmental factors	• Project destaffing plan, if applicable
• Organizational process assets	• Hazardous materials properly disposed of
• Information on work performance	• Closed contract files, project archives, project closure
	• Lessons learned documented
	• Documented project status in a final project definition file (PDF), if applicable
	• Customer feedback
	• Project is closed out in the product

Process diagram:

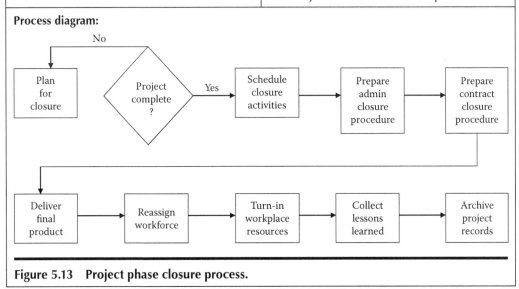

Figure 5.13   Project phase closure process.

(Continued)

Table 5.8   Project Phase Closure Process (SEP-8) (*Continued*)

Step	Action
1	**Action statement:** Update plans. Review and update plans for project closure and destaffing. The project manager (PM) updates or prepares a plan for accomplishing project closure activities to include at least the activities listed above. This planning should include making modifications and additions to a project-closure-related checklist. The PM updates or prepares a destaffing plan to account for the orderly transition of staff to new project work or other activities. The PM, as the responsible party for preparing and approving this plan, is to ensure that a balance is struck between maintaining adequate personnel to accomplish closeout activities and attending to the needs of personnel to obtain new work. **Notes and guidance:** N/A
2	**Action statement:** Determine if project is complete. Determine if the project is complete based on criteria in the agreement, tasking, or organizations' procedure. **Notes and guidance:** N/A
3	**Action statement:** Develop a schedule. Develop an event-based schedule on key events, related tasks, or relevant completion criteria for the applicable enterprise-based life cycle phase. **Notes and guidance:** N/A
4	**Action statement:** Prepare or obtain the administrative closure procedure. **Notes and guidance:** Conduct administrative closure and document the completed action on the project closure checklist. Ensure all documentation and efforts are complete to conclude the project or phase. Document final project results, progress, and status. Document lessons learned. a. Document project results to formalize acceptance of the product or termination for other reasons. b. Collect project records and ensure that they reflect final specifications and analyze project success, effectiveness, and lessons learned. Archive records, deliverables, and other information for future use.
5	**Action statement:** Prepare or obtain the contract closure procedure. **Notes and guidance:** Ensure contracts are appropriately terminated and document the completed action on the project closure checklist. The PM and contracting officer should collaborate to ensure that contracts, license agreements, and other procurement agreements are correctly closed. Informally assess the work or performance of contributing organizations (from within and outside the center) against project requirements and statement of work. a. For all contracts, ensure that work was completed correctly and satisfactorily. b. Coordinate with the contracting officer to evaluate contractor performance through formal or informal customer feedback surveys, or other mechanisms. c. Close administrative issues, update records to reflect final results, and archive this information for future use.

Table 5.8   Project Phase Closure Process (SEP-8) (*Continued*)

6	**Action statement:** Deliver the product. Deliver the final product, service, or result, and arrange for product support. **Notes and guidance:** Deliver product or service and provide user support and document the completed action on the project closure checklist. Transition the verified product to the sponsor, user, or next phase in accordance with the project requirements. Provide user training. • Build a delivery package and document all delivered work products and services. For hardware projects, this will include engineering drawings or technical documentation. For products or software projects, this will include all components, including outstanding problem reports, of the final product(s) in a version description document or similar description. Install and check the product in the user environment following the project build plan. • For final product or software deliveries, support the user's acceptance testing. Conduct or support system acceptance tests, physical configuration audits, and functional configuration audits. Obtain formal acceptance of the product by the user, sponsor, and/or customer. • Work with key stakeholders to develop and implement integrated support plans, if applicable. • Develop, acquire, and/or provide product training to users as specified in the project requirements.
7	**Action statement:** Perform workplace resources turn-in and closure. **Notes and guidance:** Plan and track the turn-in and accounting activities of sponsor-owned equipment, computers, and recyclable/reusable products, as well as the release of facilities, and the appropriate disposal of hazardous materials. Obtain receipts from turn-in of equipment and hazardous materials from the proper owner/authority.
8	**Action statement:** Perform the military force reorganization or reassignment. **Notes and guidance:** Organizational structures and staff size may need to be changed from one project phase to the next. While formal and informal organizational relationships evolve as the project progresses, there is a need for periodically publishing an updated official organizational chart to let personnel know where they fit in the new structure. Additional team structures may also be useful. Project reorganization and closure is frequently a time of increased levels of stress and anxiety for personnel. Frequent team and individual communication should be accomplished during drawdown, destaffing, and/or reassignment.
9	**Action statement:** Collect lessons learned and submit them as required. **Notes and guidance:** Collect and document lessons learned, and implement approved process improvement activities. The lessons learned and historical information are to be transferred to the organizational lessons-learned knowledge base for use by future projects using the forms and guidance. Lessons-learned documentation of the causes of variance and the reasoning behind the corrective actions chosen are archived in organizational repositories for use by all projects of the performing organization.

(Continued)

Table 5.8   Project Phase Closure Process (SEP-8) (*Continued*)

10	**Action statement:** Archive project records. Collect and archive project records as required by the agreement and/or organizational policy.
	Notes and guidance: N/A

Measures:

- Percent completion of project or phase closure checklist items
- Time and effort devoted to performance of the closure process activities

Process Task Outcomes

- Equipment returned.
- Project destaffing plan, if applicable.
- Hazardous materials properly disposed of.
- Closed contract files, project archives, project closure.
- Lessons learned documented.
- Documented project status in a final project definition file (PDF), if applicable.
- Customer feedback.
- Project is closed out in the product.

5.3 Solution Design Processes

The solution design processes are used to convert customer requirements into a set of realizable products that satisfy stakeholder requirements. These processes serve two functions: assessment of the requirements and review of the product characteristics, as illustrated in Figure 5.14.

The solution design process is a top-down, comprehensive, iterative, and recursive problem-solving process applied sequentially through all life cycle phases and stages of the product's development, as described in Chapter 3. During the stages of development, the iterative process is used to

- Transform needs and derived requirements into a set of product and process descriptions (adding value and more detail with each level of development)
- Generate information for decision makers
- Provide input for the next level of development

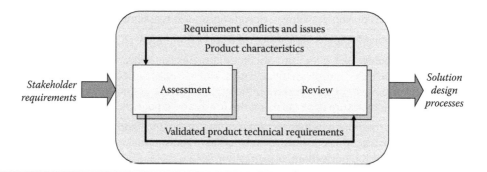

Figure 5.14   Solution design.

The fundamental design activities are stakeholder requirements definition, product technical requirements definition, logical solutions representation (functional analysis and allocation), physical solutions representation (design synthesis), and specified requirements definition. They are all balanced by other processes such as assessment, control, and product analysis. These processes are used to make decisions and track requirements, maintain technical baselines, manage interfaces, identify and manage risks, track cost and schedule, track technical performance, verify requirements are met, and review/audit the progress.

During the solution design iteration, derived requirements and architectures are generated to better describe and understand the product. The word "architecture" is used in various contexts. It is used as a general description of how the subproducts join together. It can also be a detailed description of an aspect of a product, for example, the operational, product, and technical architectures used in hardware- and software-intensive developments. Sustainability engineering recognizes three universally usable architectures that describe important aspects of the product: functional, physical, and product architecture. The functional architecture identifies and structures the allocated functional and performance requirements. The physical architecture depicts the product by showing how it is broken down into subproducts and components. The product architecture identifies all the products (including enabling products) that are necessary to support the product and, by implication, the processes necessary for development, production/construction, deployment, operations, support, disposal, training, and verification.

Life cycle phase integration is achieved through integrated development—that is, concurrent consideration of all life cycle needs during the development process. DoD policy requires integrated development to be practiced at all levels in the acquisition chain of command. Concurrent consideration of all life cycle needs can be greatly enhanced through the use of integrated product teams (IPTs). The objective of an IPT is to

- Produce a design solution that satisfies initially defined requirements and communicates that design solution clearly, effectively, and in a timely manner
- Place balanced emphasis on product and process development
- Assure early involvement of all disciplines appropriate to the team task
- Achieve concurrent technical management

Life cycle phase functions are the characteristic actions associated with the product life cycle. They are development, production and construction, deployment (fielding), operation, support, disposal, training, and verification. These activities cover the "cradle to grave" life cycle process. The customers of product design perform the life cycle functions. The product user's needs are emphasized because their needs generate the requirement for the product, but it must be remembered that all of the life cycle phase functional areas generate requirements for the product design once the user has established the basic need. Those that perform these functions also provide life cycle representation in design-level integrated teams.

The solution design effort begins with identifying, collecting, and defining acquirer and other stakeholder requirements. These requirements are transformed into a set of validated product technical requirements. The validated product technical requirements are then transformed into a design solution described by a set of specified requirements. The specified requirements take the form of specifications, drawings, models, or other design documents depending on design maturity. These are used to (1) build, code, assemble and integrate end products; (2) verify end products against requirements; (3) obtain off-the-shelf products; or (4) assign to a supplier for the development of subproducts. Requirements traceability is instituted for tracking requirements from the

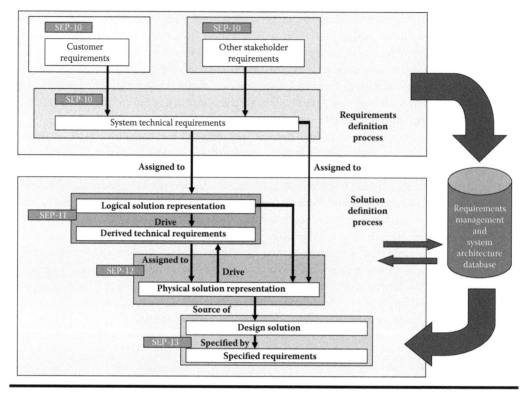

Figure 5.15   Solution design process relationships diagram.

identification of acquirer and other stakeholder requirements to the product technical requirements logical solution representations, physical solution representations, derived technical requirements, and specified requirements (see the requirements management process [SEP-18]). The relationships among the requirements involved with the solution design processes are illustrated in Figure 5.15.

Inputs to the requirements definition process shown in Figure 5.15 include (1) requirements from the customer's agreement, other documents, and individuals or groups that have a stake in the outcome of the engineering or reengineering of the product; (2) requirements in the form of outcomes from other processes, such as technical plans and decisions from technical reviews; and (3) requested or approved changes to the requirements. The requirements defined by this process come from stakeholders who have an interest in the product being engineered. Stakeholders are of two kinds: the acquirer (customer) of the products and all other stakeholders. The requirements definition process is used to transform stakeholder requirements into a set of product technical requirements. These requirements are stated in acceptable technical terms and represent a reasonably complete description of the problem that must be solved to provide a set of end products and enabling products that meet the acquirer's and other stakeholders' needs and expectations. The requirements definition process is reaccomplished, as necessary, whenever requirements in an agreement change, or when other stakeholder requirements are identified that affect the product design or otherwise constrain the technical effort required to engineer a new product, develop a derivative product, or reengineer a legacy product. Such changes could be caused by technology limitations, project schedule and cost anomalies, or new requirements. Sometimes, it is important to preserve competition when defining requirements to ensure that there will be more than one supplier that can meet the requirements. Otherwise, the cost of a single supplier can be too high since there can sometimes be little incentive to give a low-cost bid.

There are five processes associated with solution design. They are shown in Figure 5.16.

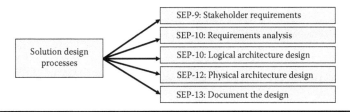

Figure 5.16 Solution design processes.

Table 5.9 Stakeholder Requirements Definition Process (SEP-9)

Process purpose: The supplier performs the stakeholder requirements definition process to provide the products and services needed by users and other stakeholders in a defined environment.	
Roles and agents: • Project leader • Sustainability engineer • Integrated product team	
Inputs: • Specifications from higher level product • Building blocks • Requirements documents • Customer requirements • Other stakeholder requirements • Effectiveness analysis report • Effectiveness models • Analysis of Alternatives technical report • Requirement statements validation revisions • System technical requirements validation revisions • Technology roadmap • Life cycle support plans	**Outputs:** All outputs should be archived • Utilization environment • Verification approach • Operational profiles • Physical and functional requirements • Mission profiles • Cycle timelines • Functional performance • Human interface requirements • Function concurrency/capacity • Technology constraints • Design constraints • Enabling products requirements • Conflicting requirements • Effectiveness analysis request • Trade options and constraints • System requirements document • System technical requirements

(Continued)

Table 5.9   Stakeholder Requirements Definition Process (SEP-9) (*Continued*)

Process diagram:

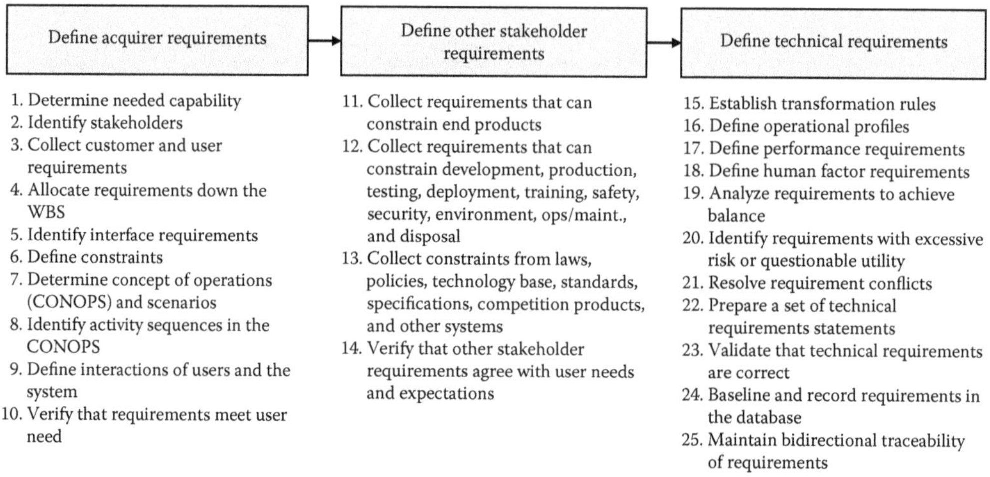

Define acquirer requirements	→	Define other stakeholder requirements	→	Define technical requirements

Define acquirer requirements	Define other stakeholder requirements	Define technical requirements
1. Determine needed capability 2. Identify stakeholders 3. Collect customer and user requirements 4. Allocate requirements down the WBS 5. Identify interface requirements 6. Define constraints 7. Determine concept of operations (CONOPS) and scenarios 8. Identify activity sequences in the CONOPS 9. Define interactions of users and the system 10. Verify that requirements meet user need	11. Collect requirements that can constrain end products 12. Collect requirements that can constrain development, production, testing, deployment, training, safety, security, environment, ops/maint., and disposal 13. Collect constraints from laws, policies, technology base, standards, specifications, competition products, and other systems 14. Verify that other stakeholder requirements agree with user needs and expectations	15. Establish transformation rules 16. Define operational profiles 17. Define performance requirements 18. Define human factor requirements 19. Analyze requirements to achieve balance 20. Identify requirements with excessive risk or questionable utility 21. Resolve requirement conflicts 22. Prepare a set of technical requirements statements 23. Validate that technical requirements are correct 24. Baseline and record requirements in the database 25. Maintain bidirectional traceability of requirements

Figure 5.17   Stakeholder requirements definition process.

Step	Action
1	**Define Customer Requirements**
1.1	**Action statement:** Determine needed capability. Determine a definition of required functionality (capability) that corresponds to operational scenarios. **Notes and guidance:** N/A
1.2	**Action statement:** Identify users, customers, and other stakeholders. Identify the individual stakeholders or classes of stakeholders who have a legitimate interest in the system or product throughout its life cycle. **Notes and guidance:** This includes, but is not limited to, users, operators, supporters, developers, producers, trainers, maintainers, disposers, customer and supplier organizations, parties responsible for external interfacing entities, regulatory bodies, and members of society. Where direct communication is not practicable (e.g., for consumer products and services), representatives or designated proxy stakeholders are selected.
1.3	**Action statement:** Collect customer and user requirements. Elicit, identify, collect, and prioritize assigned, customer, user, or operator requirements for the product or portion thereof, including and requirements for development, production, test, deployment/installation, training, operations, support/maintenance, and disposal of the systems products. **Notes and guidance:** Although the sponsor typically provides these inputs, analyses and validation are required to ensure the team has a clear understanding of the customer requirements. In cases where these documents are not provided, the team is to perform appropriate modeling, simulation, and analysis to develop comparable requirements studies. These analyses include the following: • Surveying the sponsor, fleet operators, and maintainers • Mission analysis (effectiveness analysis)

Table 5.9 Stakeholder Requirements Definition Process (SEP-9) (*Continued*)

Step	Action
	• System concept analysis (effectiveness analysis) • Operational concept analysis (effectiveness analysis) • Operational requirements analysis
1.4	**Action statement:** Allocate requirements down the work breakdown structure (WBS). Allocate requirements down the WBS to product elements and enabling product components. **Notes and guidance:** The requirements for product components of the defined solution include allocation of product performance; design constraints; and fit, form, and function to meet requirements and facilitate production. In cases where a higher level requirement specifies performance that will be the responsibility of two or more product components, the performance must be partitioned for unique allocation to each product component as a derived requirement. Refer to the logical design process for additional detail.
1.5	**Action statement:** Identify interface requirements. **Notes and guidance:** N/A
1.6	**Action statement:** Define constraints. Define the constraints on a system or product solution that are unavoidable consequences of existing agreements, management decisions, and technical decisions. **Notes and guidance:** N/A
1.7	**Action statement:** Determine concept of operations (CONOPS) and scenarios. **Notes and guidance:** N/A
1.8	**Action statement:** Identify activity sequences in the CONOPS. Define a representative set of activity sequences to identify all required services that correspond to anticipated operational and support scenarios and environments. **Notes and guidance:** Scenarios are used to analyze the operation of the product in its intended environment in order to identify requirements that may not have been formally specified by any of the stakeholders; for example, legal, regulatory, and social obligations. The context of use of the product is identified and analyzed. Include in the context analysis the activities that users perform to achieve product objectives, the relevant characteristics of the end users of the product (e.g., expected training, degree of fatigue), the physical environment (e.g., available light, temperature), and any equipment to be used (e.g., protective or communication equipment). The social and organizational influences on users that could affect product use or constrain its design are analyzed when applicable.
1.9	**Action statement:** Define interactions of users and the product. Define interactions between users and the product to account for needed skills, knowledge, and any limitations (human systems integration area). **Notes and guidance:** N/A
1.10	**Action statement:** Verify that requirements meet user needs. Ensure that the resulting set of requirements agrees with the customer needs expectations. **Notes and guidance:** Refer to the verification process.

(Continued)

Table 5.9 Stakeholder Requirements Definition Process (SEP-9) (*Continued*)

Step	Action
2	**Define Other Stakeholder Requirements**
2.1	**Action statement:** Collect requirements that can constrain end products. Elicit, identify, and collect other stakeholder requirements that can constrain the system's end products. **Notes and guidance:** Be sure to consider joint-project stakeholders requirements.
2.2	**Action statement:** Collect requirements that can constrain development, production, testing, safety, security, environment, and disposal. Elicit, identify, and collect other stakeholder requirements that can constrain development, production, test, deployment/installation, training, health, safety, security, environmental, support/maintenance, and disposal of system's end products. **Notes and guidance:** Stakeholder requirements describe the needs, wants, desires, expectations, and perceived constraints of identified stakeholders. They are expressed in terms of a model that may be textual or formal, that concentrates on product purpose and behavior, and that is described in the context of the operational environment and conditions. A product quality model and quality requirements may be useful for aiding this activity. Stakeholder requirements include the needs and requirements imposed by society, the constraints imposed by an acquiring organization, and the capabilities and operational characteristics of users and operator staff. It is useful to cite sources, including solicitation documents or agreements, and, where possible, their justification and rationale, and the assumptions of stakeholders and the value they place on the satisfaction of their requirements. For key stakeholder needs, the measures of effectiveness (MOEs) are defined so that operational performance can be measured and assessed. If significant risks are likely to arise from issues (i.e., needs, wants, constraints, limits, concerns, barriers, factors, or considerations) relating to people (users and other stakeholders) and their involvement in or interaction with a product at any time in the life cycle of that product.
2.3	**Action statement:** Collect constraints from laws, policies, standards, specifications, and competition. Identify and collect other stakeholder constraints such as applicable laws, regulations, policies, technology base, standards and specifications, competitor's product capabilities and trends, and interfaces with other evolving products or platforms. **Notes and guidance:** N/A
2.4	**Action statement:** Verify that other requirements meet user needs and expectations. Ensure that the resulting set of requirements agrees with other stakeholder needs and expectations (see the product verification process SEP-28). **Notes and guidance:** N/A
3	**Define Technical Requirements**
3.1	**Action statement:** Establish transformation rules. Establish required transformation rules, priorities, inputs, states, modes, and configurations, as appropriate to each system product.

Table 5.9   Stakeholder Requirements Definition Process (SEP-9) (*Continued*)

Step	Action
	Notes and guidance:
	1. The actual transformation of all requirements (customer, other stakeholder, technical) into a set of testable technical requirements statements is accomplished as part of activity 3.8 of this process and the requirements analysis process.
	2. Review concept of operations and elaborate where necessary on describing product behavior, starting with outputs generated by external products (modified as appropriate by passing through the natural product environment), which act as stimuli to the product, causing it to take specified actions and produce outputs that are absorbed by external products. These single threads of behavior are traced from source document statements and cover every aspect of operational performance, including logistical modes of operation, operation under designated conditions, and behavior required when experiencing mutual interference with multiobject products.
	Aggregation of these single threads of behavior is a more or less mechanical process depending on the level of sophistication of tool support supplied with the design decision database. When aggregated, the logical sum of these single threads of behavior represents a dynamic statement of what the product is required to do. In some cases, the word "scenario" is used to describe a single thread of behavior, and in other cases, it describes a superset of many single threads operating concurrently.
	In defining the requisite product behavior within the operating environment(s), transformation rules are important in characterizing a product. A transformation rule is anything that tells a product how to transform one or more inputs into one or more outputs (transform inputs into outputs) or change from one mode/state/configuration to another given certain conditions to be true (e.g., transform from state X to state Y). For example:
	a. Given inputs A and B produce output C (inputs/outputs)
	b. Do the above only when in XYZ mode (mode/state)
	c. Do the above only when in configuration LMN (configuration)
	d. Convert A to A-prime by using the JKL algorithm (transformation rule)
	e. When both A and B received at same time, process A first (priority)
	Basically the nature of these transformation rules will differ depending on the technology being used, type of product (hardware, software, facilities, etc.), or the standard methods and tools used in a particular industry or company.
	Define the various modes of operation (embedded training capability, fully operational, etc.) for the system products under development. The conditions (environmental, configuration, operational, etc.), which determine the modes of operation, are also defined.
	Identify all possible types of observable input and output events that can occur between the product and its interacting external products. Record them as input and output events in the database including information to trace the reason for their existence to prevent dilution of originating requirements.

(Continued)

Table 5.9   Stakeholder Requirements Definition Process (SEP-9) (*Continued*)

Step	Action
3.2	**Action statement:** Define operational profiles. Define operational requirements to include operational profiles, and for each operational profile, the utilization environment, events to which system's end products must respond, frequency of use, physical and functional interfaces, and product functional requirements (what system's end products must accomplish).
	Notes and guidance: At the beginning of the program, sustainability engineering is concerned primarily with operational requirements analysis—leading to the translation of user needs into a quantifiable set of performance requirements that can be translated into design requirements. These objectives are then quantified in broad terms, and basic functions are identified that could fulfill the need. The objective of operational requirement analysis is to identify and express technical requirements in measurable parameters that state user needs in appropriate terms to guide product concept development. Performing the mission analysis in a parametric manner ensures that an appropriate product sizing (of communication links, data processing throughput and capacity, number of computers and personnel, and facility space) can be performed. The context diagram serves as a useful tool to depict input/process/output requirements analysis. The total sustainability engineering process is an iterative operation, constantly refining and identifying new requirements as the concept develops and additional details are defined.
	Each operational profile should include the following items:
	1. The utilization environment and factors, natural or induced, that can affect end product performance. This task is to define the utilization environments for each of the operational scenarios. All environmental factors, natural or induced, which may affect product performance, should be identified and defined. Factors which ensure that the product minimizes the potential for human or machine errors or failures that cause injurious accidents or death and impart minimal risk of death, injury, or acute chronic illness, disability, and/or reduced job performance of the humans who support the product life cycle are identified. Specifically, weather conditions (e.g., rain, snow, sun, wind, ice, dust, and fog), temperature ranges, topologies (e.g., ocean, mountains, deserts, plains, and vegetation), biological factors (e.g., animal, insects, birds, and fungi), time (e.g., day, night, and dusk), induced factors (e.g., vibration, electromagnetic, acoustic, and chemical), or other environmental factors are defined for possible locations and conditions where the product may be operated. Effects on hardware, software, and humans should be assessed for impact on product performance and life cycle processes.
	2. If the inputs/outputs are expected to be significantly affected by the environment between the product and the external products, add concurrent functions to the context diagram to represent these transformations, and add input and output events to the database to account for the differences in event timing between when it is emitted and when it is received.

Table 5.9   Stakeholder Requirements Definition Process (SEP-9) (*Continued*)

Step	Action
	3. The events to which end products must respond. Define all external stimuli impinging on the product that elicits a response.
	4. The physical and functional interfaces (e.g., mechanical, electrical, thermal, data, and procedural) including physical interactions (e.g., form and fit), product boundaries (what is controlled by the supplier), and interactions (e.g., information flows and behaviors) of products or environments within supplier control and those products or environments outside product boundaries. Provide a detailed definition of each external interface to the product, typically documented in an information exchange requirements, interface requirements document, and interface control document.
3.3	**Action statement:** Define the performance requirements. Define the performance requirements (how well each functional requirement must be accomplished), including identification of critical performance parameters. • Define performance objectives • Define affordability objectives/constraints • Define schedule constraints • Define technical constraints
3.4	**Action statement:** Define human factor requirements. Analyze customer and other stakeholder requirements to define human factor effects and concerns, establish capacities and timing, define technology and product design constraints, define enabling product requirements, identify conflicts, and determine criteria for analysis of alternatives to resolve conflicts. **Notes and guidance:** 1. Define human system integration effects—Define the operator roles, as applicable, and the human interface requirements (ergonomic limitations, workspace, eye movement, access, cultural background, natural and induced environmental constraints, work tasks, and time constraints) associated with functional and performance requirements on potential users, operators, installers, or recipients and handlers of system's end products. Early inclusion of human interfaces in requirements definition assures a good user interface and a product that achieves the required performance by operators and control and maintenance personnel. 2. Do the required concurrency capacities (e.g., memory, storage, and flows) of end products and timing of events, states, modes, and functions related to each operational profile. Ensure that concurrent functions are clearly depicted in a timeline analysis covering the entire product. A composite picture of total demand on the product (particularly "worst case" scenarios) is essential. Add traceability information to the database to record what external products stimulate the functions, traced from functional source requirements.

(*Continued*)

Table 5.9   Stakeholder Requirements Definition Process (SEP-9) (*Continued*)

Step	Action
	3. Determine any constraints that will influence or affect end product design (e.g., materials, special skills, and automated tools), required physical characteristics (e.g., size, color, texture, weight, and buoyancy), operator safety, product security, reuse requirements, standardization of end products, open system architecture, maintainer access, handling and storage, transportability, and other attributes of end products or design processes of which trade-offs cannot be made. Design constraints recognize inherent limitations on the sizing and capabilities of the product, its interfacing systems, and its operational and physical environment. These typically include power, weight, propellant, data throughput rates, memory, and other resources within the vehicle or which it processes. These resources must be properly managed to insure mission success. Design constraints are of paramount importance in the development of derivative products. A derivative product is a product, which by mandate must retain major components of a prior product. For example, an aircraft may be modified to increase its range while retaining its fuselage or some other major components. The constraints must be firmly established: Which components must remain unmodified? What can be added? What can be modified? The key principle to be invoked in the development of derivative products is that the requirements for the product as a whole must be achieved while conforming to the imposed constraints. Within this realm of product definition, sustainability engineering personnel may also withhold a margin to accommodate unforeseen problems. The margin is held at the product level. In communication links, typically a 3 dB system margin is maintained throughout the development phase. These allocations are analyzed by engineering personnel to verify their achievability. As the design progresses, the current status of the allocations is reviewed at the control board meetings. Care must be exercised that "margins-on-margins" are not overdone, resulting in too conservative (possibly too expensive) a design.
	4. Define technical requirements for enabling products associated with processes to develop, produce, test, deploy/install, operate, support/maintain, train, and retire/dispose of end products under development or being improved. Identify and resolve requirements that have questionable utility or have unacceptable risk of not being satisfied. The above analysis is usually directed at the mission or payload requirements and does not consider the total product requirements, which include communications, command and control, security, supportability, life expectancy, and so on. It is necessary to expand the analysis to include supporting areas in order to obtain the total product requirements.
3.5	**Action statement:** Analyze requirements to achieve balance. **Notes and guidance:** Refer to the requirements analysis process for additional detail.
3.6	**Action statement:** Question requirements. Identify and resolve requirements that have questionable utility or have unacceptable risk of not being satisfied.

Table 5.9   Stakeholder Requirements Definition Process (SEP-9) (*Continued*)

Step	Action
	Notes and guidance: Examine any adverse consequences of incorporating requirements: • Is unnecessary risk being introduced? • Is the product cost within budget limitations? • Is the technology ready for production? • Are sufficient resources available for production and operation? • Is the schedule realistic and achievable?
3.7	**Action statement:** Resolve conflicting requirements. Resolve identified conflicts between sets of customer requirements and other stakeholder requirements, and among these sets (trade-off process). **Notes and guidance:** The sustainability engineer does not perform mission analysis and requirements analysis as discrete sequential operations. Rather the analyses are performed concurrently with mission needs playing the dominant role. It is essential that the sustainability engineer proceeds in this manner to assure progression toward the most cost-effective solution to the mission need. Throughout this process, the sustainability engineer makes cost/requirements trade-offs. The significant or controversial ones are formally documented and presented to the customer for review. Following mission/requirements analysis, product functional analysis proceeds leading to candidate product design(s), which is(are) evaluated in terms of performance, cost, and schedule. While this process ideally results in an optimum technical product, in actuality, limitations on cost, schedule, and risk place constraints on product design, which result in selection of a preferred product from a number of candidates, rather than the optimum technical solution. Where existing user requirements cannot be confirmed, trade studies should be performed to determine more appropriate requirements to achieve the best-balanced performance at minimum cost. Where critical resources (weight, power, memory, throughput, etc.) must be allocated, trade studies may be required to determine the proper allocation.
3.8	**Action statement:** Prepare a set of technical requirements statements. Transform a set of technical requirement statements that is well formulated in accordance with the transformation criteria in action 3.1 above. **Notes and guidance:** Technical requirements are expressed as "shall" statements that are quantitative and measurable. They are inputs to the requirements analysis process. Assess and confirm requirements as to degree of certainty of estimate, and place a "To Be Reviewed" (TBR) flag after any requirement that is not completely agreed upon or a "To Be Determined" (TBD) flag where the value is unknown. Place a list of all TBD/TBR items with responsibilities and closure dates at the back of the specification. Prioritize all requirements as to the criticality of mission success. Since resources on any program are limited, this identifies where the effort should be concentrated in refining, deriving, and flowing down requirements.

(*Continued*)

Table 5.9   Stakeholder Requirements Definition Process (SEP-9) (*Continued*)

Step	Action
3.9	**Action statement:** Validate that technical requirements are correct. Ensure that the set of customer, other stakeholder, and product technical requirements is correct in accordance with the "technical requirements validation" process. **Notes and guidance:** N/A
3.10	**Action statement:** Baseline and record requirements in the database. Record and baseline the resulting set of customer, other stakeholder, and product technical requirements in the established information database. **Notes and guidance:** The validated set of product technical requirements and associated assumptions is captured in the project's information database and maintained and controlled throughout the life of the project. Controlled maintenance of the product technical requirements in the information database allows for traceability, supports validation, and is essential for change management.
3.11	**Action statement:** Maintain bidirectional traceability. Maintain stakeholder requirements traceability to the source of stakeholder need (see the requirements management process). **Notes and guidance:** N/A

Measures:

- Percent completion of analysis and output products
- Percent of product technical requirements that have been validated

Process Task Outcomes	
Representative Tasks	*Process Task Outcomes*
a. Identify, collect, and prioritize customer's product requirements	User, customer, or assigned requirements for a product, or portion thereof, have been identified and defined in terms of needs, expectations, capabilities, and priorities, or of assigned requirements for a product, or portion thereof, as expressed in specifications. Specifically, the following have been identified, as applicable: 1. Concept of operation 2. What the customer wants the products of the product to accomplish (functional requirements) 3. How well each function must be accomplished (performance requirements) 4. Natural and induced environments in which the product must operate or be used

Table 5.9   Stakeholder Requirements Definition Process (SEP-9) (*Continued*)

Representative Tasks	Process Task Outcomes
	5. Design constraints such as use of nondevelopmental or reusable items
	6. Requirements pertaining to the availability, electromagnetic compatibility, health factors, human factors, interoperability, maintainability, reliability, safety, and security
	7. MOEs that reflect overall expectations against which satisfaction will be determined
	8. Constraints pertaining to development, production, test, deployment/installation, training, support/maintenance, and disposal
b. Ensure completeness and consistency of the set of collected customer requirements	The collected user, customer, or assigned requirements are validated. Resolution of all conflicts and variances is completed. Invoked the requirements validation process.
c. Record the set of customer requirements	Validated set of customer requirements is captured in the established information database.
d. Identify and collect other stakeholders' end product requirements	Other types of requirements that can contain the engineering of the system's end products are identified, collected, and defined, such as 1. Project plans 2. Team assignments and organization 3. Automated tools availability and approval for use 4. Required metrics 5. Decisions from management or technical reviews 6. Enterprise standards, guides, policies, and procedures 7. Enterprise technologies 8. Enterprise physical and financial resources
e. Identify and collect other stakeholders' enabling product requirements	Enabling product requirements associated with manufacturing/production, test, deployment/installation, training, support, and disposal (including disposal) processes including enterprise capacities (facilities, equipment, tools, and staff) to accomplish these processes are identified, collected, and defined.

(Continued)

Table 5.9   Stakeholder Requirements Definition Process (SEP-9) (*Continued*)

Representative Tasks	Process Task Outcomes
f. Identify and collect other stakeholders' external constraints	Other end product and development process constraints from external sources are identified, collected, and defined, such as 1. National and international standards, laws, and regulations (including environmental protection, hazardous material exclusion list, and waste disposal) 2. Technology base 3. Industry and international standards and general specifications 4. Competitor product capabilities and trends 5. Interfaces with other existing or evolving products and platforms
g. Establish required transformation rules, priorities, inputs, outputs, states, modes, and configurations	Transformation rules, priorities, inputs, outputs, states, modes, and configurations that will influence and affect the other tasks for definition of product technical requirements are identified and defined, as appropriate to each product.
h. Define operational requirements	The range of anticipated use of the end products, as identified in the concept of operations or specification, or for potential end products, is defined, including for each operational profile the definition of 1. The utilization environment and factors, natural or induced, that can affect end product performance 2. The events to which end products must respond 3. The physical and functional interfaces (e.g., mechanical, electrical, thermal, data, and procedural) including physical interactions (e.g., form and fit) 4. Product boundaries (what is controlled by the supplier) and interactions (e.g., information flows and behaviors) of products or environments within supplier control and those products or environments outside product boundaries 5. What system's end products must be able to accomplish (functional requirements) to satisfy customer identified requirements. Includes factors such as production ability, testability, transportability, installability, operability, supportability, disposability, reliability, availability, maintainability, security, and safety 6. How often end products will be used, cycle time between use, and how often each product function will be accomplished

Table 5.9   Stakeholder Requirements Definition Process (SEP-9) (*Continued*)

Representative Tasks	Process Task Outcomes
i. Define performance requirements	The following are defined: 1. The performance expectations for each functional requirement (how well the function must be accomplished) 2. The set of measure of performances (MOPs), made up of the functional and performance requirements combinations, associated with each MOE 3. The key performance parameters selected from the MOPs that will be key indicators of end product or product performance, and if not met, that will cause the associated MOE to not be satisfied and will put the project in cost, schedule, or performance risk 4. Functional and performance testability approach for each requirement statement
j. Analyze customer and other stakeholder requirements to 1. Define human factors effects 2. Establish capacities and timing 3. Define technology constraints 4. Define product design constraints	The following are identified and defined, as applicable: 1. The user or operator roles, as applicable, and the human factor effects (ergonomic limitations, work space, eye movement, access, cultural background, natural and induced environmental constraints, work tasks, and time constraints) associated with functional performance requirements on potential users, operators, installers, or recipients and handlers of the system's end products 2. Required capacities (e.g., memory, storage, and flows) of end products and timing of events, states, modes, and functions related to each operational profile 3. Any constraints or limitations from use of existing technologies and the risks associated with using any unproven technologies 4. Any constraints that will influence or affect end product design (e.g., materials, special skills, and automated tools) required physical characteristics (e.g., size, color, texture, weight, and buoyancy), operator safety, product security, reuse requirements, standardization of end products, open system architecture, maintainer access, handling and storage, transportability, and other attributes of end products or design processes for which trade-offs cannot be made

(Continued)

Table 5.9   Stakeholder Requirements Definition Process (SEP-9) (*Continued*)

Representative Tasks	Process Task Outcomes
5. Define enabling product requirements 6. Identify conflicts 7. Determine analysis of alternatives criteria	5. Technical requirements for enabling products associated with processes to develop, produce, test, deploy/install, operate, support/maintain, train, and retire/dispose of end products under development or being improved 6. Conflicts among the requirements set 7. The set of risk, cost, schedule, and performance criteria to be used in conducting trade-off analyses for conflict resolution **NOTES:** 1. Suppliers are to ensure that residual risks from constraints are not significant to harm or otherwise prevent the product from performing its functions, create unacceptable costs, or price the system's end products out of competitiveness. 2. Analyses of product requirements can necessitate consideration of existing or possible physical solutions to ensure feasibility.
k. Challenge questionable requirements	Customer and other stakeholder requirements that are of questionable utility or that have an unacceptable risk of satisfaction are identified and resolved.
l. Resolve identified conflict of requirements	Any conflicts between combinations of functional requirements, performance requirements, or constraints, as well as within respective sets of those requirements, are resolved. Invoked the system analysis process.
m. Prepare a set of acceptable product technical requirements	Associated assumptions and technical requirement statements for the product are prepared and then validated. Invoked the requirements validation process.
n. Ensure completeness and consistency of the set of product technical requirements	System technical requirements are validated. Resolution of variances is completed. Invoked the requirements validation process.
o. Record the set of product technical requirements	The validation set of product technical requirements and associated assumptions is captured in the project's information database and maintained and controlled throughout the life of the project. NOTE: Controlled maintenance of the product technical requirements in the information database allows for traceability, supports validation, and is essential for change management.

Table 5.10 Requirements Analysis Process (SEP-10)

Process purpose: The supplier is to perform requirements analysis activities to transform the other stakeholder requirements and requirements-driven views of desired capabilities into a representation of a future product that will meet stakeholder requirements and that, as far as constraints permit, does not imply any specific implementation.

Roles and Agents:

- Project leader
- Sustainability engineer
- Integrated product team

Inputs:	**Outputs:**
• Requirements summary forms	• Validated functional requirements listings
	• Derived functional requirements listing
	• Records of requirements analysis
	• Functional architecture

Process diagram:

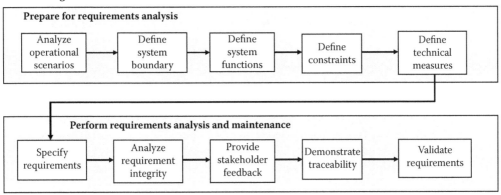

Figure 5.18 Requirements analysis process.

Step	*Action*
Define System Requirements	
1	**Action statement:** Analyze the operational concepts and scenarios
	Notes and guidance: Perform the following subactivities:
	1. Describe operational concepts and scenarios that include functionality, performance, maintenance, support, and disposal as appropriate. Identify and develop scenarios, consistent with the level of detail in the stakeholder needs, expectations, and constraints in which the proposed product or product components are expected to operate.
	2. Define the environment in which the product or product components will operate, including boundaries and constraints.

(*Continued*)

Table 5.10   Requirements Analysis Process (SEP-10) (*Continued*)

Step	Action
	3. Review operational concepts and scenarios to discover and refine requirements. This activity is performed in an iterative process as the requirements evolve.
	4. Analyze the operational concepts in terms of the interactions with the users, the product, the operational environments, maintenance support, and disposal needs.
2	**Action statement:** Define the functional boundary of the product in terms of the behavior and properties to be provided.
	Notes and guidance: This includes the product's stimuli and its responses to user and environment behavior, and an analysis and description of the required interactions between the product and its operational environment in terms of interface constraints, such as mechanical, electrical, mass, thermal, data, and procedural flows. This establishes the expected product behavior, expressed in quantitative terms, at its boundary.
3	**Action statement:** Define each function that the product is required to perform.
	Notes and guidance:
	1. This includes how well the product, including its operators, is required to perform that function, the conditions under which the product is to be capable of performing the function, the conditions under which the product is to commence performing that function and the conditions under which the product is to cease performing that function.
	2. Conditions for the performance of functions may incorporate reference to required states and modes of operation of the product. System requirements depend heavily on abstract representations of proposed product characteristics and may employ multiple modeling techniques and perspectives to give a sufficiently complete description of the desired product requirements.
4	**Action statement:** Define necessary implementation constraints that are introduced by stakeholder requirements or are unavoidable solution limitations.
	Notes and guidance: This includes the implementation decisions that are allocated from a design at higher levels in the structure of the product. Stakeholder requirements are to be analyzed from a variety of areas such as those listed below:
	1. Selected standards—these standards are identified in the requirements analysis process to meet the required quality or design considerations imposed as defined stakeholder requirements or derived to meet enterprise, industry, or domain requirements.
	2. System boundaries—clearly identify product elements under design control of the project team and/or enterprise and expected interactions with products external to that control boundary as defined in the negotiated interface control document (ICD). After this agreement, the ICD is placed under formal change control.
	3. External interfaces—functional and design interfaces to interacting products, platforms, and/or humans external to the product boundary as negotiated in the ICD.

Table 5.10   Requirements Analysis Process (SEP-10) (*Continued*)

Step	Action
	4. Utilization environment(s)—identify all environmental factors (natural or induced) that may affect product performance, impact human comfort, safety, or cause human error for each of the operational scenarios envisioned product use.
	5. Life cycle process requirements—conditions or design factors that facilitate and foster efficient and cost-effective life cycle functions (i.e., production, deployment, transition, operation, maintenance, reengineering/upgrade, and disposal).
	6. Design considerations—including human systems integration (manpower, personnel, training, human factors engineering, environment, safety, occupational health, survivability, habitability), product security requirements (e.g., information assurance, antitamper provisions), and potential environmental impact.
	7. Design constraints—including physical limitations (e.g. weight, form/fit factors), manpower, personnel, and other resource constraints on operation of the product, and defined interfaces with host platforms and interacting products external to the product boundary, including supply, maintenance, and training infrastructures.
	8. Define verification criteria—concurrent with analysis, to ensure verifiable.
5	**Action statement:** Define technical and quality in use measures that enable the assessment of technical achievement.
	Notes and guidance: This includes defining critical performance parameters associated with each effectiveness measure identified in the stakeholder requirements. The critical performance measures are analyzed and reviewed to ensure stakeholder requirements are met and to ensure identification of project cost, schedule, or performance risk associated with any noncompliance.
6	**Action statement:** Specify product requirements and functions, as justified by risk identification or criticality of the product, that relate to critical qualities, such as health, safety, security, reliability, availability, and supportability.
	Notes and guidance: This includes analysis and definition of safety considerations, including those relating to methods of operation and maintenance, environmental influences, and personnel injury. It also includes each safety-related function, and its associated safety integrity, expressed in terms of the necessary risk reduction, is specified and allocated to designated safety-related products. Applicable standards should be used concerning functional safety and environmental protection. Analyze security considerations including those related to compromise and protection of sensitive information, data, and material. The security-related risks are defined, including, but not limited to, administrative, personnel, physical, computer, communication, network, emission, and environment factors using, as appropriate, applicable security standards.
Analyze and Maintain System Requirements	
7	**Action statement:** Analyze the integrity of the product requirements to ensure that each requirement, pairs of requirements, or sets of requirements possess overall integrity.

(*Continued*)

Table 5.10   Requirements Analysis Process (SEP-10) (*Continued*)

Step	Action
	Notes and guidance: Each product requirement statement is checked to establish that it is unique, complete, unambiguous, consistent with all other requirements, implementable, and verifiable. Deficiencies, conflicts, and weaknesses are identified and resolved within the complete set of product requirements. The resulting product requirements are analyzed to confirm that they are complete, consistent, achievable (given current technologies or knowledge of technological advances), and expressed at an appropriate level of detail.
8	**Action statement:** Feedback the analyzed requirements to applicable stakeholders to ensure that the specified product requirements adequately reflect the stakeholder requirements to address the needs and expectations.
	Notes and guidance: Confirmation is made that they are a necessary and sufficient response to stakeholder requirements and a necessary and sufficient input to other processes, in particular architectural design.
9	**Action statement:** Demonstrate traceability between the product requirements and the stakeholder requirements.
	Notes and guidance: Maintain mutual traceability between the product requirements and the stakeholder requirements; that is, all achievable stakeholder requirements are met by one or more product requirements, and all product requirements meet or contribute to meeting at least one stakeholder requirement. The product requirements are held in an appropriate data repository that permits traceability to stakeholder needs and architectural design.
10	**Action statement:** Validate requirements through product analysis.
	Notes and guidance: Refer to the product analysis process for additional information on product analysis, and refer to the product validation process for additional detail on validation.
11	**Action statement:** Maintain throughout the product life cycle the set of product requirements together with the associated rationale, decisions, and assumptions.
	Notes and guidance: N/A

Measures:

- Number of identified and analyzed requirements.
- Percent of total project resources committed to requirements analysis activities.

Process Task Outcomes

As a result of the successful implementation of the requirements analysis process:

a. The required characteristics, attributes, and functional and performance requirements for a product solution are specified.

b. Constraints that will affect the architectural design of a product and the means to realize it are specified.

Table 5.10   Requirements Analysis Process (SEP-10) (*Continued*)

Process Task Outcomes
c. The integrity and traceability of product requirements to stakeholder requirements is achieved.
d. A basis for verifying that the product requirements are satisfied is defined.

Table 5.11   Logical Architecture Design Process (SEP-11)

Process purpose: The implementer shall perform the logical design process to define one or more logical solution representations (a likeness, picture, drawing, block diagram, description, or symbol that logically portrays a physical, operational, or conceptual [product] image or situation that conforms to technical requirements of the product).

Roles and agents:

- Project leader
- Sustainability engineer
- Integrated product team

Inputs:	**Outputs:**
• System technical requirements (STRs) • Operational capabilities • Physical and functional requirements • Mission areas • Cycle timelines • Measures of performance • Key performance parameter • Functional performance • Human interface requirements • Function concurrency/capacity • Enabling products requirements • Conflicting requirements • System requirements document • Effectiveness analysis report • Effectiveness models • Analysis of alternatives technical report • Requirement statement validation revisions • Logical solution representation validation revisions	• Functional analysis products: functional decomposition, timeline • Structured analysis products: context, data dictionaries, entity relationship, modes and states diagrams • Object-oriented analysis (OOA) products: classical, behavior, domain, use case analyses • Effectiveness analysis request • Trade options and constraints • Derived technical requirements • Logical solution representation

(*Continued*)

Table 5.11   Logical Architecture Design Process (SEP-11) (*Continued*)

Process diagram:

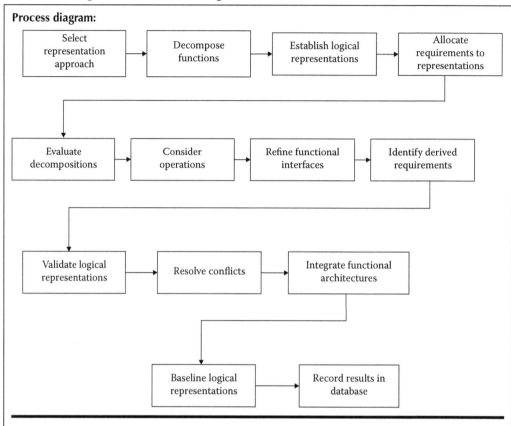

Figure 5.19   Logical architecture design process.

Step	Action
1	**Action statement:** Select an approach for abstract representation. Select one or more appropriate approaches to providing an abstract representation of the solution to the product technical requirements. **Notes and guidance:** N/A
2	**Action statement:** Decompose functions. Decompose each function to lower level functions (use functional flow block diagrams [FFBDs]), as part of the logical architecture design description. **Notes and guidance:** The approach can be a combination of various approaches tailored to the type of product at a given system level. The application of the various analyses, or a combination thereof, is dependent on many variables, such as product type (e.g., hardware, or software), size, and the functional complexity. The traditional sustainability engineering approach for developing logical solution representations has been the functional analysis. This approach is primarily supported by the development of FFBDs and the functional decomposition methods. Other types of analyses have been developed to support logical solution representations; each method favors particular product types and development activities and has advantages and disadvantages. For example, the structured analysis, which includes context

Table 5.11   Logical Architecture Design Process (SEP-11) (*Continued*)

Step	Action
	diagrams, control/data flows, data dictionaries, entity-relationships diagrams, and state transition diagrams, is typically applied in development of complex software-intensive systems (e.g., an air traffic control system). Another type, the OOA using case analysis/unified modeling language (UML), is commonly applied in the development of information systems and other software applications. The resultant output of this task is typically a logical solution analysis approach. The analyses considered for the range of system/software are shown in Figure 5.20. In the OOA task, one must establish a method/approach to the STRs. That task defines these methods in more detail including the specific procedures that should be considered for developing a logical solution.
	A combination of these may be used for a product that contains both hardware and software. One approach might be to perform a functional analysis at the system level and use OOA for the software elements. If multiple approaches are used, traceability must be maintained across methodologies.
	Analysis: Functional Structured Object-oriented ◄───► **Complexity/Functionality** Simple system Complex system Little software software intensive
	Figure 5.20 Analysis considered for product/software.
	NOTE: Functional analysis, object-oriented analysis, structured analysis, and information engineering analysis are recognized approaches found in text books and other literature to develop logical solution representations in terms of, for example, functional flows, behavioral responses, state and mode transitions, timelines, control flows, data flows, information models, object services and attributes, context diagrams, threads, data structures, and functional failure modes and effects.
3	**Action statement:** Establish logical representations. Establish sets of logical solution representations by performing 1. Analysis of alternatives (see the product analysis process) 2. Identifying and defining interfaces (see the interface management process) 3. Analyzing behaviors 4. Performing failure modes, effects, and criticality analysis (FMECA) **Notes and guidance:** *Functional analysis* • FFBD: The translation of the product operational concept into a series of time-sequenced blocks that contain a description of the product function. • Functional decomposition: The breakdown of the product functions from higher level to lower level. This approach is not time sequenced.

(*Continued*)

Table 5.11 Logical Architecture Design Process (SEP-11) (*Continued*)

Step	Action
	• Timelines and sequencing: When time is critical to the sequencing of events that a product must perform, a timeline analysis shall be conducted. A method for defining timing and sequencing is the time analysis sheet and time line analysis chart. Some of the automated sustainability engineering tools provide the capability to perform a simulation and give timeline charts.
	Structured analysis • Context diagram: A diagram that shows the product and its interfaces with external components/elements. • Control data flow diagrams: Data and control flow diagrams are used to document all data transmission, control, and processing functional requirements. • Data dictionaries: A data dictionary is an organized listing of all the data elements that are pertinent to a product. It should be used to describe data elements in both the control data flow diagrams and context diagrams. It should contain name, type, kind, and description. • Activity models: A diagram that identifies the system entities (other systems, devices, or people at the system must keep track of) connected by an arrow that is labeled with the cause/effect relationship (verbs) with other entities in the diagram. • State transition diagrams: A diagram that shows the possible modes and states that can exist between the product and the event or action under which the product can transition. Preliminary states and modes are derived from the concept of operations, and the STRs are further refined in increased detail. *Object-oriented analysis* • Classical approach: Definition of the product through categorization of things, roles, events, and interaction. • Behavior analysis: Definition of the products through the grouping of objects that exhibit similar behavior. • Domain analysis: Definition of the products based on objects, operations, and relationships that are important to the domain (technical area). • Use case analysis/UML: Definition of the product based on a particular form or example of usage/scenario. This also supports analyzing behaviors. *Logical solution trade-off analyses* An optimum logical solution representation should be developed by formulating alternative sets and down-selecting through the trade-off process. Trade studies of alternative product logical solutions must be performed by taking into account cost, customer/user requirements (fleet project team input), open system considerations, and constraints such as the customer requesting the use of a specific commercial off-the-shelf product or interface with legacy systems. After the appropriate approach is selected (functional analysis, structured analysis, or object-oriented analysis), ensure the following analytical techniques are applied in the trade-off decision process where appropriate.

Table 5.11   Logical Architecture Design Process (SEP-11) (*Continued*)

Step	Action
	• Defining interfaces (N2 charts)—Logical solution requirements shall be sequenced with input, output, and logical solution interface (internal and external) requirements defined and be traceable from beginning to end conditions and across their interfaces. A method for defining functional interfaces is the N2 chart. Description of interface is critical in taking an open systems approach to product definition.
	• Analyzing behaviors—Analyze product logical solution behavior through simulation. Some of the automated sustainability engineering tools provide the capability to perform a run-time simulation and check various product logic and threads/paths through the product logical solution definition.
	• FMECA—Analyze, define, and prioritize logical solution (functional level) failure modes and effects through an FMECA. This analysis shall be used to define fault detection, isolation, and recovery functions such as built-in-test and redundancy requirements.
4	**Action statement:** Allocate requirements to representations. Allocate product technical requirements to elements of the logical solution representations. **Notes and guidance:** Establish performance requirements for each logical solution requirement (functional area) and interface. A method for gathering requirements allocation is the requirement allocation sheet. Time requirements that are prerequisite for a logical solution or set of logical solutions shall be determined and allocated. The resulting set of requirements shall be defined in measurable terms, applicable go/no-go criteria, and in sufficient detail for use as design criteria. Performance requirements shall be traceable throughout the logical solution architecture, through the analysis by which they were allocated, to the higher level requirements they are intended to fulfill. Logical solution architecture refers to logical solution definition of the product and the allocation of performance requirements to these functions, not the hardware/software architecture. There can be product technical requirements that are neither appropriate to assign to the sets of logical solution representations nor modifiable into derived technical requirements. An example is a characteristic or constraint applicable only to the product, not to the products of the system. These product technical requirements must be analyzed and assigned during physical architecture design process. There will be additional derived technical requirements prepared to reflect product analysis results from physical architecture design process.
5	**Action statement:** Evaluate decompositions. Evaluate alternative decompositions and select one. **Notes and guidance:** Eventually, each subfunction in the lowest levels of the functional architecture is going to be allocated to hardware, software, or manual operations. In addition, each of these functions will have to be tested. The objective here is to select those decompositions that lend themselves to straightforward implementation and testing. Also, we may be able to come up with decompositions that allow a single function to be used at several places within the hierarchy, thereby simplifying development.

(Continued)

Table 5.11   Logical Architecture Design Process (SEP-11) (*Continued*)

Step	Action
	This is a task that requires sound engineering judgment. There are various ad hoc figures of merit that can be applied to evaluate alternative decompositions. The degree of interconnectivity among functions is one possible measure. There are several measures for software-intensive products that can be applied, such as high cohesion and low coupling. The sustainability engineer needs to be aware of opportunities for use of nondestructive inspection hardware and software. That means that a subfunction that has already been implemented in a compatible form on another product may be preferred to one that has not.
6	**Action statement:** Consider operators. Determine which requirements to allocate to operators. **Notes and guidance:** This determination takes account of the context of use factors and considers the following limitations of human capabilities for the most effective, efficient, and reliable human–machine interaction: • Human actions critical to safety and how the consequences of error are addressed • Integration of human performance into systems and their operation
7	**Action statement:** Refine functional interfaces. Define and refine all internal and external functional interfaces. **Notes and guidance:** N2 charts/diagrams can be used to develop interfaces. These apply to product interfaces, equipment (hardware) interfaces, or software interfaces. Alternatively, or in addition, data/control flow diagrams can be used to characterize the flow of information among functions and between functions and the outside world. As the system architecture is decomposed to lower and lower levels, it is important to make sure that the interface definitions keep pace and that interfaces are not defined that ignore lower level decompositions.
8	**Action statement:** Identify derived requirements. Identify and define derived technical requirements. **Notes and guidance:** N/A
9	**Action statement:** Resolve conflicts. Resolve derived technical requirements conflicts within each logical solution representation. **Notes and guidance:** 1. Resolve derived technical requirement conflicts within each logical analysis model and among logical analysis models: a. Model conflicts? Identify conflicts within each logical analysis model, if any b. Requirements conflicts? Identify conflicts within derived technical requirements c. Trade-off criteria: Use the established set of risk, cost, schedule, and performance criteria in planning

Table 5.11 Logical Architecture Design Process (SEP-11) (*Continued*)

Step	Action
	2. Assure derived technical requirements are necessary and sufficient
	a. Trace up? Check upward traceability of derived technical requirements from each logical analysis model to its source set of technical requirements
	b. Trace down? Check the downward traceability of technical requirements to their derived technical requirements from each set of logical analysis model
	c. Assumptions still OK? Check that any assumptions and decisions made in forming logical analysis models and the related set of derived technical requirements are consistent with the source set of technical requirements
	d. Resolve anomalies: Revolve any anomalies identified in the above subtasks
10	**Action statement:** Validate logical representations. Ensure that each set of logical solutions is correct in accordance with the validation processes.
	Notes and guidance: Refer to the product validation process.
	• Ensure that validated sets of derived technical requirements have been established
	• Ensure that validated sets of logical models have been established
	• Confirm that assumptions and decisions are valid
	• Confirm that identified voids, variances, and conflicts have been resolved
11	**Action statement:** Integrate functional architectures. Define, refine, and integrate the functional architecture.
	Notes and guidance: It may be necessary to make some final modifications to the functional definitions, FFBDs, and interfaces in order to arrive at a viable allocation. The product of this activity is a final FFBD hierarchy with each function (or subfunction) at the lowest possible level uniquely described. The functional flow diagrams, interface definitions, and allocation of requirements to functions and subfunctions constitute the functional architecture.
12	**Action statement:** Baseline logical representations. Baseline the logical analysis representations, derived and allocated technical requirements.
	Notes and guidance: Specially designated working IPTs are used to coordinate logical models and obtain feedback and guidance. Key stakeholders are to be included in these IPTs. Other methods such as quality function deployment (QFD) are also recommended.
13	**Action statement:** Record results in database. Record the results of logical representations in the established information database. Include sets of derived requirements, any unassigned technical requirements along with rational and assumptions.
	Notes and guidance: This records the structural and functional partitioning, interface, and control definitions and the design decisions and conclusions, with traceability to the requirements baseline. The architectural design baseline enables review in the event of change throughout the life cycle, as well as providing information for any subsequent reuse of the architecture. It is also the information source from which tests during integration are defined.

(Continued)

Table 5.11   Logical Architecture Design Process (SEP-11) *(Continued)*

Process Task Outcomes	
Representative Tasks	*Process Task Outcomes*
Select and implement one or more of the four approaches below, or the approach designated by enterprise policies, guides, or standards: 1. Functional analysis 2. Object-oriented analysis 3. Structured analysis 4. Information modeling 5. Other techniques	An abstract definition of the solution is provided in the form of 1. Functional flow, timelines, behaviors, data and control flows, states and modes, and functional failure modes and effects. 2. Objects encapsulating a partition and mapping of STRs and characterized by services (behaviors, functions, and operations) provided and by encapsulated attributes (values, characteristics, and data) 3. Model data and functions with algorithms derived from contextual diagrams and data flow diagrams used to decompose functions while explicitly showing the data needed for each function 4. Data structures with their functions and processing flows related to the data and associated with assigned product technical requirements 5. Outcomes from other techniques (dependent on the nature of that particular methodology)
Establish sets of logical solution representations by 1. Performing trade-off analyses 2. Identifying an defining interfaces	**NOTE:** There is no set format or form for the various definitions of logical solutions. The format or form selected is that which best defines the functional, behavior, or data flow or data structure, as appropriate, and that will allow best assignment to potential end products, manual operations, or enabling products for generating physical solution representations. One or more sets of logical solution representations that are appropriate to the engineering life cycle phase and the product being engineered or reengineered have been formed and defined and include the following: 1. Acceptable logical arrangements and sequencing, or derivative representations (e.g., subfunctions, timelines, objects, data structures, and threads) defined by invoking the system analysis process. 2. Interfaces related to logical arrangements and sequencing, or derivative representations, to include, for example, start and end of states and inputs and outputs defined. Interface attributes identified and defined that trigger, for example, a behavioral response, change of state or mode, or data flow.

Table 5.11   Logical Architecture Design Process (SEP-11) (*Continued*)

Representative Tasks	Process Task Outcomes
3. Analyzing behaviors 4. Identifying and defining state and modes 5. Identifying and defining timelines 6. Identifying and defining data and control flows 7. Analyzing failure modes and defining failure effects	3. The responses (outputs) of the subfunction, group of subfunctions, objects, and so on, to stimuli (inputs) for each operational profile identified and defined, as appropriate. Executable threads identified and defined, as appropriate, through the logical arrangements and sequencing, or derivative representations. 4. The states and modes for which subfunctions, groups of subfunctions, groups, objects, and so on, exhibit different behaviors are identified and defined. 5. Timelines associated with a sequence of functions and objects for each operational profile are defined, as appropriate. Ranges for execution time and conditions that cause normal and abnormal performance are identified and defined. 6. The following are defined, as appropriate: (a) data flows among subfunctions, groups of subfunctions, objects, and so on, for each operational profile and (b) execution controls of each subfunction, and among groups of subfunctions or objects, for each operational profile 7. The functional or behavioral consequences of any specific functional failure that represent significant safety, security, human factor, performance, or environmental hazards are determined and prioritized. Alternative actions to resolve high-priority failure consequences are determined.
Assign product technical requirements (including performance requirements and constraints)	STRs (including performance requirements of a functional requirement and constraints) assigned to appropriate subfunctions, groups of subfunctions, objects, data structures, and so on. **NOTE:** There can be unassigned product technical requirements after the tasks of the physical design architecture process are completed.
Identify, define, and validate derived technical requirement statements	Derived technical requirement statements prepared that (1) reflect the requirements associated with defined logical solution representations, (2) constitute expansion of previously defined derived technical requirements into more detailed lower level requirements, (3) represent product technical requirement statements (such as range) that are not appropriate for logical solution representations but through analysis can be made more specific (such as fuel capacity, engine efficiency, and vehicle resistance), and (4) individually and as a set, are well formulated in accordance with SP-33.
Ensure completeness and consistency of logical solution representations	Logical solution representations and assumptions are validated. Resolution of identified variances is completed. Invoked the validation process SP 33.

(*Continued*)

Table 5.11 Logical Architecture Design Process (SEP-11) (*Continued*)

Representative Tasks	Process Task Outcomes
Record logical solution representations and derived technical requirements	The following are captured in the information database: (1) the data generated, selected arrangements and sequencing, assignments of product performance requirements, and constraints; (2) the validated sets of logical solution representations; (3) the derived technical requirements, along with source rationale and assumptions; and (4) any unassigned product technical requirements (see the note in SP 26).

Table 5.12 Physical Architecture Design Process (SEP-12)

Process purpose: The implementer shall perform the physical design process to define a preferred set of physical solution representations that agree with assigned logical representations, derived technical requirements, and product technical requirements.

The tasks included are often referred as system architecture synthesis. The system architecture synthesis is part of the overall product design process, and it runs iteratively with requirements definition and functional analysis and allocation (now included in logical solution representations).

Roles and agents:

- Project leader
- Sustainability engineer
- Integrated product team

Inputs:	Outputs:
• Design constraints • Technology constraints • Functional analysis products: functional decomposition and timeline • Structured analysis products: context/quality functional deployment/data dictionaries/entity • Relationships • Object-oriented analysis products: classical/behavior/domain/use case analyses • N2/failure modes, effects, and criticality analysis (FMECA) • Logical solution representation • Effectiveness analysis report • Effectiveness models • Analysis of alternatives technical report • Risk analysis report • Requirement statements validation revisions	• Effectiveness analysis request (for alternative physical solutions) • Trade options and constraints • Risk analysis request • Physical solution options • Derived technical requirements • Selected physical solution representation (to include supporting documentation, e.g., concept description sheet, design sheet, system hierarchy definition, functional and performance allocation, system specification tree, functional flow block diagram [FFBD] and system schematic, FMECA [based on FFBD], integrated diagnostic analysis [testability], system architecture views)

Table 5.12   Physical Architecture Design Process (SEP-12) (*Continued*)

Process diagram:

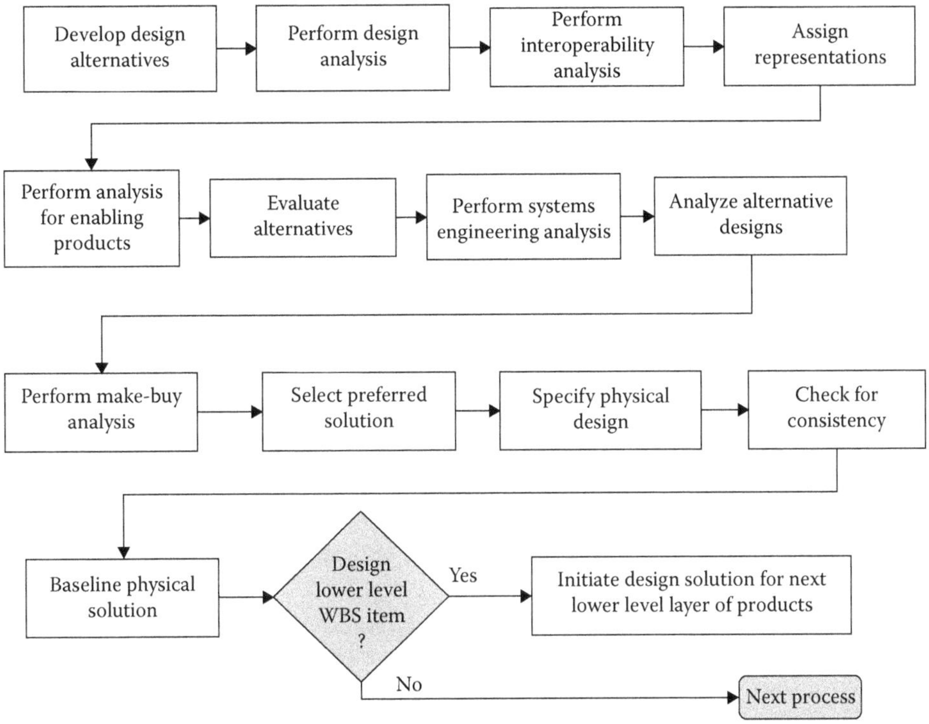

Figure 5.21   Physical architecture design process.

Step	Action
1	**Action statement:** Develop design alternatives. Develop alternative design solutions and selection criteria.
	Notes and guidance: Alternative solutions need to be identified and analyzed to enable the selection of a balanced solution across the life of the product in terms of cost, schedule, and performance. Consideration for alternative solutions and selection criteria include the following:
	• Cost of development, manufacturing, maintenance, and support, etc.
	• Performance
	• Complexity of the product components-related life cycle processes
	• Robustness to product operating and use conditions, operating modes, environments, and variations in product-related life cycle processes
	• Product expansion and growth
	• Technology limitations
	• Sensitivity to production methods and materials
	• Risk

(*Continued*)

Table 5.12 Physical Architecture Design Process (SEP-12) (*Continued*)

Step	Action
	• Evolution of requirements and technology • Capabilities and limitations of end users and operators • Characteristics of commercial off-the-shelf (COTS) products • Disposal
2	**Action statement:** Perform design analysis. Perform analysis of each alternative logical design representation, derived technical requirements, and any unassigned product technical requirements. Determine assignment of requirements to 1. The product 2. Hardware, software, firmware, humans 3. Enabling products **Notes and guidance:** See the notes under the physical architecture design process to determine which requirements are for (a) enabling products; (b) can be done best manually or by facilities, materials, data, services, or techniques; and (c) can be done best by hardware, software, or firmware products (new or existing). The developer shall initiate the physical solution representation analysis by defining alternatives of the product hierarchy. This hierarchy is described in Chapter 4. These product hierarchy alternatives create the design space for all possible choices of elements. The product hierarchy is derived from the logical solution representation, and its purpose is to create the product elements, which constitute the building blocks from which the system architecture is generated. The system elements include hardware, software, information, procedures, and people and are defined top down beginning with the system, subsystem, and configuration items. The product hierarchy can be applied in the planning process to develop the work breakdown structure (WBS) in accordance with the building block concept that consists of the breakdown of end products and enabling products.
3	**Action statement:** Perform interoperability analysis as appropriate. **Notes and guidance:** Interoperability depends on the compatibility of components of large and complex products (which may sometimes be called a system of systems or a family of systems) to work as a single entity. This feature is increasingly important as the size and complexity of products continues to grow.
4	**Action statement:** Assign representations. Assign logical representations to physical entities that make up the physical solution, to include unassigned product technical requirements and derived requirements.

Table 5.12   Physical Architecture Design Process (SEP-12) (*Continued*)

Step	Action
	Notes and guidance: The developer shall assign (requirements allocation) logical solution representation in the form of functions and product technical requirements (i.e., performance, reliability, maintainability, interfaces, environmental requirements, human systems integration, survivability, safety, security, supportability, materials, cost, and other constraints) to the physical elements in the product hierarchy, thus creating a design space and range of values for those physical elements alternatives. These allocations and design descriptions for each physical element should not be constrained by the values of other elements. Assignments (allocation) of design requirements shall be based on the mathematical formulation and representations relative to that discipline (i.e., performance models, reliability and maintainability model and schema, etc.). After requirements assignments are completed, the next step is the identification of the systems hierarchy specification tree for the various system elements alternatives. The assignment to physical entities and the generation of alternative solutions composed of these entities are tightly coupled and iterative.
5	**Action statement:** Perform analysis. Perform analysis to identify and specify enabling products: 1. Manufacturing and production ability analysis 2. Testing analysis 3. Logistics support analysis (LSA) **Notes and guidance:** 1. Production ability analysis is a key task in developing low-cost, quality products. Multidisciplinary teams work to simplify the design and stabilize the manufacturing process to reduce risk, manufacturing cost, lead time, and cycle time and to minimize strategic or critical materials use. Design simplification considers ready assembly and disassembly for ease of maintenance and preservation of material for recycling. The selection of manufacturing methods and processes is included in early decisions. 2. The capability to economically produce a product element is as essential as the ability to properly define and design it. If a designed product cannot be effectively and efficiently manufactured, this causes design rework and program delays with concomitant cost overruns. For this reason, production engineering analysis and trade studies for each design alternative form an integral part of the architectural design process. One objective is to determine if existing proven processes are satisfactory since this could be the lowest risk and most cost-effective approach. 3. Design for testability. 4. Design for logistics (supportability). Perform an LSA as described in the maintenance support process.

(*Continued*)

Table 5.12 Physical Architecture Design Process (SEP-12) (*Continued*)

Step	Action
6	**Action statement:** Evaluate alternatives. Evaluate alternative design solutions, modeling them to a level of detail that permits comparison against specifications expressed in the product requirements and the performance, costs, time scales, and risks expressed in the stakeholder requirements. **Notes and guidance:** This includes the following: • Assessing and communicating the emergence of adverse product properties resulting from the interaction of candidate product elements or from changes in a product element • Ensuring that the constraints of enabling products are taken account of in the design • Performing effectiveness assessments, trade-off analyses, and risk analyses that lead toward realizing a feasible, effective, stable, and optimized design.

7	**Action statement:** Perform sustainability engineering analysis. Perform sustainability engineering analysis and generate alternative physical solution to include applicable characteristics.
	
	Notes and guidance: Generate alternative physical solutions by sizing, configuring, and integrating of the physical product elements alternatives in relation to the logical representation options and assigned requirements range. At this point, the developer shall begin to synthesize the system architecture alternatives. This approach together with the schematic block diagram, systems view (C4ISR architecture framework), and N2 diagrams enables the generation of architectural alternatives. In developing these architectural alternatives, the developer shall consider the following:
	
	1. Identification and definition of physical interfaces to include information exchange requirements
	
	2. Identification and analysis of critical parameters (measures of effectiveness and technical performance measurements)
	
	3. Identification and assessment of physical solution options:
	
	   a. Technology requirements
	
	   b. Off-the-shelf availability and nondevelopmental items (NDIs)
	
	   c. Competitive considerations
	
	   d. Failure modes, effects, and criticality (integrated diagnostics/testability)
	
	   e. Performance assessment
	
	   f. Life cycle considerations
	
	   g. Capacity to evolve
	
	   h. Make versus buy
	
	   i. Standardization considerations (open system architecture)
	
	   j. Integration concerns

Table 5.12 Physical Architecture Design Process (SEP-12) (*Continued*)

Step	Action
8	**Action statement:** Analyze alternative designs. Analyze each alternative design solution, and identify and define derived technical requirements statements using acceptable requirements technical terminology.
	Notes and guidance: See the decision analysis process, data management process, and supplier performance management process, including performance design and parametric analyses, to optimize operating target parameters. This effort helps establish sensitivities, connects hardware requirements to mission measures, exposes thresholds and risks, and creates the range for robust design goals. The system analysis will include considerations in the design for performance, cost, reliability and maintainability, and testability (reference integrated diagnostics, supportability, manufacturability, maintainability, safety, security, and production ability). Supportability and LSA play a key role in the development of physical solution representation. This should include analyses of human systems integration engineering, electromagnetics, survivability, materials, parts, environmental, supportability design, LSA, open system, COTS/NDI, and system and performance design.
9	**Action statement:** Perform make-buy-reuse analysis. Evaluate whether the product should be developed, purchased, or reused based on established criteria.
	Notes and guidance: The make-or-buy decision can be made using formal decision analysis. Factors affecting the make-or-buy decision may include the following:
	a. Functions the products will perform and how these functions fit into the product
	b. Available project resources and skills
	c. Cost for acquiring versus developing internally
	d. Critical delivery and integration dates
	e. Strategic business alliances, including organizational policies and objectives
	f. Market research of available products, including COTS products
	g. Functionality and quality of available products
	h. Skills and capabilities of potential suppliers
	i. Impact of core competencies
	j. Propriety issues
	k. Licenses, warranties, responsibilities, and limitations of acquired products
10	**Action statement:** Select preferred solution. Select the preferred physical solution representation, using criteria, for further characterization into a design solution from the evaluation of each physical solution representation (see the product verification and product validation processes).

(Continued)

Table 5.12 Physical Architecture Design Process (SEP-12) (*Continued*)

Step	Action
	Notes and guidance: Selecting product components that best satisfy the criteria establishes the requirement allocations to product components. Lower level requirements are generated from the selected alternative and used to develop the product component design. Interface requirements among product components are described, primarily functionally. Physical interface descriptions are included in the documentation for interfaces to items and activities external to the product.
	The description of the solutions and the rationale for selection are documented. The documentation evolves throughout development as solutions and detailed designs are developed and those designs are implemented. Maintaining a record of rationale is critical to downstream decision making. Such records keep downstream stakeholders from redoing work and provide insights to apply technology as it becomes available in applicable circumstances.
11	**Action statement:** Specify the physical design. Specify the selected physical design solution as an architectural design that is consistent with assigned logical design representations and derived requirements.
	Notes and guidance: These specifications are the basis of the product solution and an origin for product element acquisition agreements, including acceptance criteria. They may be in the form of sketches, drawings, or other descriptions appropriate to the maturity of the development effort; for example, feasibility design, conceptual design, prefabrication design. They are the basis for deciding whether to produce, reuse, or acquire product elements, for verifying the product elements and for defining an integration strategy for the product.
12	**Action statement:** Check for consistency. Ensure that the selected physical design representation is consistent with the assigned logical solution representation, derived requirements, and any unassigned product technical requirements.
	Notes and guidance: N/A
13	**Action statement:** Baseline physical solution. Baseline the physical solution and record it in the technical data package (TDP) to include specifications and drawings, and place the TDP in the established information database to include selected rationale and assumptions (see the requirements management process).
	Notes and guidance: The design is recorded in a TDP that is created during preliminary design to document the architecture definition. This TDP is maintained throughout the life of the product to record essential details of the product design. The TDP provides the description of a product or product component (including product-related life cycle processes if not handled as separate product components) that supports an acquisition strategy, or the implementation, production, engineering, and logistics support phases of the product life cycle. The description includes the definition of the required design configuration and procedures to ensure adequacy of product or product component performance. It includes all applicable technical data such as drawings, associated lists, specifications, design descriptions, design databases, standards, performance requirements, quality assurance provisions, and packaging details. The TDP includes a description of the selected alternative solution that was chosen for implementation.

Table 5.12 Physical Architecture Design Process (SEP-12) (*Continued*)

Step	*Action*
	A TDP should include the following if such information is appropriate for the type of product and product component (e.g., material and manufacturing requirements may not be useful for product components associated with software services or processes): a. Product architecture description b. Allocated requirements c. Product component descriptions d. Product-related life cycle process descriptions, if not described as separate product components e. Key product characteristics f. Required physical characteristics and constraints g. Interface requirements h. Materials requirements (bills of material and material characteristics) i. Fabrication and manufacturing requirements (for both the original equipment manufacturer and field support) j. The verification and testing criteria used to ensure that requirements have been achieved k. Conditions of use (environments) and operating/usage scenarios, modes and states for operations, support, training, manufacturing, disposal, and verifications throughout the life of the product
14	**Action statement:** Determine if lower level (WBS) products are required. If there are, go to step 16 below. If not, go to the design implementation process. **Notes and guidance:** N/A
15	**Action statement:** Initiate design solution for next lower level layer of products (iterate activities above). **Notes and guidance:** N/A

Process Task Outcomes	
Representative Tasks	*Process Task Outcomes*
Analyze logical solution representation sets, assigned product, and derived technical requirements	The following are determined: 1. Which logical solution set or assigned requirement provides a requirement for an enabling product associated with development, production, test, deployment/installation, training, support/maintenance, or disposal 2. Which logical solution set or assigned requirement can best be accomplished manually or by facilities, material, or data 3. Which logical solution set or assigned requirement can best be accomplished by hardware, software, or firmware products (new or existing) Invoke the system analysis process as necessary.

(Continued)

Table 5.12   Physical Architecture Design Process (SEP-12) (*Continued*)

Representative Tasks	Process Task Outcomes
Assign representations derived from technical requirements and unassigned product technical requirements to appropriate physical entities	The appropriate sets of functions, groups of functions, objects, behaviors, derived technical requirements, and so on, are assigned to appropriate physical entities (e.g., sensor, engine, power source, storage device, structural frame, communication device, and computer) that will make up a physical solution. This assignment to physical entities and generation of alternative solutions composed of these entities is tightly coupled and iterative.
Generate and evaluate alternative physical solution representations by performing the following tasks: **NOTE:** Appropriate models (digital, hardware or software, or both, partial or complete) or prototypes are normally created to help avert risk, identify critical product characteristics and enabling product requirements, identify control requirements for product integrity, perform sensitivity analyses to establish design margins, provide quantitative performance assessments, and select preferred physical solution representation.	
Identify and define physical interfaces	Physical interfaces (human, form, fit, function, data flow, and interoperability) among specific physical entities that make up each end product physical solution alternative, among end products that make up the system, among end products and enabling products, and along with end products and other interfacing systems, are identified and defined. Physical interfaces (internal to the product and external) among specific solutions selected for each physical entity that make up the selected physical solution are designed and described.
Identify and analyze critical parameters	For each identified key performance parameter, the variability and the sensitivity of each alternative physical solution to that variability are identified and defined.
Identify and Assess Physical Solution Options	
Technology requirements	The technological needs necessary to make each alternative solution effective, the risks associated with introduction of new or advanced technologies to meet requirements, and alternative lower risk technologies that could be substituted for unacceptable higher risk technologies are identified and assessed.
Off-the-shelf availability	The availability of off-the-shelf end products (nondevelopmental hardware or reusable software) are identified and assessed.
Competitive considerations	The effect of design considerations to maintain or make a physical solution representation alternative competitive with potential or existing competitor products is identified and assessed.
Failure modes, effects, and criticality	Further design efforts are identified that will be needed to accommodate redundancy and to support graceful degradation when the results of failure modes, effects, and criticality of failure analyses have an unacceptable or high criticality rating.

Table 5.12 Physical Architecture Design Process (SEP-12) (*Continued*)

Representative Tasks	Process Task Outcomes
Performance assessment	The degree to which the performance requirements are satisfied by each alternative physical solution is identified and assessed.
Life cycle considerations	The degree to which production ability, testability, ease of deployment, installability, operability, supportability, trainability, and disposability are considered in each alternative physical solution is identified and assessed. Enabling product needs, requirements, and constraints for the associated processes are identified, assessed, and defined.
Capacity to evolve	The capacity of each alternative physical solution to evolve, or be reengineered, incorporate new technologies, enhance performance, and increase functionality, or other cost-effective or competitive improvements, once solution end products are in production or in the marketplace, are identified and assessed. Limitations that can preclude the capacity of the product to evolve are identified and documented.
Make vs. buy	The advantages and disadvantages of making the products of the solution within the enterprise or going to an established supplier are identified and assessed.
Standardization considerations	The advantages and disadvantages of using standardized end products, protocols, interfaces, and so on, for the physical solution are identified and assessed.
Integration concerns	The following are identified and assessed: (a) potential hazards to other products, operators, or the environment; (b) built-in test and fault isolation test requirements; (c) ease of access, ready disassembly, use of common tools, part count effect, advantage of modularity, standardization, and less need for cognitive skills; and (d) dynamic or static conflicts, inconsistencies, and improper functionality of the integrated products of the solution.
Perform product analysis	Which physical solution option is best for each alternative solution representation, based on each option individually or in sets.
Identify and define derived technical requirements	Derived technical requirement statements identified and defined that are (1) the consequence of design choices associated with the above tasks, (2) used to form alternative physical solution representations, as appropriate, and (3) individually and as a set (including physical interface requirements) well formulated (SP-33).
Select preferred physical solution	The preferred physical solution representation is selected, based on the results of an evaluation of each physical solution representation.
Ensure selected physical solution representation consistency	The selected physical solution representation is determined to be consistent with assigned logical solution representations, derived technical requirements, and the identified subset of unassigned product technical requirements.

(Continued)

Table 5.12   Physical Architecture Design Process (SEP-12) (*Continued*)

Representative Tasks	Process Task Outcomes
Record the outcomes of all tasks	The following are captured in the information database: selected physical solution representation, along with selection rationale, assumptions, and outcomes from all tasks in this process.

Table 5.13   Document the Design (SEP-13)

Process purpose: The provider shall document the design in specifications.	

Roles and agents:

- Project leader
- Sustainability engineer
- Integrated product team

Inputs:	Outputs:
<ul><li>System configuration items structure</li><li>Deficiencies and discrepancies</li><li>Selected physical solution representation</li><li>Requirement statements with validation revisions</li><li>Design solution deficiency and discrepancy reports</li><li>End product deficiency and discrepancy reports</li><li>Operational test/ follow-on test and evaluation (OT/ FOT&E) report</li></ul>	<ul><li>A specification tree</li><li>Specified requirements (system, subsystem, and interface specifications that describe the specified requirements [see below]) in the form of an interface control document or detailed design specification<ul><li>System/subsystem specification</li><li>External physical interfaces</li><li>Internal physical interfaces</li><li>Hardware configuration item</li><li>System architecture design</li><li>Hardware interface design description</li><li>Product descriptions</li><li>Software requirements specification</li><li>Software interface requirements</li><li>Software design description</li><li>Database design description</li><li>Software product specifications</li><li>User version description</li></ul></li><li>Specified requirements products<ul><li>Parts lists</li><li>Procedural manuals</li><li>Data and other applicable design descriptions</li><li>Verified design solution</li><li>Drawings/schematics</li><li>Supportability product specs/descriptions</li></ul></li><li>Enabling products development</li></ul>

Table 5.13   Document the Design (SEP-13) (*Continued*)

Process diagram:

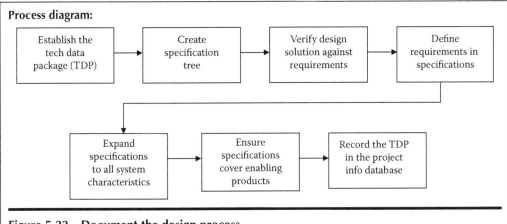

Figure 5.22   Document the design process.

Step	Action
1	**Action statement:** Establish the technical data package (TDP). The TDB includes many types of documentation, including specifications.
	Notes and guidance:
	An integrated data package includes, but is not limited to, the following items:
	A. Hardware
	1. *Arrangement drawings*—documents the relationship of the major subsystems or components of the product.
	2. *Assembly drawings*—documents the relationship of a combination of parts and subassemblies required to form the next higher indenture level of equipment or product.
	3. *Connection drawings*—documents the electrical connections of an installation or of its component devices or parts.
	4. *Construction drawings*—documents the design of buildings or structures.
	5. *Product drawings*—an engineering drawing that documents configuration and configuration limitations, performance and test requirements, weight and space requirements, access clearances, pipe and cable attachments, support requirements, and so on, to the extent necessary that an item may be developed or procured on the commercial market to meet the stated requirements.
	6. *Detail drawings*—documents complete end-item requirements for the subcomponent(s) delineated in the drawing.
	7. *Elevation drawings*—documents vertical projections of buildings and structures or profiles of equipment.
	8. *Engineering drawings*—an engineering document that discloses, by means of pictorial or textual presentations, or a combination of both, the design and functional end product requirements or design of an item.

(*Continued*)

Table 5.13   Document the Design (SEP-13) (*Continued*)

Step	Action
	9. *Installation drawings*—documents general configuration and complete information necessary to install an item relative to its supporting structure or associated items.
	10. *Logic diagrams*—documents, by means of graphic symbols or notations, the sequence and functions of logic circuitry and flows of sequences for operations, maintenance, test, and repair.
	11. *Numerical control drawings*—documents complete design and functional engineering and product requirements of an item to facilitate production by tape control means.
	12. *Piping diagrams*—documents the interconnection of components by piping, tubing, or hose, and when desired, the sequential flow of hydraulic fluids or pneumatic air in the product.
	13. *Wire lists*—documents a book-form drawing consisting of tabular data and instructions required to establish wiring connections within or between items.
	14. *Schematic diagrams*—documents, by means of graphical symbols, the electrical connections and functions of a specific circuit arrangement.
	15. *Wiring and cable harness drawings*—documents the path of a group of wires laced together in a specific configuration, so formed to simplify installation.
	16. *Models, simulations, or design databases*—provides a physical, analytical, or digital representation of any of the items listed above.
	B. Software
	1. *Software design documentation*—documents the software items architecture, design requirements, implementation logic, and data structures that provide a means of support.
	2. *Software source code listings*—documents the actual source code instructions that represented the "as-built" implementation.
	C. Human—documents cognitive, physical, and sensory characteristics of the humans who operate, maintain, and support the product throughout its life cycle that directly contribute to, or constrain, product performance and impact human–machine interfaces.
	1. *Manpower, personnel, and training documentation*—documents the knowledge, skills, and abilities; training requirements; and availability of the humans who operate, maintain, and support the product throughout its life cycle.
	2. *Workspace arrangement drawings*—documents the relationship of humans that support each phase of the life cycle with major subsystems or components of the system.
	3. *Interface design specifications and drawings*—documents all interfaces between the humans that support the product life cycle and any aspect of the product, including human–human interfaces, human–hardware interfaces, and human–software interfaces.

Table 5.13   Document the Design (SEP-13) (*Continued*)

Step	Action
	4. *Operational sequence diagrams*—a graphical representation that documents the interaction between humans and other subsystems or components of the system in the performance of a task over time.
	5. *Procedures*—documents the actions that humans who support each phase of the life cycle perform to develop, produce, test, distribute, operate, support, and dispose of the system or its products or to train humans to accomplish these actions. These procedures may be in the form of operational sequence diagrams, lists, or tables.
	6. *Safety specifications*—documents the equipment/product design features, performance specifications, and training that reduces the potential for human or machine errors or failures that cause injury or death within the constraints of operational effectiveness, time, and cost throughout the equipment/product life cycle.
	D. Life cycle processes
	Process product design architecture—documents the design architecture for the life cycle process products related to development (sustainability engineering and integration), manufacturing, verification, distribution, support, training, and disposal. Products include equipment, software, people, facilities, processes, and services integral to a specific life cycle process.
2	**Action statement:** Create a specification tree. Create a specification outline, from a standard template, for each specification in a specification tree.
	Notes and guidance: The hierarchy within a product should be a balanced hierarchy with appropriate fan-out and span of control. A level of design with too few entities likely does not have distinct design activity, and both design and testing activities contain redundancy. The hardware configuration items can have software configuration items subordinate to them. For example, a display screen on a mobile phone is dominantly hardware, but it must have embedded software to function. Also the operating system in the mobile phone is dominantly a computer software configuration item, but software defines subordinate hardware requirements in order to meet the higher level software requirements (e.g., capacity and speed).
	Developing a specification tree is one element of product design, whereby the product is decomposed into its constituent parts. This process has major ramifications on the development of the product in that it essentially determines the items to be purchased versus those to be developed and establishes the framework for the integration and test program. The objective in the design is to achieve the most cost-effective solution to the customer's requirements with all factors considered. Generally, this is achieved by identifying existing or implementation units as early as possible in the tree development. At each element or node of the tree, a specification is written, and later on in the project, a corresponding individual test will be performed. When identifying elements, it is useful to consider the element both from a design and a test perspective. The element should be appropriate from both perspectives.

(Continued)

Table 5.13 Document the Design (SEP-13) (*Continued*)

Step	Action
3	**Action statement:** Verify design solutions against requirements. Ensure that the design solution is consistent with its source requirements (selected physical solution representation [drawing] requirements, associated product technical requirements, and derived technical requirements). Refer to the product verification process.
	Notes and guidance: Craft requirements for each specification, fulfilling all flow-down and accommodating derived requirements emerging from the definitions of each configuration item.
	A specification represents a design entity and a test entity. The specification should represent appropriate complexity from both the design and the test perspective. Many factors contribute to the appropriate selection of elements. However, as a measure of complexity, a requirements specification should not have too many or too few requirements. As a "rule of thumb," 50–250 functional/performance requirements in a specification are appropriate. Requirements in the physical or environmental areas would be in addition to the functional/performance variety.
4	**Action statement:** Show requirements in specifications. Specify requirements (including functional and performance requirements, physical characteristics, and test requirements) for the system, system end products, and subsystems of each end product, as applicable to the engineering life cycle phase.
	Notes and guidance: N/A
5	**Action statement:** Expand specifications to cover all product design characteristics. Fully characterize the design solution in a specification tree from the system architecture configuration to include all configuration items.
	Notes and guidance: N/A
6	**Action statement:** Specify enabling product data in specifications or the TDP. Establish projects to engineer enabling products that require development, or to procure those that are off-the-shelf or will be reused, that will satisfy identified requirements for associated processes (production, test, deployment/installation, training, support or maintenance, and retirement or disposal) related to the system's end products.
	Notes and guidance: N/A
7	**Action statement:** Record the TDP and specifications in the information database. Record the design solution work products, including specified requirements, in the established information database. Include all analysis of alternatives results, design rationale, assumptions, and key decisions to provide traceability of requirements up and down the product structure.
	Notes and guidance: N/A

Table 5.13   Document the Design (SEP-13) (*Continued*)

Process Task Outcomes	
Representative Tasks	*Process Task Outcomes*
Fully characterized design solution	For each specific physical entity of the selected physical solution, hardware drawings and schematics, software design documents, parts lists, interface descriptions, procedural manuals, data, or other applicable design descriptions, based on the requirements assigned to the selected physical solution and engineering life cycle phase exit criteria, are completed, as applicable.
Ensure design solution consistency	The defined design solution is verified as being consistent with the selected physical solution representations as described by its encapsulated requirements for the assigned logical solution representations, associated product technical requirements, and derived technical requirements. Invoked the product verification process.
Specify requirements	System, subsystem, and interface specifications that describe the specified requirements (functional and performance requirements, and physical characteristics) are documented. Test requirements to ensure that end products satisfy their specified requirements are determined and included in the related specification, as appropriate to the engineering life cycle phase.
Establish projects for development of enabling products	A project is established to engineer the enabling products associated with the processes for development, production, test, deployment/installation, training, support/maintenance, and retirement/disposal. **NOTE:** The requirements for enabling products come from (1) user or customer or assigned requirements and other stakeholder requirements for the product and (2) derived technical requirements for end products and their subsystems generated by tasks of the solution definition process. Thus, initiation of enabling product development is dependent on the completion of the design solution for the product (building block) being engineered or reengineered.
Record design solution and related specified requirements	The design solution work products, including the specified requirements, are captured and recorded in the established information database, along with all trade-off analyses, design rationale, assumptions, and key decisions to provide traceability of requirements up and down the product structure.

5.4 Technical Management Processes

The technical management processes are to be used to plan, assess, and control the technical work efforts of the project. The relationship of these three functions is illustrated in Figure 5.23. There are twelve processes associated with these three functions. They are shown in Figure 5.24.

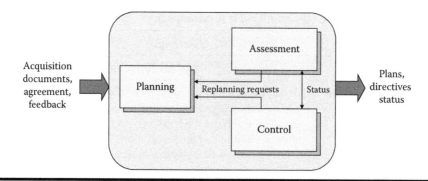

Figure 5.23 Planning, assessing, and controlling the technical efforts of the project.

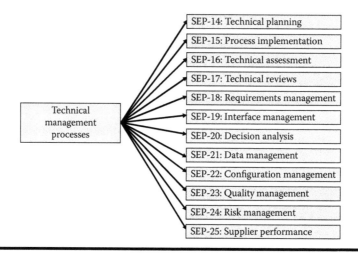

Figure 5.24 Technical management processes.

Table 5.14 Technical Planning Process (SEP-14)

Process purpose: The provider shall perform the technical planning process to produce and communicate effective and workable project technical plans.
This process determines the scope of the project technical activities; identifies process outputs, project tasks, and deliverables; and establishes schedules for project technical tasks performance, including achievement criteria and required resources to accomplish project technical tasks.
This process is used to support enterprise and project decision making and to prepare necessary technical plans that support and complement project plans to (a) arrive at a decision to supply services according to an external solicitation; (b) determine whether to proceed with an internal enterprise project for a new product or a product improvement; (c) guide the work efforts that will meet the requirements of an established agreement; or (d) replan applicable processes for engineering a product. Replanning is normally initiated (1) when required by an agreement, (2) when significant variations or anomalies are identified from other technical management process outcomes, or (3) before implementation of the next enterprise-based life cycle phase.

Table 5.14   Technical Planning Process (SEP-14) (*Continued*)

Roles and agents:

- Project leader
- Sustainability engineer
- Integrated product team

Inputs:	Outputs:

• Process implementation strategy
• Organizational structure
• Integrated master schedule
• Measures of effectiveness
• Stakeholder requirements
• Data management plans
• Configuration management plans
• Acquisition strategy
• Cost, schedule, and performance constraints
• Solicitation (request for proposal, statement of work, or statement of objectives with cost/schedule requirements)
	• Project work breakdown structure (WBS)
	• Project staffing requirements
	• Project technical plans for
	   • Sustainability engineering
	   • Test and evaluation
	   • Risk management
	   • Configuration management
	   • Data management
	   • Quality assurance
	   • Security management
	   • Verification and validation
	   • Product life cycle support

Process diagram:

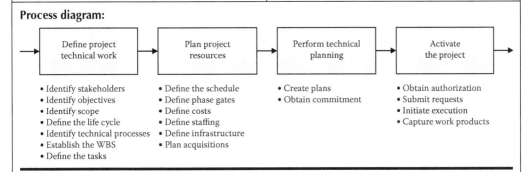

Figure 5.25   Technical planning process.

Step	Action
1	**Define Project Technical Work**
1.1	**Action statement:** Identify stakeholders. Identify stakeholders who will have an interest or stake in the outcomes of the project technical effort. **Notes and guidance:** Describe how the project will manage the involvement of stakeholders in project implementation.

(*Continued*)

Table 5.14 Technical Planning Process (SEP-14) (*Continued*)

Step	Action
1.2	**Action statement:** Identify objectives. Identify the project technical objectives and constraints. **Notes and guidance:** Objectives and constraints include performance and other quality aspects, cost, time, and stakeholder satisfaction. Each objective is identified with a level of detail that permits selection, tailoring, and implementation of the appropriate processes and activities.
1.3	**Action statement:** Identify scope. Define the project scope as established in the agreement. **Notes and guidance:** The project includes all the relevant activities required to satisfy business decision criteria and complete the project successfully. A project can have responsibility for one or more stages in the complete product life cycle. Planning includes appropriate actions for maintaining project plans, performing assessments, and controlling the project.
1.4	**Action statement:** Define the life cycle. Define and maintain a life cycle model that is comprised of stages using the defined life cycle models of the organization. **Notes and guidance:** Refer to the life cycle management process for details on creating and maintaining a project life cycle. For purposes of this technical planning process, the project technical plan should contain a description of the project life cycle and describe how it relates to technical processes discussed in activity 1.5.
1.5	**Action statement:** Identify technical processes. The project is to define how it will select, tailor, and utilize the processes in this chapter as part of planning and estimating the technical effort. **Notes and guidance:** Details on the selection and tailoring of technical processes are provided in the sustainability engineering technical reviews process. One of the important technical processes involves the establishment of a project information database that will allow the capture and secure storage of project data. More information on this topic is provided in the data and information management process. In addition, the technical plan needs to describe how technical performance measurement (TPM) will be implemented by the project. The technical measurement process will provide detail in this area. NOTE: A TPM program provides an early warning of the adequacy of a design in terms of satisfying selected key performance parameter (KPP) requirements of a system end product. TPM also examines marginal cost benefit of performance in excess of requirements. It also includes sensitivity analysis. A KPP is one that characterizes a significant total product qualifier. In addition, it must be possible to project the evolution of the parameter as a function of time toward the desired value at the completion of development. The projection can be based on verification validation planning or historical data.

Table 5.14 Technical Planning Process (SEP-14) (*Continued*)

Step	Action
1.6	**Action statement:** Establish a WBS. Establish a WBS based on the evolving system architecture.
	Notes and guidance: The WBS should address each element of the system architecture and appropriate processes and activities that are described with a level of detail that is consistent with identified risks. Related tasks in the WBS are grouped into project tasks according to organizational responsibilities. Project tasks identify every work item being developed or produced and its associated tasks.
1.7	**Action statement:** Define the tasks. Identify applicable technical tasks based on an analysis of processes of this guide and key phases of the project life cycle and entry and exit criteria for each phase.
	Notes and guidance: N/A
2	**Plan Project Resources**
2.1	**Action statement:** Define the schedule. Define and maintain a project schedule based on project objectives and work estimates.
	Notes and guidance: This includes definition of the duration, relationship, dependencies and sequence of project activities, achievement milestones, resources employed, and reviews and schedule reserves for risk management necessary to achieve timely completion of the project.
2.2	**Action statement:** Define the phase gates. Define project achievement criteria for the life cycle stage decision gates, delivery dates, and major dependencies on external inputs or outputs.
	Notes and guidance: The time intervals between internal project reviews are defined in accordance with organizational policy on issues such as business and system criticality, schedule, and technical risks.
2.3	**Action statement:** Define staffing. Establish the structure of authorities and responsibilities for project work.
	Notes and guidance: This includes defining the project organization, staff acquisitions, the development of staff skills, and the methods of team working. Responsibilities include the effective use of human resources, drawing on organizational functions that contribute to all stages of the product life cycle. The structure of authority is designated, including, as appropriate, the legally responsible roles and individuals, and so on, design authorization, safety authorization, and award of certification or accreditation.
2.4	**Action statement:** Define the infrastructure. Define the infrastructure and services required by the project.
	Notes and guidance: This includes defining the capacity needed, its availability, and its allocation to project tasks. Also included are facilities, tools, communications, and information technology assets. The requirements for enabling products for each life cycle stage within the scope of the project are also specified.

(Continued)

Table 5.14   Technical Planning Process (SEP-14) (*Continued*)

Step	Action
2.5	**Action statement:** Plan acquisitions. Plan the acquisition of materials, goods, and enabling product services supplied from outside the project.
	Notes and guidance: This includes, as necessary, plans for solicitation, supplier selection, acceptance, contract administration, and contract closure. The agreement processes are used for the planned acquisitions.
2.6	**Action statement:** Define costs. Define the project costs and plan a budget.
	Notes and guidance: Costs are based on, e.g., the project schedule, labor estimates, infrastructure costs, procurement-items-acquired service and enabling product estimates, and budget reserves for risk management.
3	**Perform Technical Planning**
3.1	**Action statement:** Create and prepare appropriate technical plans. Sustainability engineering planning addresses the scope of the technical effort required to develop the system or product. The basic questions of "who will do what" and "when" must be answered. A technical plan describes what must be accomplished, how sustainability engineering will be done, how the effort will be scheduled, what resources will be needed, and how the effort will be monitored and controlled. The number and type of plans will vary depending on the project scope, life cycle phase, and other factors.
	Notes and guidance: Examples of project technical plans include some of the following:
	1. *Engineering plan.* For most programs, this implies a sustainability engineering management plan (SEMP). On major programs, the SEMP is a contract deliverable and is prepared by the prime contractor. The software development plan is the equivalent of a SEMP when the product under development is purely software and for the software component of a product. On programs that are procuring software-intensive systems, the planning information should be incorporated into the corresponding sections of documents, such as the acquisition plan and the SEMP.
	2. *Risk management plan.* The development of the risk management plan supports the risk and opportunity management process. The risk management plan should address the elements of risk management including risk identification, risk analysis, risk assessment, and risk handling. Plans for a risk management board and risk reporting should be defined.
	3. *Technical review plan.* A review plan shall identify any significant technical reviews required, when they will occur, and the purpose of the review. Typically the review plan is not a stand-alone document but is incorporated in the SEMP and in other program documentation. The normal sequence of reviews for a typical system is system requirements review; system functional review/software specification review; preliminary design review; and critical design review/test readiness review. The nomenclature and acronyms for these reviews are often modified for specific programs, but the purpose of the reviews should not change. Additional guidance can be found in the sustainability engineering technical review process.

Table 5.14   Technical Planning Process (SEP-14) (*Continued*)

Step	Action
	4. *Verifications plans.* Verification plans take many forms depending on the life cycle phase and program content. The product verification process requires verification plans that are often very informal and consist only of a verification matrix. A verification matrix shows how every requirement will be verified such as by analysis, modeling and simulation, lab test, or full-scale test.
	5. *Validation plans.* Planning for validation should be encompassed in the SEMP.
3.2	**Action statement:** Obtain commitment. Obtain stakeholder commitments to technical plans. **Notes and guidance:** Perform the following subactivities: a. Review and coordinate plans that affect the technical effort to determine work, authority, responsibility, accountability, and conflict control to ensure a common understanding of the scope, objectives, roles, and relationships required to successfully complete the technical effort. b. Identify and reconcile any differences between estimates and available resources as reflected in project technical plans and funding profiles. c. Obtain stakeholder commitment on technical plans. Commitment should reflect confidence that the work can be performed within cost, schedule, and performance constraints and within acceptable levels of risk.
4	**Activate the Project**
4.1	**Action statement:** Obtain authorization. Obtain authorization for the project to start plan execution. **Notes and guidance:** Refer to the project definition process for additional detail.
4.2	**Action statement:** Submit requests. Submit requests and obtain commitments for necessary resources to perform the project. **Notes and guidance:** Resources includes facilities, testing labs, computer resources, and skilled people. Technical plans need to describe staffing profiles that show the build-up, steady state, and eventual decline of required personnel.
4.3	**Action statement:** Initiate execution. Initiate the implementation of the project technical plans to satisfy the objectives, criteria, and exercising project control. **Notes and guidance:** Perform the following subtasks: a. Develop work assignment packages for teams and individuals that describe the work to be done b. Generate work authorizations that provide approval for applicable teams and/or line organizations to complete their work packages (see the work directives process for more detail). c. Generate work or resource orders to release applicable resources to the appropriate teams or organization (see the work directives process for more detail). d. Obtain official approval of work authorizations and work orders e. Distribute approved work authorizations and work orders to the intended teams and/or the organizational elements

(*Continued*)

Table 5.14 Technical Planning Process (SEP-14) (*Continued*)

Step	Action
4.4	**Action statement:** Capture work products. Capture the technical planning work products. **Notes and guidance:** Assemble and assess the technical planning work products. Insure that technical planning work products are stored in the project information database (see SP-18) and made available to appropriate stakeholders.

Measures

- Number of project with project technical plans approved by appropriate management level
- Time required to draft, coordinate, and obtain approval of project technical plans

Process Task Outcomes

As a result of the successful implementation of the project planning process:

1. Users and stakeholders have been identified, and project involvement with them has been documented.

2. Stakeholder expectations concerning cost, schedule, and product technical performance are understood.

3. Organizational policies, guidance, priorities, and constraints on funding, personnel, capabilities, infrastructure, and other resources that can affect the technical effort are understood.

4. Applicable technical processes, standards, and specifications; technical risks; net cost targets; methods of resource allocation; how work and changes will be authorized; how information will be captured and stored; how work packages will be formed and controlled; scope and procedures for sustainability engineering analysis and risk management will be captured and documented.

5. Specific technical effort task requirements including work product outputs and data to be produced have been identified and budgeted for.

6. Potential conflicts between the work agreement, the technical effort, organizational policies and procedures, and available funding have been identified, addressed, and resolved.

7. Key events, phases, and review gates associated with the technical effort have been planned, coordinated, and committed to.

8. Project plans are available.

9. Roles, responsibilities, accountabilities, and authorities are defined.

10. Resources and services necessary to achieve the project objectives are formally requested and committed.

11. Project staff are directed in accordance with the project plans.

12. Plans for the execution of the project are activated.

Table 5.15   Process Implementation Strategy (SEP-15)

Process purpose: The provider shall define a detailed strategy for adopting standard processes of this guide and their implementation as the basis for technical planning and that is in accordance with the agreement. The intent is to provide enough information for the user to determine whether a given process activity is appropriate in supporting the objectives of the program or project they support and how to go about implementing the process activity.

Note that the act of planning should not be carried out in a vacuum. It is interactive and iterative and thus will require inputs regarding the technical effort, schedule, technical plans, and work directives.

Roles and agents:

- Project leader
- Sustainability engineer
- Integrated product team

Inputs:	**Outputs:**
• Requirements-related documents	• List of stakeholders and roles
	• Associated process approaches
	• Life cycle phase chart (milestones)
	• Work products and outputs—WBS
	• Work product reviews
	• Life cycle phase exit criteria
	• List of applicable tasks
	• Program metrics and reporting requirements
	• Project library
	• Process implementation strategy

Process diagram:

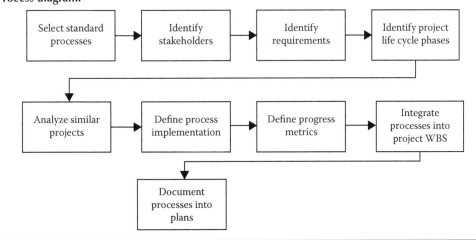

Figure 5.26   Process implementation strategy.

(*Continued*)

Table 5.15   Process Implementation Strategy (SEP-15) (*Continued*)

Step	Action
1	**Action statement:** Select processes. Select and use the organizational standard process assets for planning the project. **Notes and guidance:** N/A
2	**Action statement:** Identify stakeholders. Identify stakeholders who will have an interest or stake in the outcomes of the project. **Notes and guidance:** Consider stakeholders in both the funding chain and beneficiary chain (other stakeholders, primary users, etc.).
3	**Action statement:** Identify requirements. Identify and acquire requirements-related documents that will effect process selection. **Notes and guidance:** This will ensure the current and accurate documentation of the engineering baseline. The sustainability engineer is responsible for the implementation of, and adherence to, approved policies and processes. Making the applicable documents available in a project library enables the project's personnel to easily access the same baseline information as they perform their work. At a minimum, list the document name, version, and date for historical purposes. This information should be stored in the Enterprise Data Repository established in the configuration management process.
4	**Action statement:** Identify project life cycle phases. Identify the project life cycle phases, phase exit criteria, and related technical reviews. **Notes and guidance:** N/A
5	**Action statement:** Analyze similar projects. Identify and analyze like and similar projects along with their approaches to process selection and application. **Notes and guidance:** • Check that the similar project appears to have applied a reasonable approach to applying this process. • Include essential rationale elements in the process, such as risk management, training, testing, modeling and simulation, open systems, cost as an independent variable, environment considerations, and source of support.
6	**Action statement:** Define the processes that will be implemented. What processes are applicable to this project? **Notes and guidance:** 1. Read through all of this chapter to get the overall interrelationship of the processes and the document's philosophy and approach. 2. Take into account the phase and scope of your program/project using the available documents. Do this early in a program, since fewer guiding documents will be available later in the program/project. 3. Identify an initial list of which inputs and outputs are required to execute the program.

Table 5.15   Process Implementation Strategy (SEP-15) (*Continued*)

Step	Action
	4. Tracing the inputs and outputs through processes will reveal a number of things: a. Determine the level of process applicability and tailoring required. b. Additional inputs required. c. Support resources required and where these resources are available. 5. Check to see what outputs are produced by each process to see if all apply to the program considering its phase and scope. The descriptive portion of the tasks of a process contains clarifications of these outputs. This portion also gives guidance on developing the output by identifying the tools and organizations that are involved and detailing some interrelationships between the organizations. 6. Create a tailored version of this sustainability engineering process for your project. Creating a top-level plan can be accomplished by developing a Gantt chart using the schedule and tasking information in the inputs and the tailored process list. Consult with those responsible for the technical effort, schedule, technical plans, and work directives to determine how the details will be filled in.
7	**Action statement:** Define progress metrics. Identify and define progress assessment metrics and reporting requirements. **Notes and guidance:** The frequency and format of progress reports will impact the effort calculations in the technical planning process and the establishment of schedules in that process. The decision on whether or not to use an EVMS will also have impacts in technical planning. Projects should select meaningful metrics and measures specific to the program and add them to the generic list. Acknowledge that someone else is responsible for executing the process. That person will be responsible for defining and collecting metrics for both the process itself and the products that are produced. Without measuring the process itself, there is no way to tell that a change to the process was actually an improvement.
8	**Action statement:** Integrate processes into the WBS. Integrate applicable processes into the project WBS. **Notes and guidance:** The WBS is used to estimate the scope of the project, and it evolves with the project. Develop a WBS that reflects the produce and process architecture. The WBS should consider product and process work package outputs and effort-related activities that support project cost estimation.
9	**Action statement:** Document processes in plans. Document the selected process activities and selection rationale in appropriate planning documents. **Notes and guidance:** This documentation should also include details for modifications to the process implementation strategy.

Measures:

- Percentage of project technical plans that show that this process was implemented
- Number of man hours required to implement this process by a project

(*Continued*)

Table 5.15   Process Implementation Strategy (SEP-15) (*Continued*)

Process Task Outcomes	
Representative Tasks	*Process Task Outcomes*
Identify stakeholders	Intended users or customers and other stakeholders who will have an interest or stake in the outcome of the project are established.
Identify applicable documents	Applicable source and technical documents and the requirements therein that could affect the project effort are identified and acquired, including 1. The scope and purpose of both the project and products to be developed or reengineered 2. Stated purpose of the products, expectations of the stakeholders, expected benefits to stakeholders, as well as the goals and objectives of the product, or portion thereof, to be developed or reengineered 3. Enterprise policies, priorities, and constraints on funding, personnel, facilities, manufacturing capability and capacity, and critical resources that will affect accomplishing the requirements and goals of the source and technical documents 4. (a) Applicable processes, standards, and specifications; (b) core enterprise technologies; (c) risks to business growth by new project; (d) must-win criteria; (e) net cost targets; (f) methods of resource allocation; (g) how work and changes will be authorized; (h) how information will be captured; (i) how work packages will be formed and controlled; and (j) scope and procedures for trade-off analyses, effectiveness analyses, and risk management based on enterprise goals and planning baselines.
Identify associated process approaches	How development of enabling products associated with production, test, deployment, or installation, and logistics processes will be implemented is determined.
Identify applicable life cycle phases	Applicable enterprise-based life cycle phases, the expected work product outputs and management reviews, and the relevant exit criteria for each applicable enterprise-based life cycle phase, including level of product maturity expected, level of acceptable risk, management review concerns, and documentation requirements, are determined.
Identify and define technical process and project integration	How the applicable processes of this chapter will be integrated with each other and with other processes specified in enterprise and agreement documents, and which internal and external projects that will be involved and how they will be integrated are determined.

Table 5.15   Process Implementation Strategy (SEP-15) (*Continued*)

Representative Tasks	Process Task Outcomes
Identify and define metrics for measuring technical requirements progress in an assessment	Required reporting requirements, specific product, and process metrics to be used, how and when metrics will be collected and by whom, and how progress will be assessed are determined.
Prepare the process implementation strategy	A process implementation strategy document based on the integrated results of the outcomes of the above tasks is prepared.

Table 5.16   Technical Assessment Process (SEP-16)

Process purpose: The provider shall perform the project assessment and control process to determine the status of the project and direct project plan execution to ensure that the project performs according to plans and schedules, within projected budgets, to satisfy technical objectives.

This process evaluates, periodically and at major events, the progress and achievements against requirements, plans, and overall business objectives. Information is communicated for management action when significant variances are detected. This process also includes redirecting the project activities and tasks, as appropriate, to correct identified deviations and variations from other project management or technical processes. Redirection may include replanning as appropriate.

The technical assessment process is used to (1) determine progress of the technical effort against both plans and requirements; (2) review progress during technical reviews; and (3) support control for the engineering of a system. The product and process metrics selected for assessing progress should provide information for risk aversion, meaningful financial and nonfinancial performance, and support of project management.

NOTE: When variations are sufficiently significant or cannot be corrected by reaccomplishment of the process tasks that generated the outcome data, the technical planning process is reinitiated in order to implement appropriate corrective actions.

This process uses metrics (see the technical planning process) to track the progress of the processes. Product technical requirements essential to the product being acquired are also tracked. The technical assessment process uses metrics to track the progress against the program plans and schedules used to manage the program, while the project phase closure process tracks the progress in meeting product-related technical requirements. The sustainability engineering technical reviews process provides a status of design maturity and requirement satisfaction, identifies risks and issues to be resolved, and determines whether the product is ready for the next engineering phase. Cost, schedule, and performance variances reflected in the metrics are fed into a risk management system (see the risk and opportunity management process), which produces a risk management system with risk mitigations identified, the effect of which can be observed and adjusted. A program, which does not employ a closed loop to feed system variances into the risk management system, cannot be effective in making positive changes in the management of the system.

(Continued)

Table 5.16   Technical Assessment Process (SEP-16) (*Continued*)

Roles and agents:
• Project leader
• Sustainability engineer
• Integrated product team

Inputs:	Outputs:
• Technical performance measurements (TPMs)	• List of appropriate events, tasks, and process metrics
• Work breakdown structure	• Process metrics data
• Inputs to earned value management system	• Program metrics data
• Program metrics	• Deficiency reports
• Process metrics	• Discrepancy reports
• Integrated master schedule	• Status of the technical effort
• Sustainability engineering management plan or software development plan	• Degree of satisfaction of technical requirements
• Computer resources life cycle management plan	• Records of recommended corrective actions
• Configuration management plan	
• Analysis of alternatives technical report	
• Design solution discrepancy reports	
• Product deficiency reports	

Process diagram:

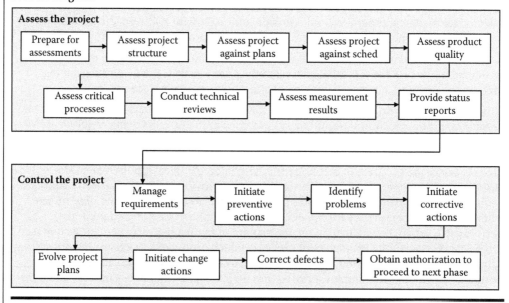

Figure 5.27   Technical assessment process.

Table 5.16   Technical Assessment Process (SEP-16) (*Continued*)

Step	Action
1	**Action statement:** Assess the project. **Notes and guidance:** N/A
1.1	**Action statement:** Prepare for technical assessments. Document the strategy and integrate with other related plans. **Notes and guidance:** Integrate technical plans to include the following: • Periodically determining the status and achievements against technical plans • Periodically determining the quality of the product under development • Conducting technical reviews to assess, redirect, or approve progress of the technical effort toward meeting criteria for entering the next project phase.
1.2	**Action statement:** Assess the effectiveness of project team structure, roles, responsibilities, accountabilities, and authorities. **Notes and guidance:** This includes assessment of the adequacy of team member competencies to perform project roles and accomplish project tasks. Use objective measures wherever possible; for example, efficiency of resource use, project achievement.
1.3	**Action statement:** Assess the adequacy and availability of the project's supporting infrastructure. **Notes and guidance:** This includes confirming that intraorganizational commitments are satisfied.
1.4	**Action statement:** Assess project status against appropriate project plans to determine actual and projected cost, schedule, and quality variations. **Notes and guidance:** Perform earned value management (EVM) to measure productivity based on technical cost and schedule performance. Use EVM to track completion of work packages and actual expenditures to complete the work over a given period of time. The time period is usually one project phase or stage of the project life cycle.
1.5	**Action statement:** Assess project progress using measured achievement and milestone completion. **Notes and guidance:** Collect and evaluate, at planned times, the actual or estimated labor, material, and service costs. Compare against defined project measures of achievement. This includes conducting effectiveness assessments to determine the adequacy of the evolving product against requirements. It also includes the readiness of enabling products to deliver their services when needed.

(*Continued*)

Table 5.16 Technical Assessment Process (SEP-16) (*Continued*)

Step	Action
1.6	**Action statement:** Assess technical work product quality. Perform quality assurance (QA) in accordance with project plans. Assess project adherence to performing processes in this guide as outlined in the approved project technical plans.
	Notes and guidance: Refer to the quality management process for details on the QA process.
	Assess technical product quality by performing the following subtasks:
	a. System product development progress and achievement against technical requirements using established events, methods, tools, and product measures.
	b. Measure expected values at established events where measurement data will be made available.
	c. Collect and analyze product data that describes the currently defined system product or service characteristics against requirements.
	d. Record decisions, rational for decisions, and assumptions made with respect to collected data.
	e. Compare results against system product or services requirements to determine degree of requirement satisfaction and progress toward satisfying expected maturity of the system products or services for the applicable project phase.
	f. Identify needed changes and recommendations for implementing changes.
	g. Report recommendations and the results of assessments and analysis to the appropriate technical managers for action.
1.7	**Action statement:** Monitor critical processes and new technologies.
	Notes and guidance: N/A
1.8	**Action statement:** Assess technical work productivity (efficiency).
	Notes and guidance: Refer to TPM as part of the technical measurement process.
1.9	**Action statement:** Conduct required management and technical reviews, audits, and inspections to determine readiness to proceed to the next stage of the product life cycle or project milestone. Also identify deficiencies and discrepancies to specifications and configuration baselines.
	Notes and guidance: This includes identifying and evaluating technology insertion according to project plans.
1.9	**Action statement:** Analyze measurement results to identify deviations or variations from planned values or status and make appropriate recommendations for corrections.
	Notes and guidance: This includes, where appropriate, statistical analysis of measures that indicates trends; for example, fault density to indicate quality of outputs and distribution of measured parameters that indicate process repeatability.

Table 5.16 Technical Assessment Process (SEP-16) (*Continued*)

Step	Action
1.10	**Action statement:** Provide periodic status reports and required deviation reports as designated in the agreement, policies, and procedures. **Notes and guidance:** N/A
2	**Control the Project**
2.1	**Action statement:** Manage project requirements and changes to requirements in accordance with the project plans. **Notes and guidance:** N/A
2.2	**Action statement:** Initiate the corrective actions needed to achieve the goals and outputs of project tasks that have deviated outside acceptable or defined limits. **Notes and guidance:** Corrective action may include replanning or redeployment and reassignment of personnel, tools, and project infrastructure assets when inadequacy or unavailability has been detected.
2.3	**Action statement:** Initiate preventive actions, as appropriate, to ensure achievement of the goals and outputs of the project. **Notes and guidance:** N/A
2.4	**Action statement:** Initiate problem resolution actions to correct nonconformances. **Notes and guidance:** This includes performing corrective actions to the implementation and execution of the life cycle processes when nonconformances are traced to them. Actions are documented and reviewed to confirm their adequacy and timeliness.
2.5	**Action statement:** Evolve with time the scope, definition, and the related breakdown of the work to be carried out by the project in response to the corrective action decisions taken and the estimated changes they introduce. **Notes and guidance:** N/A
2.6	**Action statement:** Initiate change actions when there is a contractual change to cost, time, or quality due to the impact of a customer or supplier request. **Notes and guidance:** N/A
2.7	**Action statement:** Act to correct defective provision of acquired goods and services through constructive interaction with the supplier. **Notes and guidance:** This may include consideration of modified terms and conditions for supply or initiating new supplier selection.
2.8	**Action statement:** Obtain authorization to proceed toward the next milestone or event if justified. **Notes and guidance:** N/A

(*Continued*)

Table 5.16 Technical Assessment Process (SEP-16) (*Continued*)

Process Task Outcomes
As a result of the successful implementation of the project assessment and control process:
a. Project performance measures or assessment results are available.
b. Adequacy of roles, responsibilities, accountabilities, authorities, and resources and services necessary to achieve the project is assessed.
c. Deviations in project performance indicators are analyzed.
d. Affected parties are informed of project status.
e. Corrective action is defined and directed when project achievement is not meeting planned targets.
f. Project replanning is initiated when project objectives or constraints have changed or when planning assumptions are shown to be invalid.
g. Project action to progress (or not) from one scheduled milestone or event to the next is authorized.
h. Project objectives are achieved.

Table 5.17 Sustainability Engineering Technical Reviews (SEP-17)

Process purpose: The supplier shall conduct technical reviews of progress and accomplishments in accordance with appropriate technical plans and organizational policies.	
Roles and agents: • Project leader • Sustainability engineer • Integrated product team	
Inputs: • Testing metrics • Technical performance measurements • Integrated master schedule • Validation plan • Sustainability engineering master plan • Test and evaluation master plan • Plans and schedules trend analysis • Requirement trend analysis • Deficiencies and discrepancies • System requirements document • System technical requirements • Specified requirements • Design solution deficiency and discrepancy reports • End product deficiency and discrepancy reports	**Outputs:** All outputs should be archived • List of issues needed corrective actions • Technical review report

Table 5.17   Sustainability Engineering Technical Reviews (SEP-17) (*Continued*)

Process diagram:

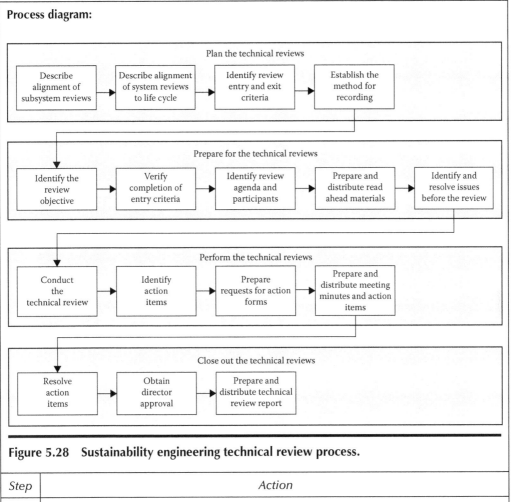

Figure 5.28   Sustainability engineering technical review process.

Step	Action
	Plan the Technical Review
1	**Action statement:** Align system, subsystem, and configuration item (CI) reviews.
	Notes and guidance: Actions 1 through 5 are to be described in the project technical plan (sustainability engineering plan). Start with the individual CIs and subsystem reviews and audit schedules, then allowing some reserve time for any required major fixes to CIs, and schedule the summary system-level reviews.
2	**Action statement:** Align product-level reviews with the product life cycle.
	Notes and guidance: Align the technical reviews with a project life cycle.

(*Continued*)

Table 5.17 Sustainability Engineering Technical Reviews (SEP-17) (*Continued*)

Step	Action
3	**Action statement:** Identify entry criteria and exit criteria for each review. **Notes and guidance:** After the number, type, and schedule for the reviews and audits has been established, write down the criteria for entering and successfully exiting each review or audit. These criteria must be distributed to all who will have a role in preparing for or participating in the reviews and audits. • Identify the anticipated completion at the stage of product maturity (technical performance measures [TPMs] and drawings) evaluated against the anticipated status/requirements. • Identify criteria to confirm that necessary subsystem reviews, inspections, deliveries, fabrication, and coding have been completed properly as specified/required.
4	**Action statement:** Identify review roles. **Notes and guidance:** Identify the core review board, decision authority for review completion. Also identify speakers, demonstration presenters, facility tour leaders, and recorders of activity minutes and action items. As each review/audit time approaches, a chairman and recording secretary must be named, along with all other representatives of the review team. Select from experts knowledgeable in the areas being reviewed, both from within and outside the project.
5	**Action statement:** Establish the method of recording, status monitoring, and closing action items. **Notes and guidance:** Establish a method for recording, reviewing, status monitoring, and closing action items assigned during the review or audit, including a formal signoff by the review chairman, as appropriate.
6	**Action statement:** Identify the review objectives and requirements cited in the sustainability engineering management plan (SEMP); enterprise policies and procedures; and agreement, as applicable. **Notes and guidance:** N/A
7	**Action statement:** Verify completion of the technical review entry requirements. **Notes and guidance:** 1. Identify the anticipated completion at that stage of maturity (TPMs, drawings) evaluated against the anticipated status/requirements. 2. Confirm that necessary reviews, inspections, tests, processes, deliveries, and coding were completed properly as specified/required.
8	**Action statement:** Establish the technical review board, agenda, and speakers. **Notes and guidance:** N/A
9	**Action statement:** Prepare the appropriate materials to include in the read-ahead technical review package and presentation package. **Notes and guidance:** N/A

Table 5.17   Sustainability Engineering Technical Reviews (SEP-17) (*Continued*)

Step	Action
10	**Action statement:** Analyze technical issues. Analyze the issues and action items to identify corrective actions necessary to address the issues. **Notes and guidance:** Place highest priority on issues that if left unresolved, may prevent the project from meeting its objectives.
11	**Action statement:** Facilitate and support identification and resolution of emerging issues prior to the review. **Notes and guidance:** N/A
12	**Action statement:** Conduct the technical review according to the SEMP, identifying and documenting action items required to meet the review objectives. **Notes and guidance:** 1. Evaluate the design for compliance with known technical requirements. 2. Verify interfaces compatibility. 3. Determine what issues remain to be resolved. 4. Verify that the emerging design is ready to enter the next stage of development. 5. Verify that the product is testable, buildable, usable, safe, and reliable. 6. Verify that the product exhibits the characteristics necessary to prove effective and suitable during operational evaluation throughout the development phase. 7. Challenge the design and related processes for optimization. 8. Communicate requirements, design concepts, and descriptions to other departments.
13	**Action statement:** Track action items to closure. Prepare and distribute minutes and action items. Take the appropriate corrective actions and track them to closure. **Notes and guidance:** Perform the following subactivities: 1. Determine and document the appropriate actions needed to address the identified issues. 2. Review and get agreement with relevant stakeholders on the actions to be taken. 3. Negotiate changes to internal and external commitments. 4. Monitor action results. 5. Analyze results of corrective actions to determine the effectiveness of the corrective actions. 6. Determine and document appropriate changes to approved actions needed to achieve needed results in solving issues.
14	**Action statement:** Close out the review after minutes have been prepared, approved, and distributed; action items have been resolved; and the review has been signed off by the director. **Notes and guidance:** Prepare the technical review report using the organizational standard technical review report template.

(Continued)

Table 5.17 Sustainability Engineering Technical Reviews (SEP-17) (*Continued*)

Measures
a. Percent completion of technical review action items
b. Percent of achievement of cost, schedule, and performance key performance parameters
c. Resources expended on performing the technical review (preparation and review)

Process Task Outcomes	
Representative Tasks	*Process Task Outcomes*
Identify technical review object are requirements	The following are identified and documented: (1) purpose and objectives of the review, (2) agenda requirements, (3) tasks to be completed at each required review, (4) entrance and exit requirements, (5) documentation requirements, (6) distribution requirements, and (7) responsibilities of the review participants.
Determine progress against event-based plan	The satisfaction of entrance requirements to the review is determined and documented.
Establish technical review board, agenda, and speakers	For each review, the following are established: (1) persons who will participate in the review, (2) chairpersons, (3) secretary, (4) reviewers of the presentation, (5) agenda that meets review requirements and ensures that all required tasks are completed, and (6) members of the design team that will prepare the data package, prepare the presentation, prepare material for distribution at the review, make presentations, answer questions, and accomplish task close out action items.
Prepare technical review package and presentation materials	Comprehensive read-ahead material is prepared that includes sufficient information so that technical board members can understand the design and participate effectively in the review. Review team responsibilities, agendas, plans, and expectations from the review are defined and documented. A comprehensive set of presentation materials that describe the assigned design topics and that satisfy review objectives is prepared.
Facilitate resolution of emerging issues	Emerging issues identified and resolved prior to the review.
Conduct technical review	The following are assessed by the review: (1) maturity of the product, or portion thereof, being engineered, (2) progress according to plans and requirements, (3) risks and variances in cost schedule, and performance, and (4) readiness to proceed with the next phase of development. Action items required to meet review objectives are generated, recorded, and assigned.

Table 5.17 Sustainability Engineering Technical Reviews (SEP-17) (*Continued*)

Representative Tasks	Process Task Outcomes
Closeout review	The following are completed for review closeout: (1) preparation and distribution of minutes that include purpose, time, place, attendees, decisions, action items, due date, and persons responsible for resolving action items, (2) resolution of action items, and (3) sign off by chairperson.

Table 5.18 Requirements Management Process (SEP-18)

Process purpose: The supplier is to perform requirements management process activities to control the requirements of the project's products and product components and to identify any inconsistencies between those requirements and the projects plans and work products.

Roles and agents:

- Project leader
- Sustainability engineer
- Integrated product team

Inputs:	Outputs:
a. New or changed requirements b. Customer/sponsor requirements c. Validated requirements from earlier phase iterations	a. Lists of criteria for distinguishing appropriate requirements providers b. Criteria for evaluation and acceptance of requirements c. Results of analysis of requirements using criteria d. Requirements impact assessment reports e. Documented commitments to approved requirements and requirements changes f. Requirements status reports g. Requirements database h. Requirements decision database i. An approved requirements verification and traceability matrix j. Requirements tracking system k. Documentation of inconsistencies, including sources, conditions, and rationale l. Documentation of corrections of inconsistencies

(*Continued*)

Table 5.18   Requirements Management Process (SEP-18) (*Continued*)

Process diagram:

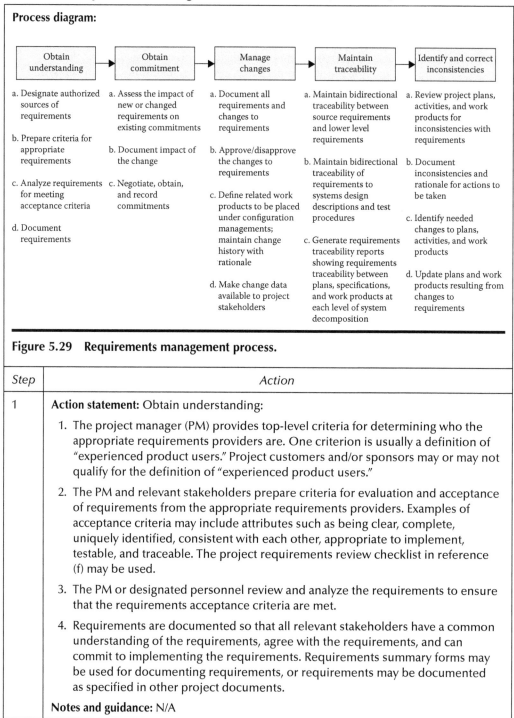

Obtain understanding	Obtain commitment	Manage changes	Maintain traceability	Identify and correct inconsistencies
a. Designate authorized sources of requirements	a. Assess the impact of new or changed requirements on existing commitments	a. Document all requirements and changes to requirements	a. Maintain bidirectional traceability between source requirements and lower level requirements	a. Review project plans, activities, and work products for inconsistencies with requirements
b. Prepare criteria for appropriate requirements	b. Document impact of the change	b. Approve/disapprove the changes to requirements	b. Maintain bidirectional traceability of requirements to systems design descriptions and test procedures	b. Document inconsistencies and rationale for actions to be taken
c. Analyze requirements for meeting acceptance criteria	c. Negotiate, obtain, and record commitments	c. Define related work products to be placed under configuration managements; maintain change history with rationale	c. Generate requirements traceability reports showing requirements traceability between plans, specifications, and work products at each level of system decomposition	c. Identify needed changes to plans, activities, and work products
d. Document requirements		d. Make change data available to project stakeholders		d. Update plans and work products resulting from changes to requirements

Figure 5.29   Requirements management process.

Step	Action
1	**Action statement:** Obtain understanding: 1. The project manager (PM) provides top-level criteria for determining who the appropriate requirements providers are. One criterion is usually a definition of "experienced product users." Project customers and/or sponsors may or may not qualify for the definition of "experienced product users." 2. The PM and relevant stakeholders prepare criteria for evaluation and acceptance of requirements from the appropriate requirements providers. Examples of acceptance criteria may include attributes such as being clear, complete, uniquely identified, consistent with each other, appropriate to implement, testable, and traceable. The project requirements review checklist in reference (f) may be used. 3. The PM or designated personnel review and analyze the requirements to ensure that the requirements acceptance criteria are met. 4. Requirements are documented so that all relevant stakeholders have a common understanding of the requirements, agree with the requirements, and can commit to implementing the requirements. Requirements summary forms may be used for documenting requirements, or requirements may be documented as specified in other project documents. **Notes and guidance:** N/A

Table 5.18 Requirements Management Process (SEP-18) (*Continued*)

Step	Action
2	**Action statement:** Obtain commitment: 1. Relevant stakeholders (who implement the requirements) assess the impact of new or changed requirements on existing commitments to requirements, project plans, project activities, and project work products. 2. The PM or designated stakeholders negotiate, obtain, and document commitments to requirements or requirements changes. Commitments should be negotiated before project participants commit to the requirement or requirement change. **Notes and guidance:** N/A
3	**Action statement:** Manage changes: 1. The PM or designated personnel document all requirements changes that are given to or generated by the project. The PM, or designated personnel, collect requirement changes requested externally or generated by the project. 2. The PM approves/disapproves the requirement changes. 3. The PM defines what products of this process are to be maintained under configuration control. Configuration management (CM) maintains a requirements change history with impact analysis and accompanying rationale for changes. 4. CM makes approved requirements and changes data available to the project's relevant stakeholders. **Notes and guidance:** N/A
4	**Action statement:** Maintain traceability: 1. CM or designated personnel maintain bidirectional traceability from source requirements to lower level requirements as the product or task is decomposed across the life cycle. This may be accomplished using the guidance in the Project Requirements Traceability Matrix Guide, a software tool such as the tool template in the Project Requirements Traceability Matrix Template, or as defined in the project plan. 2. CM or designated personnel maintain bidirectional traceability of requirements to their allocation of functions, objects, processes, and work products (as appropriate). To verify requirements, bidirectional traceability to test procedures and reports is also maintained. 3. CM or designated personnel maintain a method of generating traceability reports showing traceability of requirements to and from project plans and work products at each applicable level of product or task decomposition. The reporting method is to be defined in the project plans. **Notes and guidance:** N/A

(*Continued*)

Table 5.18   Requirements Management Process (SEP-18) (*Continued*)

Step	Action
5	**Action statement:** Identify and correct inconsistencies: 1. The PM and relevant stakeholders review project plans, activities, and work products for consistency with requirements and requirement changes and document inconsistencies. 2. The PM, relevant stakeholder, or designated personnel identify the source of any inconsistencies and the rationale for actions to be taken. 3. The PM, relevant stakeholder, or designated personnel identify the changes that need to be made to plans and work products resulting from any inconsistencies in or changes to the requirements baseline. 4. The PM initiates corrective action as needed to update plans and work products resulting from any inconsistencies in or changes to the requirements baseline. **Notes and guidance:** N/A

Measures:
a. The number of new or changed requirements during the reporting period (requirements volatility)
b. Planned versus actual requirements process task costs

Table 5.19   Interface Management Process (SEP-19)

Process purpose: The provider shall perform interface management activities to ensure interface definition and compliance among the elements that compose the product, as well as with other products with which the product or system elements will interoperate (i.e., system of systems), and to ensure that all internal and external interface requirement changes are properly documented and controlled in accordance with the configuration management plan and communicated to all affected stakeholders.

Roles and agents:
• Project leader
• Sustainability engineer
• Integrated product team (IPT)

Inputs:	**Outputs:**
• Interface management plan	• Interface specifications
• Interface requirements	• Interface control documents (ICDs) /drawings
• Interface requirements changes	• Interface action control sheets

Table 5.19   Interface Management Process (SEP-19) (*Continued*)

Process diagram:

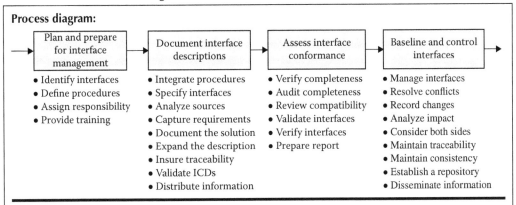

Figure 5.30   Interface management process.

Step	Activity
1	**Activity statement:** Plan and prepare for interface management. **Notes and guidance:** Interface management control measures ensure that all internal and external interface requirement changes are properly documented in accordance with the configuration management plan and communicated to all affected configuration items. Interface management deals with the following: • Defining and establishing interface specifications • Assessing compliance of interfaces among configuration items comprising systems or system of systems • Monitoring the viability and integrity of interfaces within a product • Establishing an interface management plan to assess existing and emerging interface standards and profiles, update interfaces, and abandon obsolete architectures An interface management plan is a part of a configuration management plan that • Documents a product's internal and external interfaces and their requirement specifications • Identifies preferred and discretionary interface standards and their profiles • Provides justification for selection and procedure for upgrading interface standards • Describes the certifications and tests applicable to each interface or standard
1.1	**Activity statement:** Identify Interfaces. Describe procedures for identifying interface requirements to be managed, interface requirement sources, and traceability of interface requirements. **Notes and guidance:** N/A
1.2	**Activity statement:** Define procedures. Establish formal procedures for the initiation, assessment, review, approval, and disposition of engineering change proposals (engineering change proposals involving interfaces). **Notes and guidance:** N/A

(*Continued*)

Table 5.19   Interface Management Process (SEP-19) (*Continued*)

Step	Activity
1.3	**Activity statement:** Establish the interface control working group (ICWG), and manage responsibilities for those interfaces that are part of agreement boundaries, as well as other interfaces that need to be managed.
	Notes and guidance: The ICWG is a specialized technical working group composed of appropriate technical representatives from the interfacing activities and other interested participating organizations. The ICWG serves as a forum to develop and provide interface requirements, as well as to focus on interface detail definition and timely resolution of issues. The ICWG requires collaboration with external program offices and contractors in a product or system of systems environment.
1.4	**Activity statement:** Provide training. Train IPT and other technical personnel in the established interface management procedures.
	Notes and guidance: N/A
2	**Activity statement:** Document the interface descriptions.
	Notes and guidance: N/A
2.1	**Activity statement:** Integrate procedures. Integrate interface management procedures with the requirements management procedures and configuration management procedures.
	Notes and guidance: N/A
2.2	**Activity statement:** Specify interfaces. Identify, capture, and document interfaces, both internal and external to the system model and internal to the system model products.
	Notes and guidance: N/A
2.3	**Activity statement:** Analyze sources. Analyze the concept of operations and similar documents to identify any interfaces not included in the original set of stakeholder requirements.
	Notes and guidance: N/A
2.4	**Activity statement:** Capture requirements. Identify and capture the requirements for the identified interfaces including origin, destination, stimulus, and special characteristics based on the type of interface (internal, external, and enabling product).
	Notes and guidance: N/A
2.5	**Activity statement:** Document the solution. Document the design solution interfaces for the system model under development.
	Notes and guidance: N/A

Table 5.19   Interface Management Process (SEP-19) (*Continued*)

Step	Activity
2.6	**Activity statement:** Expand the description. Ensure that the design solution for the end product includes the interface requirements defined during stakeholder requirements definition (SP-23), and requirements analysis (SP-24), and includes the requirement origin, destination, stimulus, and special characteristics. **Notes and guidance:** N/A
2.7	**Activity statement:** Insure traceability. Maintain traceability of interface requirements across interfaces and document status in the established information database. **Notes and guidance:** N/A
2.8	**Activity statement:** Validate ICDs. Ensure that ICD or drawing that is established has been validated with representatives from both sides of the interface. **Notes and guidance:** N/A
2.9	**Activity statement:** Distribute information. Provide authorized users with the needed interface information for integration into technical efforts and for external interface control. **Notes and guidance:** N/A
3	**Activity statement:** Assess interface conformance. **Notes and guidance:** N/A
3.1	**Activity statement:** Verify completeness. Review interface documentation for completeness; and identify and report any discrepancies. **Notes and guidance:** N/A
3.2	**Activity statement:** Audit completeness. Perform a visual inspection to check on all physical interfaces before connecting products together. **Notes and guidance:** N/A
3.3	**Activity statement:** Review compatibility. Evaluate implemented and assembled components for interface compatibility. **Notes and guidance:** N/A
3.4	**Activity statement:** Validate interfaces. Confirm that product validation plans, approaches, and procedures include confirmation of defined interfaces of each implemented product to be integrated. **Notes and guidance:** N/A
3.5	**Activity statement:** Verify interfaces. Confirm that verification plans, approaches, and procedures include confirmation of both external and internal specified interfaces. **Notes and guidance:** N/A
3.6	**Activity statement:** Prepare report. Prepare an interface evaluation report upon the completed verifications and validations. **Notes and guidance:** N/A

(*Continued*)

Table 5.19   Interface Management Process (SEP-19) (*Continued*)

Step	Activity
4	**Activity statement:** Baseline and control interfaces.
	Notes and guidance: Refinement of the interfaces is achieved through iteration. As more is learned about the product during the design phases, lower level, verifiable requirements and interfaces are defined and refined. Impacts of the original defined capabilities and interfaces, performance parameter thresholds and objectives, and the products are evaluated when defining and modifying interfaces.
4.1	**Activity statement:** Manage interfaces. Monitor proposed changes to the system model requirements, looking for those affecting established interfaces.
	Notes and guidance: N/A
4.2	**Activity statement:** Resolve conflicts. Identify and resolve conflicts, noncompliance, and change issues; propose changes to resolve discrepancies.
	Notes and guidance: N/A
4.3	**Activity statement:** Record changes. Identify and record proposed and directed changes to stakeholder or design solution interface requirements/specifications and ICDs/drawings.
	Notes and guidance: N/A
4.4	**Activity statement:** Analyze impacts. Analyze the cost, schedule, performance, and risks associated with making a proposed or directed interface change within planned time limits and resource availability.
	Notes and guidance: N/A
4.5	**Activity statement:** Consider both sides. Ensure that the interface issues are analyzed and resolved when a change affects products on both sides of the interface.
	Notes and guidance: N/A
4.6	**Activity statement:** Maintain traceability. Maintain and control traceability of changes including source of the change, processing methods, and approvals in accordance with the interface management plan.
	Notes and guidance: N/A
4.7	**Activity statement:** Maintain consistence. Ensure the consistency of the interfaces throughout the life cycle of the product.
	Notes and guidance: N/A
4.8	**Activity statement:** Establish a repository. Establish and maintain a repository for interface data. The project "Integrated Digital Data Environment" (IDE) is one tool that can be used for this data.
	Notes and guidance: N/A
4.9	**Activity statement:** Disseminate information. Distribute approved interface change information/data for implementation at every level of the project and for integration into technical efforts.
	Notes and guidance: N/A

Table 5.19   Interface Management Process (SEP-19) (*Continued*)

Measures:
• Number of product interfaces documented and baselined
• Dollar amount of resources spent on interface management per reporting period

Process Task Outcomes
• The procedures for performing interface management activities are established and made available to IPTs and technical managers.
• Qualified technical personnel capable of conducting established interface management procedures are identified.
• Appropriate ICWGs are established.
• Stakeholder interface requirements are captured in appropriate interface specifications, ICDs, or interface control drawings.
• Assurance that applicable internal and external interfaces (including human) are identified, defined, assigned, documented, and managed for the system model.
• Mapping of product-to-product interfaces and product-to-product integration environment has been completed.
• Interface evaluation reports have been documented.
• Information about the relationships among different products of a system model, between models throughout the system structure, and with the external environment is available to stakeholders.
• ICWG meeting reports/minutes with action items status are available.
• Traceability of interface requirement changes as to source and changes to requirements baselines and interface specifications and ICDs/drawings is maintained.

Table 5.20   Decision Analysis and Management Process (SEP-20)

Process purpose: The implementer shall perform the decision analysis process to analyze decision alternatives using a formal evaluation process that evaluates and documents identified alternatives against established criteria.
Roles and agents: • Project leader   • Sustainability engineer   • Integrated product team

(*Continued*)

Table 5.20   Decision Analysis and Management Process (SEP-20) (*Continued*)

Inputs:	Outputs:
• Request for formal decision support	• Guidelines for when to apply a formal evaluation process • Documented evaluation criteria • Rankings of criteria importance • Identified alternatives • Selected evaluation methods • Evaluation results • Recommended solutions

Process diagram:

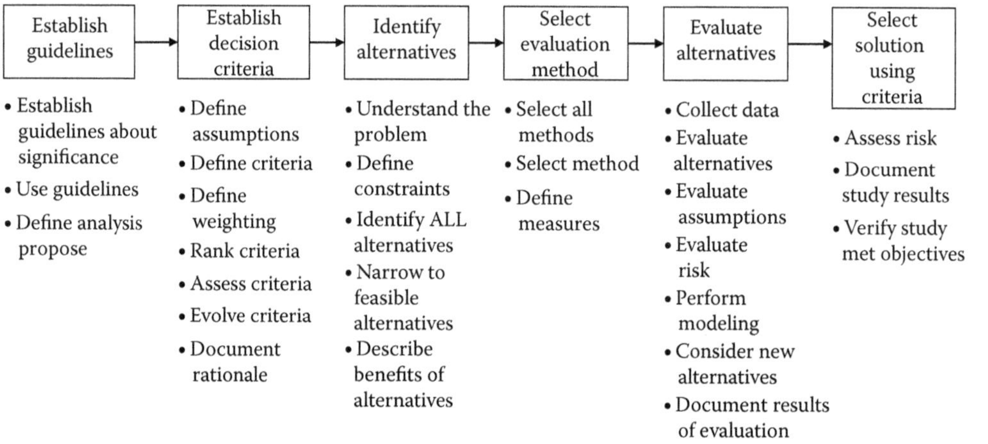

Figure 5.31   Decision analysis and management process.

Step	Activity
1	**Activity statement:** Establish decision analysis guidelines. Issues requiring a formal evaluation process may be identified at any time. The objective should be to identify issues as early as possible to maximize the time available to resolve them. **Notes and guidance:** Guidelines are created for deciding when to use formal evaluation processes to address unplanned issues. Guidelines often suggest using formal evaluation processes when issues are associated with medium to high risks or when issues affect the ability to achieve project objectives.
1.1	**Activity statement:** Establish guidelines. Establish guidelines to determine if the decision is significant enough to warrant formal analysis. **Notes and guidance:** A decision management strategy includes the identification and allocation of responsibility for, and authority to make, decisions and the identification of decision categories and a prioritization scheme. Decisions may arise as a result of an effectiveness assessment (technical assessment process), a technical trade-off (product analysis process), a problem needing to be solved, an action needed as a response to risk exceeding the acceptable threshold, a new opportunity, or an approval for project progression to the next life cycle stage.

Table 5.20   Decision Analysis and Management Process (SEP-20) (*Continued*)

Step	Activity
1.2	**Activity statement:** Incorporate guidelines. Incorporate the use of these guidelines into other applicable processes where appropriate. **Notes and guidance:** Not every decision is significant enough to require a formal evaluation process. The choice between the trivial and the truly important will be unclear without explicit guidance. Whether a decision is significant or not is dependent on the project and circumstances, and is determined by the established guidelines. Typical guidelines for determining when to require a formal evaluation process include the following: • When a decision is directly related to topics assessed as being of medium or high risk • When a decision is related to changing work products under configuration management • When a decision would cause schedule delays over a certain percentage or specific amount of time • When a decision affects the ability to achieve project objectives • When the costs of the formal evaluation process are reasonable • When compared to the decision's impact • When a legal obligation exists during a solicitation
1.3	**Activity statement:** Define analysis purpose. Define the purpose of the analysis/decision. **Notes and guidance:** N/A
2	**Activity statement:** Establish evaluation criteria. The evaluation criteria provide the basis for evaluating alternative solutions. The criteria are ranked so that the highest ranked criteria exert the most influence on the evaluation. This process is referenced by many other process areas in this guide, and there are many contexts in which a formal evaluation process can be used. Therefore, in some situations, you may find that criteria have already been defined as part of another process. This specific practice does not suggest that a second development of criteria be conducted. Document the evaluation criteria to minimize the possibility that decisions will be second-guessed or that the reason for making the decision will be forgotten. Decisions based on criteria that are explicitly defined and established remove barriers to stakeholder buy-in.
2.1	**Activity statement:** Define ground rules and assumptions. **Notes and guidance:** N/A

(*Continued*)

Table 5.20 Decision Analysis and Management Process (SEP-20) (*Continued*)

Step	Activity
2.2	**Activity statement:** Define criteria. Define the criteria for evaluating the alternatives. **Notes and guidance:** Criteria should be traceable to requirements, scenarios, business case assumptions, business objectives, or other documented sources. Types of criteria to consider include the following: • Technology limitations • Environmental impact • Risks • Total ownership and life cycle costs
2.3	**Activity statement:** Define scale for weighting. Define the range and scale for ranking the evaluation criteria. **Notes and guidance:** Scales of relative importance for evaluation criteria can be established with nonnumeric values or with formulas that relate the evaluation parameter to a numeric weight.
2.4	**Activity statement:** Rank the criteria. **Notes and guidance:** The criteria are ranked according to the defined range and scale to reflect the needs, objectives, and priorities of the relevant stakeholders.
2.5	**Activity statement:** Assess the criteria. Assess the criteria and their relative importance. **Notes and guidance:** N/A
2.6	**Activity statement:** Evolve criteria. Evolve the evaluation criteria to improve their validity. **Notes and guidance:** N/A
2.7	**Activity statement:** Document rationale. Document the rationale for the selection or rejection of evaluation criteria. **Notes and guidance:** Documentation of selection criteria and rationale may be needed to justify solutions or for future reference and use.
3	**Activity statement:** Identify alternative solutions. A wider range of alternatives can surface by soliciting as many stakeholders as practical for input. Input from stakeholders with diverse skills and backgrounds can help teams identify and address assumptions, constraints, and biases. Brainstorming sessions may stimulate innovative alternatives through rapid interaction and feedback. Sufficient candidate solutions may not be furnished for analysis. As the analysis proceeds, other alternatives should be added to the list of potential candidate solutions. The generation and consideration of multiple alternatives early in a decision analysis and resolution process increase the likelihood that an acceptable decision will be made and that consequences of the decision will be understood.

Table 5.20   Decision Analysis and Management Process (SEP-20) (*Continued*)

Step	Activity
3.1	**Activity statement:** Understand the problem. Perform a literature search to better understand the problem.
	Notes and guidance: A literature search can uncover what others have done both inside and outside the organization. It may provide a deeper understanding of the problem, alternatives to consider, barriers to implementation, existing trade studies, and lessons learned from similar decisions.
3.2	**Activity statement:** Define constraints. Define the requirements and constraints applicable to the decision and alternatives.
	Notes and guidance: N/A
3.3	**Activity statement:** Identify all alternatives. Identify all potential alternatives for consideration.
	Notes and guidance: Evaluation criteria are an effective starting point for identifying alternatives. The evaluation criteria identify the priorities of the relevant stakeholders and the importance of technical, logistical, or other challenges.
	Combining key attributes of existing alternatives can generate additional and sometimes stronger alternatives.
	Solicit alternatives from relevant stakeholders. Brainstorming sessions, interviews, and working groups can be used effectively to uncover alternatives.
3.4	**Activity statement:** Narrow to feasible alternatives. Document the proposed, feasible alternatives.
	Notes and guidance: N/A
3.5	**Activity statement:** Describe benefits of alternatives. Describe the advantages and disadvantages of each alternative.
	Notes and guidance: N/A
4	**Activity statement:** Select evaluation methods.
	Methods for evaluating alternative solutions against established criteria can range from simulations to the use of probabilistic models and decision theory. These methods need to be carefully selected. The level of detail of a method should be commensurate with cost, schedule, performance, and risk impacts.
	While many problems may need only one evaluation method, some problems may require multiple methods. For instance, simulations may augment a trade study to determine which design alternative best meets a given criterion.
4.1	**Activity statement:** Select the methods. Select the analysis methods based on the purpose of the analysis and the economical availability on information.
	Notes and guidance:
	For example, the methods used for evaluating a solution when requirements are weakly defined may be different from the methods used when the requirements are well defined. Typical evaluation methods include the following:
	1. Modeling and simulation
	2. Engineering studies

(*Continued*)

Table 5.20   Decision Analysis and Management Process (SEP-20) (*Continued*)

Step	Activity
	3. Manufacturing studies 4. Cost studies 5. Business opportunity studies 6. Surveys 7. Extrapolations based on field experience and prototypes 8. User review and comment 9. Testing 10. Judgment provided by an expert or group of experts (e.g., Delphi method)
4.2	**Activity statement:** Select methods based on their ability to focus on the issues. **Notes and guidance:** Results of simulations can be skewed by random activities in the solution that are not directly related to the issues at hand.
4.3	**Activity statement:** Define measures. Define the measures needed to support the evaluation method. **Notes and guidance:** Consider the impact on cost, schedule, performance, and risks.
5	**Activity statement:** Evaluate alternatives. Evaluating alternative solutions involves analysis, discussion, and review. Iterative cycles of analysis are sometimes necessary. Supporting analyses, experimentation, prototyping, piloting, or simulations may be needed to substantiate scoring and conclusions. Often, the relative importance of criteria is imprecise, and the total effect on a solution is not apparent until after the analysis is performed. In cases where the resulting scores differ by relatively small amounts, the best selection among alternative solutions may not be clear cut. Challenges to criteria and assumptions should be encouraged.
5.1	**Activity statement:** Collect data. Collect needed supporting data (to populate the analysis method). **Notes and guidance:** N/A
5.2	**Activity statement:** Evaluate alternatives. Evaluate the alternatives using evaluation criteria. **Notes and guidance:** N/A
5.3	**Activity statement:** Evaluate assumptions. Evaluate the ground rules and assumptions relative to the criteria. **Notes and guidance:** N/A
5.4	**Activity statement:** Evaluate risk. Evaluate risk and uncertainty concerning ranking values and input data. **Notes and guidance:** For instance, if the score can vary between two values, is the difference significant enough to make a difference in the final solution set? Does the variation in score represent a high risk? To address these concerns, simulations may be run, further studies may be performed, or evaluation criteria may be modified, among other things.

Table 5.20   Decision Analysis and Management Process (SEP-20) (*Continued*)

Step	Activity
5.5	**Activity statement:** Perform modeling. Perform modeling and simulations, prototypes, and pilots as necessary to gather data in areas of evaluation criteria. **Notes and guidance:** Untested criteria, their relative importance, and supporting data or functions may cause the validity of solutions to be questioned. Criteria and their relative priorities and scales can be tested with trial runs against a set of alternatives. These trial runs of a select set of criteria allow for the evaluation of the cumulative impact of the criteria on a solution. If the trials reveal problems, different criteria or alternatives might be considered to avoid biases.
5.6	**Activity statement:** Consider new alternatives. Consider new alternatives, criteria, of methods if the proposed solutions do not test well. **Notes and guidance:** N/A
5.7	**Activity statement:** Document results. Document the results of the evaluation for reporting to stakeholders. **Notes and guidance:** Document the rationale for the addition of new alternatives or methods and changes to criteria, as well as the results of interim evaluations.
6	**Activity statement:** Select solution based on evaluation criteria Selecting solutions involves weighing the results from the evaluation of alternatives. Risks associated with implementation of the solutions must be assessed.
6.1	**Activity statement:** Assess the risk. Assess the risks associated with implementing the recommended solution. **Notes and guidance:** Decisions must often be made with incomplete information. There can be substantial risk associated with the decision because of having incomplete information. When decisions must be made according to a specific schedule, time and resources may not be available for gathering complete information. Consequently, risky decisions made with incomplete information may require reanalysis later. Identified risks should be monitored.
6.2	**Activity statement:** Document study results. Document the results and rational for the recommended solution. **Notes and guidance:** It is important to record both why a solution is selected and why another solution was rejected.
6.3	**Activity statement:** Verify study met objectives. Verify that the study met the objectives and expectations of sponsors and stakeholders. **Notes and guidance:** N/A
	Measures: • Number of project decision support records per reporting period • Time to document, process, and record decisions (recorded on decision report form)

(*Continued*)

Table 5.20   Decision Analysis and Management Process (SEP-20) (*Continued*)

Step	Activity
	Process Task Outcomes

As a result of the successful implementation of the decision management process:

1. A decision management strategy is defined.
2. Alternative courses of action are defined.
3. A preferred course of action is selected.
4. The resolution, decision rationale, and assumptions are captured and reported.

Table 5.21   Data Management Process (SEP-21)

Process purpose: The provide shall perform the data and information management process plan to acquire, access, manage, protect, share, and use data of a technical nature to support the total life cycle of the system or product.

Data management in this guide will focus on technical data, which is any data other than computer software that is of a scientific or technical nature. This includes data associated with product development, configuration management, test and evaluation, installation, parts, spares, repairs, and product sustainment. Data specifically not included would be data relating to tactical operations information; sensor or communications information; financial transactions; personnel data; and other data of a purely business nature. Technical data can exist in many forms: paper or electronic documents, specifications, drawings, lists, records, repositories, standards, models, correspondence, and other descriptions of a product.

The purpose of this process is to ensure that required and requested information is properly disseminated so that necessary communications within the project and enterprise, and with the customer and other stakeholder community, are efficiently and effectively completed throughout the product life cycle. Project risks are increased when information is not available for decision making in a timely manner or if the information provided is of insufficient quality (e.g., too much, incomplete, not relevant, or inaccurate).

Roles and agents:

- Project leader
- Sustainability engineer
- Integrated product team

Inputs:	Outputs:
• Data management plan	• Technical data management procedures
• Data products to be managed	• Forms of data representations
• Data requests	• Data exchange formats
	• Requested data-delivered

Table 5.21   Data Management Process (SEP-21) (*Continued*)

Process diagram:

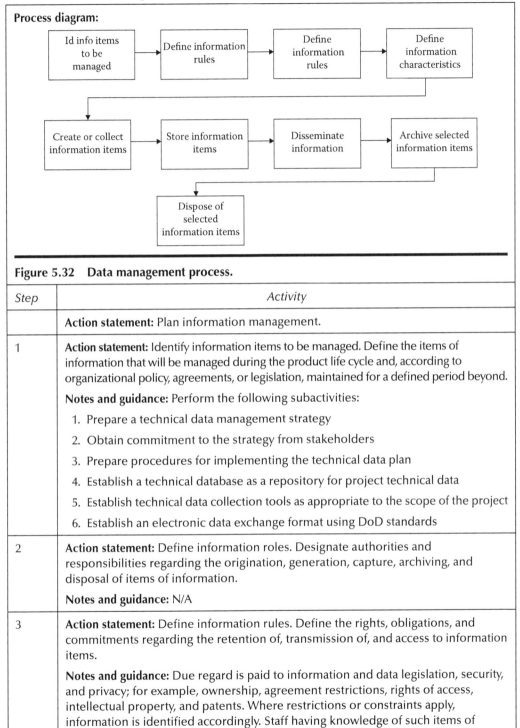

Figure 5.32   Data management process.

Step	Activity
	Action statement: Plan information management.
1	**Action statement:** Identify information items to be managed. Define the items of information that will be managed during the product life cycle and, according to organizational policy, agreements, or legislation, maintained for a defined period beyond. **Notes and guidance:** Perform the following subactivities: 1. Prepare a technical data management strategy 2. Obtain commitment to the strategy from stakeholders 3. Prepare procedures for implementing the technical data plan 4. Establish a technical database as a repository for project technical data 5. Establish technical data collection tools as appropriate to the scope of the project 6. Establish an electronic data exchange format using DoD standards
2	**Action statement:** Define information roles. Designate authorities and responsibilities regarding the origination, generation, capture, archiving, and disposal of items of information. **Notes and guidance:** N/A
3	**Action statement:** Define information rules. Define the rights, obligations, and commitments regarding the retention of, transmission of, and access to information items. **Notes and guidance:** Due regard is paid to information and data legislation, security, and privacy; for example, ownership, agreement restrictions, rights of access, intellectual property, and patents. Where restrictions or constraints apply, information is identified accordingly. Staff having knowledge of such items of information are informed of their obligations and responsibilities.

(*Continued*)

Table 5.21   Data Management Process (SEP-21) (*Continued*)

Step	Activity
4	**Action statement:** Define information characteristics. Define the content, semantics, formats and medium for the representation, retention, transmission, and retrieval of information.
	Notes and guidance: The information may originate and may terminate in any form (e.g., verbal, textual, graphical, numerical) and may be stored, processed, replicated, and transmitted using any medium (e.g., electronic, printed, magnetic, optical). Pay due regard to organization constraints; for example, infrastructure, interorganizational communications, distributed project working. Relevant information storage, transformation, transmission, and presentation standards and conventions are used according to policy, agreements, and legislation constraints.
5	**Action statement:** Define information maintenance actions.
	Notes and guidance: This includes status reviews of stored information for integrity, validity, and availability and any needs for replication or transformation to an alternative medium. Consider the need to either retain infrastructure as technology changes so that archived media can be read or the need to rerecord archived media using newer technology.
6	**Action statement:** Perform information management.
7	**Action statement:** Produce and collect information items. Obtain the identified items of information.
	Notes and guidance: This may include generating the information or collecting it from appropriate sources.
8	**Action statement:** Store information items. Maintain information items and their storage records according to integrity, security, and privacy requirements.
	Notes and guidance: Record the status of information items; for example, version description, record of distribution, and security classification. Information should be legible and stored and retained in such a way that it is readily retrievable in facilities that provide a suitable environment and that prevent damage, deterioration, and loss.
9	**Action statement:** Retrieve and distribute information to designated parties as required by agreed schedules or defined circumstances.
	Notes and guidance: Information is provided to designated parties in an appropriate form.
10	**Action statement:** Distribute information. Provide official documentation to stakeholders as required.
	Notes and guidance: Examples of official documentation are certification, accreditation, license, and assessment ratings.
11	**Action statement:** Archive selected information. Archive designated information, in accordance with the audit, knowledge retention, and project closure purposes.
	Notes and guidance: Select the media, location, and protection of the information in accordance with the specified storage and retrieval periods and with organization policy, agreements, and legislation. Ensure arrangements are in place to retain necessary documentation after project closure.

Table 5.21   Data Management Process (SEP-21) (*Continued*)

Step	Activity
12	**Action statement:** Dispose of information items. Dispose of unwanted, invalid, or unverifiable information according to organization policy and security and privacy requirements. **Notes and guidance:** N/A

Measures:

- Percent of data requests filled within organizational guidelines
- Amount of resources expended per measurement period in maintaining the data management system

Process Task Outcomes

1. Information to be managed is identified.
2. The forms of the information representations are defined.
3. Information is transformed and disposed of as required.
4. The status of information is recorded.
5. Information is current, complete, and valid.
6. Information is made available to designated parties.

Table 5.22   Configuration Management Process (SEP-22)

Process purpose: The supplier is to perform configuration management (CM) to establish and maintain the integrity of work products by using configuration identification, configuration control, configuration status accounting (CSA), and configuration audits.	
Roles and agents: - Project leader - Sustainability engineer - Integrated product team	
Inputs: - Programmatic data: project plans and schedules (e.g., project management plan [PMP]), reports, review results - Change requests (CRs), and waivers and deviations that need to be controlled - Technical data: specifications, requirements, designs, code, documentation	**Outputs:** - CM plans (CMPs) - CM procedures - CM review and audit reports - Baseline and controlled work products - Measurements of CM process and audits - CSA reports

(*Continued*)

Table 5.22   Configuration Management Process (SEP-22) (*Continued*)

Process diagram:

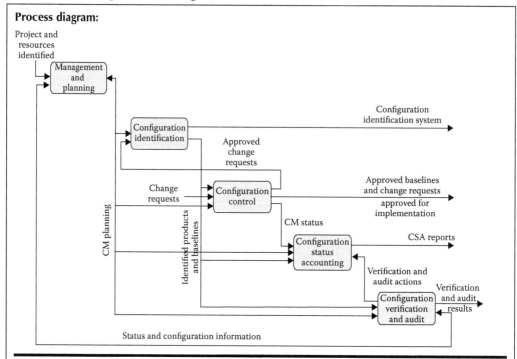

Figure 5.33   Configuration management process.

Step	Activity
1	**Action statement:** Create and maintain project CMP
	1. The CM group documents plans in a PMP and/or in a CMP. Plans include purpose of CM; governing standards; CM organization, roles, and responsibilities; configuration items (CIs); control boards; and CM functions, activities, tools, and procedures.
	2. The PM and Configuration Control Board (CCB) review and approve the CMP and provide commitment to the plan.
	Notes and guidance: Include the following topics in the CMP:
	1. Describe CM requirements and activities
	2. Describe the roles and responsibilities for performing CM
	3. Describe procedures and schedules for performing CM activities
	4. Define criteria or events for commencing CM and baselining of evolving configurations
	5. Define the creation of, disposition of, and control of information concerning CIs.
2	**Action statement:** Perform configuration identification.
	1. The CM group, in agreement with the PM, identifies the items to be placed under CM.

Table 5.22 Configuration Management Process (SEP-22) (*Continued*)

Step	Activity
	2. The CM group assigns unique identifiers (including the associated baseline) to CIs and related technical documentation and data.
	3. The CM group assigns tracking numbers to CRs and maintains a CR database.
	4. The CM group establishes CM libraries.
	Notes and guidance: Perform the following subactivities:
	1. Select the CIs and work products that compose them, based on documented criteria
	2. Assign unique identifiers to CIs
	3. Mark CIs in accordance with relevant standards and conventions so that CIs are traceable to their specifications and/or documentation
	4. Identify the documentation that serves as the baseline for each CI
	5. Specify the important characteristics of each CI
	6. Specify when each CI is to be placed under CM
	7. Identify the "owner" responsible for each CI
3	**Action statement:** Perform configuration control.
	1. The CM group places CIs and applicable technical artifacts in CM libraries to maintain the integrity of the products throughout the life cycle.
	2. The CCB authorizes baselines (formally approved versions designated to be fixed at a specific time during the life cycle) and reviews and approves CRs.
	3. The CM group establishes baselines and delivers releases and associated changes to authorized baselines.
	Notes and guidance: Perform the following subactivities:
	1. Establish the control authority
	2. Create or change configuration baselines
	3. Document the list of CIs that are contained in each baseline.
	4. Identify and record CRs
	5. Analyze and evaluate CRs
	6. Review deferred CRs with stakeholders
	7. Approve or disapprove CRs
	8. Record changes
	9. Ensure that an audit trail exists concerning the rational and approval of modifications to include coverage of specialty engineering and enabling products

(*Continued*)

Table 5.22   Configuration Management Process (SEP-22) (*Continued*)

Step	Activity
4	**Action statement:** Perform CSA.
	The CM group produces CSA reports to provide visibility into the status of baselines. CSA reports are developed periodically to address status and history of controlled products, approved identification numbers, library and baseline contents, CR implementation status, CCB decisions, and deficiencies.
	Notes and guidance: Perform the following subactivities:
	1. Track the status of CRs to closure
	2. Manage records and status reports that show status and change history of CRs
	3. Issue status reports that describe approved changes and the latest version of baselines
	4. Describe versions of CIs, baselines, approved changes, and status of changes being considered
	5. Confirm CM information correctness, timeliness, integrity, and security
	6. Ensure that stakeholders have access to configuration status
5	**Action statement:** Perform configuration verification, audits, and reviews.
	1. QA audits the functional characteristics of the products to verify they have achieved the requirements specified in the functional and allocated configuration documentation.
	2. QA audits the as-built product configurations against the technical documentation to establish or verify the product baseline.
	3. The CM group supports the functional and physical audits, provides requested data, and performs periodic informal review of CM tasks, procedures, CSA reports, and products.
	4. The CM manager oversees resolution of reported deficiencies against CM activities.
	Notes and guidance: Perform the following subactivities:
	1. Verify that CIs satisfy requirements
	2. Verify that CIs match approved configuration description information
	3. Assess the integrity of the baselines
	4. Review the structure and integrity of the data and baselines in the CM information system
	5. Confirm the completeness and correctness of the information in the CM information system
	6. Confirm the compliance with applicable CM policies, standards, processes, and procedures
	7. Track action items from the audit to closure

Table 5.22   Configuration Management Process (SEP-22) (*Continued*)

Measures:
• Effort and funds expended for CM tasks (planned vs. actual)
• Impact of requirements changes on project cost and schedule
• Number of configuration audits completed (planned vs. actual)
• Numbers of CIs, CRs (by status), trouble reports (by status), and waivers and deviations
Process Task Outcomes
As a result of the successful implementation of the CM process:
• A CM strategy is defined.
• Items requiring CM are defined.
• Configuration baselines are established.
• Changes to items under CM are controlled.
• The configuration of released items is controlled.
• The status of items under CM is made available throughout the life cycle.

Table 5.23   Quality Management Process (SEP-23)

Process purpose: The supplier shall set up a quality management process to assure that products, services, and implementations of life cycle processes meet organization quality objectives and achieve customer satisfaction and shall perform quality assessments of processes and associated work products, including the tracking of noncompliance issues to closure.	
Roles and agents: • Project leader • Sustainability engineer • Integrated product team	
Inputs: • Products and/or processes undergoing quality assurance (QA) (such as plans, requirements, specifications, deliverable items, reports, and review results)	**Outputs:** • Evaluation reports • Noncompliance reports • Corrective action artifacts • Status reports of corrective actions • Reports of quality trends

(*Continued*)

Table 5.23   Quality Management Process (SEP-23) (*Continued*)

Process diagram:

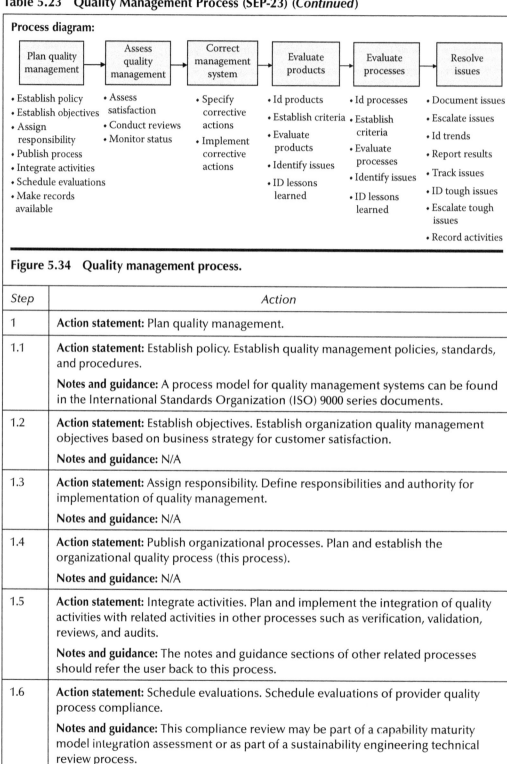

Plan quality management	Assess quality management	Correct management system	Evaluate products	Evaluate processes	Resolve issues
• Establish policy • Establish objectives • Assign responsibility • Publish process • Integrate activities • Schedule evaluations • Make records available	• Assess satisfaction • Conduct reviews • Monitor status	• Specify corrective actions • Implement corrective actions	• Id products • Establish criteria • Evaluate products • Identify issues • ID lessons learned	• Id processes • Establish criteria • Evaluate processes • Identify issues • ID lessons learned	• Document issues • Escalate issues • Id trends • Report results • Track issues • ID tough issues • Escalate tough issues • Record activities

Figure 5.34   Quality management process.

Step	Action
1	**Action statement:** Plan quality management.
1.1	**Action statement:** Establish policy. Establish quality management policies, standards, and procedures. **Notes and guidance:** A process model for quality management systems can be found in the International Standards Organization (ISO) 9000 series documents.
1.2	**Action statement:** Establish objectives. Establish organization quality management objectives based on business strategy for customer satisfaction. **Notes and guidance:** N/A
1.3	**Action statement:** Assign responsibility. Define responsibilities and authority for implementation of quality management. **Notes and guidance:** N/A
1.4	**Action statement:** Publish organizational processes. Plan and establish the organizational quality process (this process). **Notes and guidance:** N/A
1.5	**Action statement:** Integrate activities. Plan and implement the integration of quality activities with related activities in other processes such as verification, validation, reviews, and audits. **Notes and guidance:** The notes and guidance sections of other related processes should refer the user back to this process.
1.6	**Action statement:** Schedule evaluations. Schedule evaluations of provider quality process compliance. **Notes and guidance:** This compliance review may be part of a capability maturity model integration assessment or as part of a sustainability engineering technical review process.

Table 5.23   Quality Management Process (SEP-23) (*Continued*)

Step	Action
1.7	**Action statement:** Make records available. Plan and implement a method of making quality records available to sponsors, customers, and relevant stakeholders.
	Notes and guidance: This information may be posted in a sustainability engineering collaborative environment or other web-based resource as described in organizational policies and procedures.
2	**Action statement:** Assess quality management system.
2.1	**Action statement:** Assess satisfaction. Assess customer satisfaction and report.
	Notes and guidance: Use surveys or the preferred organizational approach to achieving customer satisfaction.
2.2	**Action statement:** Conduct reviews. Conduct periodic reviews of project quality plans.
	Notes and guidance: Assure that quality objectives based on the stakeholder requirements are established for each project.
2.3	**Action statement:** Monitor status. The status of quality improvements on products and services is monitored.
	Notes and guidance: N/A
3	**Action statement:** Correct the management system.
3.1	**Action statement:** Specify corrective actions. Plan corrective actions when quality management goals are not achieved.
	Notes and guidance: N/A
3.2	**Action statement:** Implement corrective actions. Implement corrective actions, and communicate results through the organization.
	Notes and guidance: N/A
4	**Action statement:** Evaluate work products.
4.1	**Action statement:** Identify products. Select work products to be evaluated, based on documented sampling criteria if sampling is used.
	Notes and guidance: N/A
4.2	**Action statement:** Establish criteria. Establish and maintain clearly stated criteria for the evaluation of work products.
	Notes and guidance: The intent of this activity is to provide criteria, based on business needs, such as the following:
	1. What will be evaluated during the evaluation of a work product
	2. When or how often a work product will be evaluated
	3. How the evaluation will be conducted
	4. Who must be involved in the evaluation

(*Continued*)

Table 5.23 Quality Management Process (SEP-23) (*Continued*)

Step	Action
4.3	**Action statement:** Evaluate products. Use the stated criteria during the evaluations of work products. **Notes and guidance:** 1. Evaluate before products are delivered to the customer 2. Evaluate work products at selected milestones in their development 3. Evaluate at in-progress reviews of work products and services against process descriptions, standards, and procedures
4.4	**Action statement:** Identify issues. Identify each case of noncompliance found during evaluations. **Notes and guidance:** N/A
4.5	**Action statement:** Identify lessons learned. Identify lessons learned that could improve processes for future products and services. **Notes and guidance:** N/A
5	**Action statement:** Evaluate processes
5.1	**Action statement:** Identify processes. Select the processes to be evaluated, based on documented process improvement schedules, based on process issues identified in process improvement proposals, or as directed. **Notes and guidance:** Organizational process improvement plans and measurement plans should include an annual schedule for process improvement events. These will likely be integrated with the "Lean Six-Sigma" program of the organization.
5.2	**Action statement:** Identify criteria. Establish and maintain clearly stated criteria for the evaluations. **Notes and guidance:** The intent of this activity is to provide criteria, based on business needs, such as the following: 1. What will be evaluated 2. When or how often a process will be evaluated 3. How the evaluation will be conducted 4. Who must be involved in the evaluation
5.3	**Action statement:** Evaluate processes. Use the stated criteria to evaluate performed processes for adherence to process descriptions, standards, and procedures. **Notes and guidance:** N/A
5.4	**Action statement:** Identify issues. Identify each noncompliance found during the evaluation. **Notes and guidance:** N/A
5.5	**Action statement:** Identify lessons learned. Identify lessons learned that could improve processes for future products and services. **Notes and guidance:** N/A

Table 5.23 Quality Management Process (SEP-23) (*Continued*)

Step	Action
6	**Action statement:** Resolve issues Noncompliance issues are problems identified in evaluations that reflect a lack of adherence to applicable standards, process descriptions, or procedures. The status of noncompliance issues provides an indication of quality trends. Quality issues include noncompliance issues and results of trend analysis. When local resolution of noncompliance issues cannot be obtained, use established escalation mechanisms to ensure that the appropriate level of management can resolve the issue. Track noncompliance issues to resolution.
6.1	**Action statement:** Document issues. Document noncompliance issues when they cannot be resolved within the project. **Notes and guidance:** Examples of ways to resolve noncompliance within the project include the following: 1. Fixing the noncompliance 2. Changing the process descriptions, standards, or procedures that were violated 3. Obtaining a waiver to cover the noncompliance issue
6.2	**Action statement:** Escalate issues. Escalate noncompliance issues that cannot be resolved within the project to the appropriate level of management designated to receive and act on noncompliance issues. **Notes and guidance:** N/A
6.3	**Action statement:** Identify trends. Analyze the noncompliance issues to see if there are any quality trends that can be identified and addressed. **Notes and guidance:** N/A
6.4	**Action statement:** Report results. Ensure that relevant stakeholders are aware of the results of evaluations and the quality trends in a timely manner. **Notes and guidance:** N/A
6.5	**Action statement:** Track issues. Track noncompliance issues to resolution. **Notes and guidance:** N/A
6.6	**Action statement:** Identify tough issues. Document noncompliance issues when they cannot be resolved within the immediate organizational level. **Notes and guidance:** N/A
6.7	**Action statement:** Escalate tough issues. Escalate noncompliance issues that cannot be resolved within the immediate organizational level to the appropriate upper level of management designated to receive and act on noncompliance issues. **Notes and guidance:** N/A

(Continued)

Table 5.23 Quality Management Process (SEP-23) (*Continued*)

Step	Action
6.8	**Action statement:** Record activities. Record process and product quality assurance activities in sufficient detail such that status and results are known. **Notes and guidance:** Revise the status and history of the quality assurance activities as necessary.

Measures:

- Number of qualify assessments performed as planned during the reporting period
- Percentage of total project resources/budget devoted to QA activities
- Number of noncompliance issues identified and resolved during the reporting period

Process Task Outcomes

As a result of the successful implementation of the quality management process:

1. Organization quality management policies and procedures are defined.
2. Organization quality objectives are defined.
3. Accountability and authority for quality management are defined.
4. The status of customer satisfaction is monitored.
5. Appropriate action is taken when quality objectives are not achieved.

Table 5.24 Risk and Opportunity Management Process (SEP-24)

Process purpose: The provider shall perform the risk management process to identify, analyze, treat, and monitor the risks all during the project life cycle.

The risk management process is a continuous process for systematically addressing risk throughout the life cycle of a system product or service. It can be applied to risks related to the acquisition, development, maintenance, or operation of a product. Risk management is a continuous process that is accomplished throughout all phases of the acquisition life cycle. It is an organized method for identifying and analyzing uncertainties (i.e., their occurrences and consequences); developing mitigation options; and selecting, planning, and implementing appropriate mitigation efforts. Risk management is not a stand-alone effort. It is supported and integrated by a number of sustainability engineering technical management processes. The reason risk is addressed as an integrated and formal, disciplined process during product development is to ensure that the technical cost, schedule, and performance thresholds and objectives defined in the program's acquisition program baseline are achieved within acceptable levels of risk to the program. In order to accomplish this, it is necessary to communicate to program-level decision makers the technical risks status and the risk reduction actions that are taken.

Roles and agents:

- Project leader
- Sustainability engineer
- Integrated product team (IPT)

Table 5.24   Risk and Opportunity Management Process (SEP-24) (*Continued*)

Inputs:	Outputs:
• Program risk management plan • Risk status measures • Risk concerns	• Technical risk management procedures • Approved risk statements • Rank-ordered risk list(s) • Risk mitigation plans • Risk contingency plan(s) • List of people with authority and responsibility for tracking and addressing risks • Risk status reports

Process diagram:

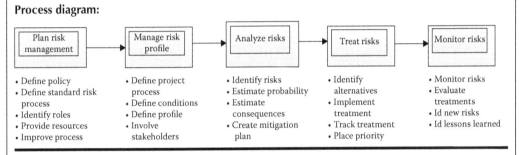

Figure 5.35   Risk and opportunity management process.

Step	Action
1	**Plan Risk Management.**
1.1	**Action statement:** Define policy. Define risk management policies. **Notes and guidance:** N/A
1.2	**Action statement:** Define risk process. Document the risk management process to be implemented. **Notes and guidance:** This standard process is intended to fulfill this requirement.
1.3	**Action statement:** Identify roles. Identify the responsible parties and their roles and responsibilities. **Notes and guidance:** N/A
1.4	**Action statement:** Provide resources. Provide the responsible parties with adequate resources to perform risk management. **Notes and guidance:** N/A

(*Continued*)

Table 5.24   Risk and Opportunity Management Process (SEP-24) *(Continued)*

Step	Action
1.5	**Action statement:** Improve risk process. Define the process for evaluating and improving the risk management process. **Notes and guidance:** Perform the following subtasks: a. Throughout the life cycle, collect risk information for purposes of improving the risk management process and generating lessons learned. The risk information includes the risks identified, their sources, their causes, their treatment, and the success of the treatments selected. b. Periodically review the risk management process for its effectiveness and efficiency. c. Periodically review risk information on the risks identified, their treatment, and the success of the treatments for the purposes of identifying systemic project and organizational risks.
1.6	**Action statement:** Perform training. Train IPT and other technical personnel in the established risk management strategy/plan and procedures. **Notes and guidance:** N/A
2	**Manage the Risk Profile**
2.1	**Action statement:** Define project process. Define and document the context of the risk management process at the project level. **Notes and guidance:** This includes a description of stakeholders' perspectives; risk categories; and a description (perhaps by reference) of the technical and managerial objectives, assumptions, and constraints. A comprehensive risk management strategy/plan should include descriptions of the following: 1. The scope of the technical risk management process activities 2. Technical risk management control, reporting, and approval levels 3. Risk measures to be used within the technical effort including the following: a. How to describe the likelihood of occurrence (e.g., levels 1 through 5, or Low, Moderate, High) b. How to describe the consequences or impacts of a risk occurrence (e.g., in terms of cost, schedule, performance) c. Thresholds in terms of risk measures taking action on identified risks 4. How identified risks (root causes) are to be (1) defined/stated (standard format), (2) organized (e.g., by criticality, cost impact, priority), (3) categorized (e.g., by phase/time, process, type product, type of risk, impact), (4) compared (e.g., priority, timing, relevancy, cost impact), and (5) consolidated (e.g., forming risk action plans, standard risk reporting models) 5. Standardized definition of risk measures to monitor the status of risks 6. Detailed approach for conducting the risk management activities including the following: a. Methods and tools to be used for risk management process activities—risk identification

Table 5.24   Risk and Opportunity Management Process (SEP-24) (*Continued*)

Step	Action
	b. Risk analysis, risk mitigation planning, mitigation plan implementation, and risk tracking c. Time intervals for risk monitoring and reporting
2.2	**Action statement:** Define risk conditions. Define and document the risks, risk thresholds, and conditions under which a level of risk may be accepted. **Notes and guidance:** N/A
2.3	**Action statement:** Define profile. Establish and maintain a risk profile. **Notes and guidance:** The risk profile records the risk management context; a record of each risk's state including its probability, consequences, and risk thresholds; the priority of each risk based on risk criteria supplied by the stakeholders; and the risk action requests along with the status of their treatment. The risk profile is updated when there are changes in an individual risk's state. The priority in the risk profile is used to determine the application of resources for treatment.
2.4	**Action statement:** Communicate with stakeholders. Periodically communicate the relevant risk profile to stakeholders based upon their needs. **Notes and guidance:** Identify and coordinate with each relevant stakeholder associated with each risk.
3	**Analyze Risks**
3.1	**Action statement:** Identify current risks. Identify risks in the categories described in the risk management context. **Notes and guidance:** Risk identification is the risk management process activity of examining each element of the technical effort to help identify associated risk root causes and begin documenting them. Risk identification sets the stage for their successful management. Risk identification begins in early phases of the acquisition life cycle and continues throughout the program with regular reviews and analyses of information from sources such as technical performance measurements, schedule, resource data, cost information, earned value management data, progress against critical path, technical baseline maturity, and other technical information available to IPT members. The risk identification activity is structured to help answer the question: "What can go wrong?" In other words, what could affect the technical effort and potentially the success of the program? It is also important to recognize that risk identification is part of everyone's job, not just that of the sustainability engineer or program manager. Examples of risk identification methods include the following: • Analyze negative trends • Look at assigned staffing, requirements, technology needs and availability, design approach, and potential changes

Table 5.24   Risk and Opportunity Management Process (SEP-24) (*Continued*)

Step	Action
	• Review previous test results, especially test failures • Review potential shortfalls in resources or technologies against expectations • Interview subject matter experts • Review lessons learned from previous, similar technical efforts • Use control charts, affinity diagrams, and interrelationship diagrams • Examine stakeholder requirements and/or specified requirements • Review verification and validation plans • Use the system model (work breakdown structure) to help pinpoint potential risk areas
3.2	**Action statement:** Estimate probability. Estimate the probability of occurrence of each identified risk. Analyze and document the likelihood (probability) associated with each risk event to quantify or qualify the probability of risk occurrence, using the established program standards. **Notes and guidance:** Assess the risk level as a function of its likelihood and consequence scores. Determine the risk exposure by considering performance, schedule, and cost consequences. Consequence scores are assigned using a tool such as "Risk Radar" scores, which are from 1 through some number such as 5, with 1 being low impact and 5 being critical impact.
3.3	**Action statement:** Estimate impact. Estimate the consequences of each identified risk. Analyze and document the consequence severity, using standards that have been established for the program. **Notes and guidance:** Prioritize risks for mitigation using a risk analysis that identifies a measure that combines the probability of occurrence with the consequence score. Determine if the risk is a "show-stopper." Determine whether the risk is such that it could stop the technical effort or affect program success by considering the following: 1. The risk exposure and whether the risk is a near-term concern 2. Options for risk mitigation 3. Whether there is any coupling, whereby one risk affects the characteristics of other risks 4. How risk occurrence can be detected, assessed, and monitored 5. The influence of other factors such as quality, safety, security, survivability, interoperability, customer/stakeholder satisfaction, and program approval
3.4	**Action statement:** Create mitigation plan. For each risk that is above its risk threshold, define and document recommended treatment strategies and measures indicating the effectiveness of the treatment alternatives. **Notes and guidance:** Risk treatment strategies include, but are not limited to, eliminating the risk, reducing its probability of occurrence or severity of consequence, or accepting the risk.

Table 5.24   Risk and Opportunity Management Process (SEP-24) (*Continued*)

Step	Action
	Select a risk-handling approach to mitigate the probability of risk occurrence or the risk consequence or each selected risk. Determine risk levels and thresholds that define when the probability of each risk occurrence becomes unacceptable and triggers execution of risk mitigation implementation.
	IPTs must develop proactive plans to mitigate and reduce risks identified during risk assessment.
	Risks are lowered by reducing the probability that the root-cause event will occur, or the consequence should the root-cause event occur, or some combination of the two.
	Techniques or options for risk mitigation include avoidance, control, transfer, and assumption strategies. These four fundamental risk mitigation strategies are described as avoidance, control, transfer, and assumption.
	Avoidance: Risk avoidance is the use of an alternate path or solution that eliminates the risk event. An example of risk avoidance would be the use of a nontoxic, benign substance that still meets product requirements in place of one that had toxic or environmentally damaging properties. A specific example of this is replacing halon gas (an ozone-depleting compound) in a fire suppression system with nitrogen (an inert gas with no environmental issues). Changing or relaxing requirements to reduce or avoid the potential for a risk event occurrence is yet another example of risk avoidance.
	Control: Risk control is a proactive action that reduces either the probability of occurrence and/or the consequence of occurrence. Risk control activities include design activities such as redundant or back-up systems, back-up designs, and fail-safe design; use of modeling and simulation or prototypes to gain early information on the potential for a risk event to occur; robust designs; use of open systems; and targeted test and evaluation. Risk control is the most widely used risk mitigation technique done during early product development acquisition life cycle phases.
	Transfer: Risk transfer is the shifting of the risk to another organization. An example of risk transfer is a contractor who subcontracts work to other contractors, who may have greater capability maturity in an area such as software. Another example is the government requesting a warranty from the contractor on the product, transferring the risk of product defect costs to the contractor through the warranty period.
	Assumption: Risk assumption is the monitoring of a risk while taking no specific action at the current time. This is the least desirable of the risk mitigation techniques. However, since risk mitigation requires resources, there may be insufficient resources to actively mitigate all risks. Some risks that are categorized as moderate or low may have to be assumed so that resources are focused on more pressing, high-risk areas. Risk assumption includes monitoring risk assessment and budgeting resources to mitigate assumed risks should they occur or begin to move toward the high-risk category. If a proactive, effective technical risk management process is not in place, then all risks, whether explicitly identified or not, are by default handled as "assumed" risks but without the necessary monitoring step. This default mode of risk management, whether it occurs at the technical risk management level or at the overall program level, is a sure recipe for disaster.

(Continued)

Table 5.24   Risk and Opportunity Management Process (SEP-24) (*Continued*)

Step	Action
4	Treat Risks
4.1	**Action statement:** Identify alternatives. Provide stakeholders with recommended alternatives for risk treatment in risk action requests. **Notes and guidance:** For each risk, determine the risk level and threshold that define when an unacceptable risk consequence triggers the execution of a contingency plan. Prepare specific technical risk mitigation and contingency action plans, and assign responsibilities and authority.
4.2	**Action statement:** Implement treatment. Implement risk treatment alternatives for which the stakeholders determine that actions should be taken to make a risk acceptable. **Notes and guidance:** 1. Work the plan: Implement the procedure for risk decision making, and use selection criteria included in the risk management plan. 2. Track: Monitor and track risk status to determine whether expectations have changed, the situation has changed, new risks have surfaced, or things are getting better or worse; determine when it is time to implement risk mitigation plans. 3. Threshold: Monitor and compare risk management mitigation thresholds to status. 4. Trigger: Monitor and compare risk triggers to risk status. 5. Status: Report technical risk status to decision authorities when triggers are activated. 6. Get help: Communicate upward to the program level when risks are beyond the control of the technical effort or when they represent a significant threat exposure to the technical effort and/or program success.
4.3	**Action statement:** Track treatment. Once a risk treatment is selected, ensure management actions in accordance with the assessment and control activities in the technical assessment process. **Notes and guidance:** Risk-tracking activities are integral to a successful technical effort. Results from the sustainability engineering technical assessment process can provide much of the information used to identify any performance, schedule, and cost barriers to meeting technical objectives and to satisfying acquisition life cycle phase exit criteria. Risk changes over time. Regularly ask the following: 1. Have expectations/situations changed? 2. Are there new risks? 3. Are things getting better or worse? 4. Is it time to take action? 5. Are risk management action plans still valid?

Table 5.24 Risk and Opportunity Management Process (SEP-24) (*Continued*)

Step	Action
4.4	**Action statement:** Prioritize treatment. When the stakeholders accept a risk that exceeds its threshold, consider it a high priority and monitor it continuously to determine if any future risk treatment actions are necessary. **Notes and guidance:** N/A
5	Monitor Risks
5.1	**Action statement:** Monitor risks. Continuously monitor all risks and the risk management context for changes and evaluate the risks when their state has changed. **Notes and guidance:** Monitor and compare trigger thresholds to risk status to provide early warning. High or critical risks may warrant special tracking, or even immediate implementation of the risk mitigation action plan. Report risk status when a trigger is activated.
5.2	**Action statement:** Evaluate treatments. Implement and monitor measures to evaluate the effectiveness of risk treatments. **Notes and guidance:** N/A
5.3	**Action statement:** Identify new risks. Continuously monitor for new risks and sources throughout the life cycle. **Notes and guidance:** N/A
5.4	**Action statement:** Identify lessons learned. Periodically review risk information on the risks identified, their treatment, and the success of the treatments for the purposes of identifying systemic project and organizational risks. **Notes and guidance:** N/A
	Process Task Outcomes

As a result of the successful implementation of the risk management process:

a. The scope of risk management to be performed is determined.

b. Appropriate risk management strategies, plans, and procedures are defined and implemented.

c. Technical personnel are qualified and capable of conducting established risk management procedures.

d. Risks are identified as they develop and during the conduct of the project.

e. Risks are analyzed, and the priority in which to apply resources to treatment of these risks is determined.

f. Clear and understandable risk statements are coordinated and understood by all relevant stakeholders.

g. Risk measures are defined, applied, and assessed to determine changes in the status of risk and the progress of the treatment activities.

h. Appropriate treatment is taken to correct or avoid the impact of risk based on its priority, probability, and consequence or other defined risk threshold.

Table 5.25 Supplier Performance Management Process (SEP-25)

Process purpose: The provider shall perform contractor performance management to ensure that acquired products are delivered in according with the formal agreement. The focus of this task is to manage supplier performance by monitoring the supplier against key product and process metrics that can include periodic reviews (i.e., incoming and final inspection, facility capability audits, and process capability studies). Process 3 is invoked whenever subsystem products are acquired from suppliers or lower tier developers outside the enterprise, as well as when the supplier is an organizational entity within the developer's own enterprise. **Scope:** The sustainability engineering process uses a requirements loop and a design loop in a progressive, iterative analytical approach to make operational requirements and design decisions at successively lower levels. As this process iterates, requirements are planned, documented, developed, identified, controlled, tracked, and verified within the configuration management process. Configuration management provides the common approach necessary to minimize variation and improve information integrity. Many projects will need to perform contract closure at the end of several life cycle phases. Refer to SSEG Sections 5 and 6 for additional detail on process usage context and foundation concepts.

Roles and agents:

- Project leader
- Sustainability engineer
- Integrated product team

Inputs:	**Outputs:**
• Customer agreements, signed • Task work statements, signed	• Supplier progress reports • Supplier review materials and reports • Action items tracked to closure • Documentation of product and document deliveries • Lists of processes selected for monitoring • Supplier activity reports • Supplier performance reports • Discrepancy reports • Lists of work products selected for monitoring • Acceptance test procedures • Acceptance test results • Corrective action reports • Transition plans • Training reports • Support and maintenance reports

Table 5.25 Supplier Performance Management Process (SEP-25) (*Continued*)

Process diagram:

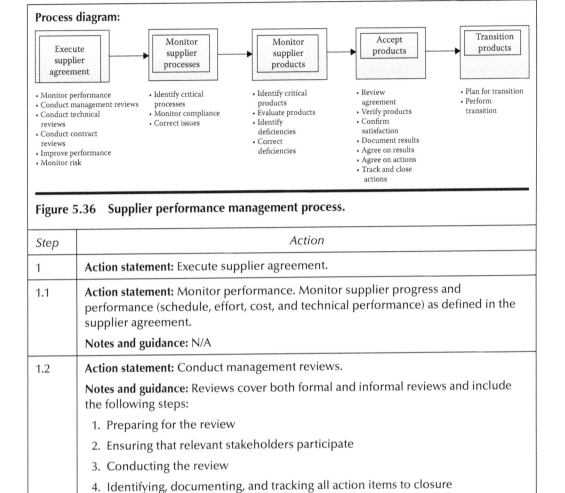

Figure 5.36 Supplier performance management process.

Step	Action
1	**Action statement:** Execute supplier agreement.
1.1	**Action statement:** Monitor performance. Monitor supplier progress and performance (schedule, effort, cost, and technical performance) as defined in the supplier agreement. **Notes and guidance:** N/A

1.2	**Action statement:** Conduct management reviews.
	
	Notes and guidance: Reviews cover both formal and informal reviews and include the following steps:
	
	1. Preparing for the review
	2. Ensuring that relevant stakeholders participate
	3. Conducting the review
	4. Identifying, documenting, and tracking all action items to closure
	5. Preparing and distributing to the relevant stakeholders a summary report of the review
	6. Management reviews typically include the following:
	  a. Reviewing critical dependencies
	  b. Reviewing project risks involving the supplier
	  c. Reviewing schedule and budget

1.3	**Action statement:** Conduct technical reviews. Conduct technical reviews with the supplier, as defined in the supplier agreement. **Notes and guidance:** Technical reviews typically include the following: 1. Providing the supplier with visibility into the needs and desires of the project's customers and end users, as appropriate

(Continued)

Table 5.25   Supplier Performance Management Process (SEP-25) (*Continued*)

Step	Action
	2. Reviewing the supplier's technical activities and verifying that the supplier's interpretation and implementation of the requirements are consistent with the project's interpretation
	3. Ensuring that technical commitments are being met and that technical issues are communicated and resolved in a timely manner
	4. Obtaining technical information about the supplier's products
	5. Providing appropriate technical information and support to the supplier
1.4	**Action statement:** Conduct contract reviews. **Notes and guidance:** The focus of this review is on status of contract deliverables and progress payments.
1.5	**Action statement:** Improve performance. Use the results of reviews to improve the supplier's performance and to establish and nurture long-term relationships with preferred suppliers. **Notes and guidance:** N/A
1.6	**Action statement:** Monitor risks. Monitor risks involving the supplier and take corrective action as necessary. **Notes and guidance:** N/A
2	**Action statement:** Monitor supplier processes. In situations where there must be tight alignment between some of the processes implemented by the supplier and those of the project, monitoring these processes will help prevent interface problems. The selection must consider the impact of the supplier's processes on the project. On larger projects with significant subcontracts for development of critical components, monitoring of key processes is expected. For most vendor agreements where a product is not being developed or for smaller, less critical components, the selection process may determine that monitoring is not appropriate. Between these extremes, the overall risk should be considered in selecting processes to be monitored. The processes selected for monitoring should include engineering, project management (including contracting), and support processes critical to successful project performance. Monitoring, if not performed with adequate care, can at one extreme be invasive and burdensome or at the other extreme be uninformative and ineffective. There should be sufficient monitoring to detect issues, as early as possible, that may affect the supplier's ability to satisfy the requirements of the supplier agreement. Analyzing selected processes involves taking the data obtained from monitoring selected supplier processes and analyzing it to determine whether there are serious issues.
2.1	**Action statement:** Identify critical processes. Identify the supplier processes that are critical to the success of the project. **Notes and guidance:** N/A

Table 5.25 Supplier Performance Management Process (SEP-25) (*Continued*)

Step	Action
2.2	**Action statement:** Monitor process compliance. Monitor the selected supplier's processes for compliance with requirements of the agreement. **Notes and guidance:** N/A
2.3	**Action statement:** Correct issues. Analyze the results of monitoring the selected processes to detect issues as early as possible that may affect the supplier's ability to satisfy the requirements of the agreement. **Notes and guidance:** Trend analysis can rely on internal and external data. Refer to assessment and control SP-13, for more detail on corrective actions.
3	**Action statement:** Monitor supplier products. The scope of this specific practice is limited to suppliers providing the project with custom-made products, particularly those that present some risk to the program due to complexity or criticality. The intent of this specific practice is to evaluate selected work products produced by the supplier to help detect issues as early as possible that may affect the supplier's ability to satisfy the requirements of the agreement. The work products selected for evaluation should include critical products, product components, and work products that provide insight into quality issues as early as possible.
3.1	**Action statement:** Identify critical products. Identify those work products that are critical to the success of the project and that should be evaluated to help detect issues early. **Notes and guidance:** Examples of work products that may be critical to the success of the project include the following: 1. Requirements 2. Analyses 3. Architecture 4. Documentation
3.2	**Action statement:** Evaluate products. Evaluate the selected work products. **Notes and guidance:** Work products are evaluated to ensure the following: 1. Derived requirements are traceable to higher level requirements. 2. The architecture is feasible and will satisfy future product growth and reuse needs. 3. Documentation that will be used to operate and to support the product is adequate. 4. Work products are consistent with one another. 5. Products and product components (e.g., custom-made, off-the-shelf, and customer-supplied products) can be integrated.

(Continued)

Table 5.25 Supplier Performance Management Process (SEP-25) (*Continued*)

Step	Action
3.3	**Action statement:** Identify deficiencies. Determine and document actions needed to address deficiencies identified in the evaluations. **Notes and guidance:** Refer to the technical assessments process for more information on corrective actions.
4	**Action statement:** Accept products. Ensure that the supplier agreement is satisfied before accepting the acquired product. Acceptance reviews and tests and configuration audits should be completed before accepting the product as defined in the supplier agreement.
4.1	**Action statement:** Define procedures. Define the acceptance procedures. **Notes and guidance:** N/A
4.2	**Action statement:** Obtain agreement. Review and obtain agreement with relevant stakeholders on the acceptance procedures before the acceptance review or test. **Notes and guidance:** N/A
4.3	**Action statement:** Verify products. Verify that the acquired products satisfy their requirements. **Notes and guidance:** Refer to the product verification process for more detail on the verification.
4.4	**Action statement:** Confirm satisfaction. Confirm that the nontechnical commitments associated with the acquired work product are satisfied. **Notes and guidance:** This may include confirming that the appropriate license, warranty, ownership, usage, and support or maintenance agreements are in place and that all supporting materials are received.
4.5	**Action statement:** Document results. Document the results of the acceptance review or test. **Notes and guidance:** N/A
4.6	**Action statement:** Get consensus on actions. Document the results of the acceptance review or test. **Notes and guidance:** N/A
4.7	**Action statement:** Track actions. Identify, document, and track action items to closure. **Notes and guidance:** Refer to assessments and control SP-13, for more detail on action closure.
5	**Action statement:** Transition products. **Notes and guidance:** 1. Before the acquired product is transferred to the project for integration, appropriate planning and evaluation should occur to ensure a smooth transition. 2. Refer to the product integration process for more information about integrating the acquired products.

Table 5.25   Supplier Performance Management Process (SEP-25) (*Continued*)

Process institutionalization: The project management plan(s) shall describe how policy, planning, resources, responsibility assignment, stakeholder involvement, monitoring, control, status reviews, feedback, quality assurance, evaluations, and audits are to be used to ensure institutionalization of this process.
Process tailoring: • The following process elements may *not* be tailored: process owner and process objective.   • The following process elements may only be tailored by obtaining a process waiver: process activities.   • The following process elements may be tailored: guidance and notes, process roles, process implementation assets, and process-related measures.

5.5 Product Realization Processes

The product realization processes, as shown in Figure 5.37, are used to (1) convert the specified requirements and other design solution characterizations into either a verified end product or a set of end products in accordance with the agreement and other stakeholder requirements; (2) deliver these to designated operating, customer, or storage sites; (3) install these at designated operating sites or into designated platforms; and (4) provide in-service support, as called for in an agreement.

There are eight processes associated with product realization, as shown in Figure 5.38.

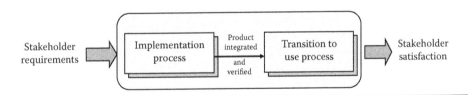

Figure 5.37   Product realization process.

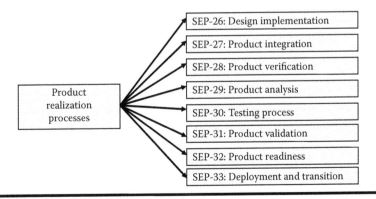

Figure 5.38   Product realization processes.

Table 5.26 Design Implementation (SEP-26)

Process purpose: The supplier shall implement (build, produce, code) the design (preliminary or final) in accordance with the specified requirements to obtain a verified end product. This process transforms specified behavior, interfaces, and implementation constraints into fabrication actions that create a product element according to the practices of the selected implementation technology. The product element is constructed or adapted by processing the materials and/or information appropriate to the selected implementation technology and by employing appropriate technical specialties or disciplines. This process results in a product element that satisfies specified design requirements through verification and stakeholder requirements through validation.

Roles and agents:

- Project leader
- Sustainability engineer
- Integrated product team

Inputs:	**Outputs:** All outputs should be archived.
• End products from supplier • Enabling products from supplier • Manufacturing plans • Quality assurance (QA) program plan • Specified requirements • Operational test/follow-on test and evaluation report	• Assembled end product(s) or enabling product(s) • Manufacturing process and personnel system • Verified and validated integrated end product or enabling product report

Process diagram:

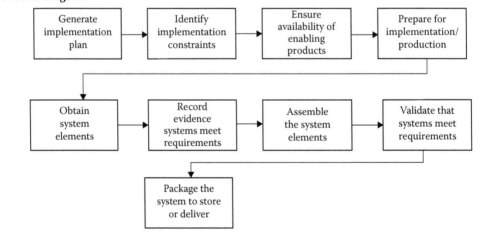

Figure 5.39 Design implementation process.

Step	Action
	Action statement: Plan the implementation.
1	**Action statement:** Generate an implementation strategy, plan, and procedures.

Table 5.26 Design Implementation (SEP-26) (*Continued*)

Step	Action
	Notes and guidance: This includes implementation procedures, fabrication processes, tools and equipment, implementation tolerances, and verification uncertainties. In the case of repeated product element implementation (e.g., mass production and replacement product elements), the implementation procedures and fabrication processes are defined to achieve consistent and repeatable production ability.
2	**Action statement:** Identify the constraints that the implementation strategy and implementation technology impose on the design solution.
	Notes and guidance: This includes current or anticipated limitations of the chosen implementation technology, customer-furnished materials, or product elements for adaptation and limitations resulting from the use of required implementation-enabling products.
3	**Action statement:** Ensure availability of enabling products. Ensure that the enabling products for each associated process will be ready and available to perform their intended support functions required by the system's end products. An area often missed is confirming dedication of fleet assets. This should include notification for fleet testing and other assets.
	Notes and guidance: The relevant end products for enabling products are verified and validated as necessary during the development of the building block related to the enabling product (see Section 6). All essential sustainability engineering technical reviews should be completed to ensure that enabling processes and resources are ready and available. A major production readiness review (PRR) is conducted near the end of the physical design period to ensure that the program is ready to proceed into low-rate initial production (LRIP). This review will validate the production facility, equipment, manufacturing processes, and personnel and help ensure that the program will enter low-rate production at a low risk. A subsequent PRR is usually conducted in LRIP to ensure the program is ready to transition from low-rate to full-rate production in production and deployment and operations and support phases.
	Action statement: Prepare to perform the implementation.
4	**Action statement:** Perform preparation for implementation

	Notes and guidance: Perform the following subtasks:
	
	  a. Perform or verify a make-buy analysis and decision
	
	  b. Ensure readiness of the enabling products for production
	
	  c. Obtain the technical data package to include drawings and specifications

	Action statement: Perform the implementation.
5	**Action statement:** Realize or adapt product elements using the implementation-enabling products and specified materials according to the defined implementation procedures for hardware fabrication, software creation, and/or operator training.

(*Continued*)

Table 5.26 Design Implementation (SEP-26) (*Continued*)

Step	Action
	Notes and guidance: Adaptation includes configuration of hardware and software elements that are reused or acquired. Realization or adaptation is conducted with regard to standards that govern applicable safety, security, privacy, and environmental guidelines or legislation and the practices of the relevant implementation technology.
	1. *Hardware fabrication*: Fabricate hardware elements using the conditioning, forming, and fabrication techniques relevant to the physical implementation technology and materials selected. As appropriate, hardware elements are tested to confirm specified product quality characteristics.
	2. *Software creation*: Develop software elements and, as appropriate, compile, inspect, and test to assure their conformance to the design criteria. ISO/IEC 12207:2008 applies to product elements realized in software.
	3. *Operator training*: Deliver appropriate training to prepare operators for performing tasks in accordance with required performance standards and operational procedures and, as appropriate, confirm that the specified range and level of competence has been attained. This may include awareness of the operational environment, including appropriate failure detection and isolation instruction.
	Supplier performance is invoked whenever subsystem products are acquired from suppliers or lower tier suppliers outside the enterprise, as well as when the supplier is an organizational entity within the supplier's own enterprise (refer to the stakeholder requirements definition process concerning supplier performance monitoring).
6	**Action statement:** Record evidence that product elements meet supplier agreements and other stakeholder requirements.
	Notes and guidance: Validate the subsystem products received or reused against their customer requirements (input requirements to the subsystem product development) using the end products validation process, process 33, unless (1) the supplier validated the products prior to delivery as required in the agreement, or (2) the reused products have already been validated. Proof of validation is needed for both conditions. Approval of suppliers' products is obtained through compliance to product specifications. This could be ascertained at suppliers' facilities, receiving incoming or via receipt inspection, first article validation, and/or test/demonstration. See ISO 9001 Section 4.6.2 for vendor management.
7	**Action statement:** Realize, assemble, make, or adapt product elements using enabling resources and specified materials. Fabricate hardware and code software.
	Notes and guidance: This should be accomplished through already approved manufacturing and QA program plans.
8	**Action statement:** Capture, verify, and validate each end product or test article against its specified requirements.

Table 5.26   Design Implementation (SEP-26) (*Continued*)

Step	Action
	Notes and guidance: Input requirements to system end product development prior to delivery, if required, by the agreement. Operational test and evaluation accomplishes such a task, and this information is incorporated into the end product or enabling product report.
9	**Action statement:** Package the product element and store or deliver it as appropriate. **Notes and guidance:** Contain the product element in order to achieve continuance of its characteristics. Conveyance and storage media, and their durations, influence the specified containment.
Process Task Outcomes	

1. An implementation strategy is defined.
2. Implementation technology constraints on the design are identified.
3. A product element is realized.
4. A product element is packaged and stored in accordance with an agreement for its supply.

Table 5.27   Product Integration (SEP-27)

Process purpose: The supplier is to perform product integration activities to assemble a product or system consistent with the product design and ensure that the integrated product meets its acceptance criteria.

Roles and agents:
- Project leader
- Sustainability engineer
- Integrated product team

Inputs:	Outputs:
• Product components • Product and component requirements	• A documented integration plan and procedures • Product integration sequence • Product integration environment established • Individual and aggregated products and components that have satisfied their allocated requirements • A list of interfaces mapped to components

(*Continued*)

Table 5.27   Product Integration (SEP-27) (*Continued*)

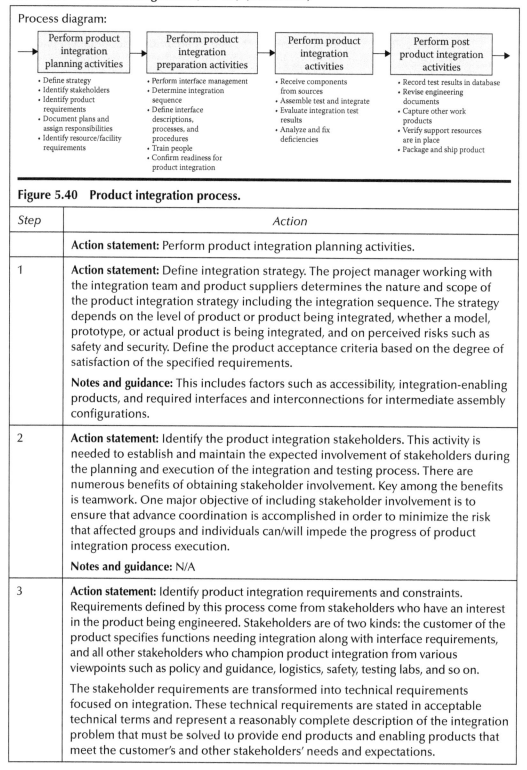

Process diagram:

Perform product integration planning activities	Perform product integration preparation activities	Perform product integration activities	Perform post product integration activities
• Define strategy • Identify stakeholders • Identify product requirements • Document plans and assign responsibilities • Identify resource/facility requirements	• Perform interface management • Determine integration sequence • Define interface descriptions, processes, and procedures • Train people • Confirm readiness for product integration	• Receive components from sources • Assemble test and integrate • Evaluate integration test results • Analyze and fix deficiencies	• Record test results in database • Revise engineering documents • Capture other work products • Verify support resources are in place • Package and ship product

Figure 5.40   Product integration process.

Step	Action
	Action statement: Perform product integration planning activities.
1	**Action statement:** Define integration strategy. The project manager working with the integration team and product suppliers determines the nature and scope of the product integration strategy including the integration sequence. The strategy depends on the level of product or product being integrated, whether a model, prototype, or actual product is being integrated, and on perceived risks such as safety and security. Define the product acceptance criteria based on the degree of satisfaction of the specified requirements. **Notes and guidance:** This includes factors such as accessibility, integration-enabling products, and required interfaces and interconnections for intermediate assembly configurations.
2	**Action statement:** Identify the product integration stakeholders. This activity is needed to establish and maintain the expected involvement of stakeholders during the planning and execution of the integration and testing process. There are numerous benefits of obtaining stakeholder involvement. Key among the benefits is teamwork. One major objective of including stakeholder involvement is to ensure that advance coordination is accomplished in order to minimize the risk that affected groups and individuals can/will impede the progress of product integration process execution. **Notes and guidance:** N/A
3	**Action statement:** Identify product integration requirements and constraints. Requirements defined by this process come from stakeholders who have an interest in the product being engineered. Stakeholders are of two kinds: the customer of the product specifies functions needing integration along with interface requirements, and all other stakeholders who champion product integration from various viewpoints such as policy and guidance, logistics, safety, testing labs, and so on. The stakeholder requirements are transformed into technical requirements focused on integration. These technical requirements are stated in acceptable technical terms and represent a reasonably complete description of the integration problem that must be solved to provide end products and enabling products that meet the customer's and other stakeholders' needs and expectations.

Table 5.27 **Product Integration (SEP-27)** (*Continued*)

Step	Action
	The product integration requirements identification process is reaccomplished, as necessary, whenever requirements in a tasking/agreement change. Such changes could be caused by technology discoveries, evolving knowledge concerning requirements by users and sponsors or by anomalies in program schedules or costs. **Notes and guidance:** N/A
4	**Action statement:** Document plans for product integration. An implementation plan for the product integration process must be developed and typically is tailored from the enterprise standard processes and documented to determine the activities that will be required to accomplish the process for the individual project/task. The plan should include the assignment of responsibility and authority for performing the product integration process, developing the work products, and providing the services of the product integration process. The project/task leader typically calls upon the expertise of the engineering and support disciplines (sustainability engineering, software engineering, test engineering, quality, and configuration management) to assist with planning and implementing the product integration process in each particular project/task. **Notes and guidance:** N/A
5	**Action statement:** Establish the product integration environment. Determine facility and resource requirements for product integration. Product integration facilities may either be acquired or developed. The product integration facilities may include reuse of existing organizational facilities and resources. The facility environment required for each step of the integration process may include assembly areas and fixtures, testing equipment, items of "real equipment," recording devices, and simulators, which take the place of nonavailable system or product components. **Notes and guidance:** N/A
	Action statement: Perform product integration preparation activities.
6	**Action statement:** Perform interface management. During this activity, interface compatibility is monitored, and conflict, noncompliance, and change issues are resolved. This effort continues throughout the life of the product. Additionally, interface data is maintained in a repository that is accessible to project/task participants. **Notes and guidance:** Consult the interface management process for additional details.
7	**Action statement:** Determine integration sequence. During this activity, the product components to be integrated are identified, along with identifying the product verifications to be performed using the definitions of the interfaces between the product components. Alternative product–component integration sequences are identified, and the best integration sequence is selected. The product integration sequence is reviewed periodically and revised as needed based on variations in production and delivery schedules that may have had an impact on the sequence. Lastly, the rational for decisions made and deferred is recorded.

(*Continued*)

Table 5.27   Product Integration (SEP-27) (*Continued*)

Step	Action
	Notes and guidance: This strategy may permit verification against a sequence of progressively more complete system element configurations. It is dependent on system element availability and is consistent with a fault isolation and diagnosis strategy. Wherever possible, an integrated configuration includes its human operators. Successive applications of the integration process and the verification process, and when appropriate the validation process, are repeated for systems at successive levels until the system of interest has been realized.
8	**Action statement:** Obtain and review interface descriptions, procedures, criteria, and enabling products. During this activity, the interface compatibility and other data are reviewed to ensure complete coverage of all interfaces. Product components and interfaces are checked to see that they are marked to ensure easy and correct connection to the joining product. Additionally, the interface descriptions are reviewed periodically to ensure they are up to date and adequate.
	Notes and guidance: The enabling product for integration may include integration facilities, jigs, conditioning facilities, and assembly equipment. Integration-enabling product requirements, constraints, and other limitations are defined.
9	**Action statement:** Train people performing and supporting integration activities. Train the people who will be performing or supporting the product integration process as needed. Training may include information, materials, and practical exercises on topics, which include system/product operation, operating instruction procedures and criteria, assembly instructions, testing procedures, and product-packaging methods.
	Notes and guidance: N/A
10	**Action statement:** Confirm product integration readiness. Review the requirements and the verification criteria and procedures for the product integration environment. A decision is made to either make or acquire the needed product integration environment. If a suitable environment cannot be acquired, one is developed and maintained throughout the project/task. Dispose of portions of the environment when they are no longer needed.
	Notes and guidance: N/A
	Action statement: Perform product integration activities.
11	**Action statement:** Receive components from source. During this activity a variety of components to be integrated may be received from a number of sources. Components are either received from commercial suppliers, reused from off-the-shelf supply, or received from the government supplier (e.g., customer-furnished items). The building block components make up the system's end products, or, as appropriate, code or build the end products (hardware or software) according to the specified requirements and detailed drawings or other design documentation. Tools for this task include the following: parts lists, parts management plans, configuration item lists, make/buy analysis, government-furnished equipment listings, and so on.

Table 5.27   Product Integration (SEP-27) (*Continued*)

Step	Action
	Notes and guidance: System elements can be received from suppliers or be withdrawn from storage. System elements are handled in accordance with relevant health, safety, security, and privacy considerations.
12	**Action statement:** Assemble, test, and integrate. This is the first of two very key activities in the product integration process. During this activity, the product components are assembled according to the product integration sequence using the available procedures. The product integration environment is checked to ensure that it is ready, and the assembly sequence is monitored to ensure it is properly performed. Based on experience gained as the activity is performed, the product integration sequence and procedures may be revised when appropriate. **Notes and guidance:** N/A
13	**Action statement:** Evaluate integration test results. This is the second of the two very key activities. It provides the performance assessment for the product integration process. This performance assessment includes monitoring and controlling the product integration process and collecting work products, measures, measurement results, and improvement information derived from planning and performing the product integration process to support the future use and improvement of the enterprise processes and process assets. This activity also provides the objective evaluation of the performed product integration process, work products, and services against the applicable process description, standards, and procedures. Finally, this activity provides for corrective actions and reviews with senior management. **Notes and guidance:** N/A
14	**Action statement:** Analyze and fix deficiencies. Analyze failures and deficiencies identified during the assembly, integration, and testing activities. Identify corrective action alternatives and select the appropriate (cost/effective) corrective action/fix for the deficiency. Apply the corrective action/fix and perform retesting of the product to verify the corrective action. **Notes and guidance:** This includes resolution of problems due to the integration strategy, the integration-enabling products, or manual assembly errors. The data is analyzed to enable corrective or improvement actions to the integration strategy and its execution. Lessons learned should also be recorded.
	Action statement: Perform post product integration activities.
15	**Action statement:** Record test results in a database. Record integration results in an appropriate information database. This includes corrective actions for system components demonstrating discrepancies/deficiencies/defects or nonconformance with any specified requirement. The database entries should also include any nonconformance with the integration strategy, the integration, the integration-enabling products, or any errors in manuals or procedures. **Notes and guidance:** N/A

(*Continued*)

Table 5.27   Product Integration (SEP-27) (*Continued*)

Step	Action
16	**Action statement:** Revise engineering documents. Update all specifications, drawing packages, technical data, parts lists, baseline descriptions, and production procedures to reflect the modifications to product elements as described in the design implementation process. **Notes and guidance:** N/A
17	**Action statement:** Capture, package, and deliver integration-related work products. **Notes and guidance:** Perform the following subactivities: 1. *Capture key information*: Capture work products and related decisions, rationale, and assumptions resulting from conducting integration process activities in the established technical data management database. 2. *Record results*: Capture the evaluation results of the assembled and integrated end product in the established technical data management database. 3. *Record lessons learned*: Capture and record the lessons learned from applying the integration process in the established technical data management database.
18	**Action statement:** Verify support is ready. At the appropriate time in the product life cycle, as the product or system is being integrated into the "system-of-systems" platform, it is appropriate to verify that all required support and enabling products are ready to perform their supporting functions. **Notes and guidance:** N/A
19	**Action statement:** Package and deliver the product. During this activity, the requirements, design, product, verification results, and documentation are reviewed to ensure that issues affecting the packaging and delivery of the product are identified and resolved. The assembled product is packaged and delivered in accordance with the applicable requirements and standards. The operational site is prepared for installation of the product. The product is delivered with the appropriate documentation, installed, and its correct operation confirmed. **Notes and guidance:** N/A

Measures:

- Percent of product development effort spent on product integration
- Number of problem reports resulting from product integration
- Effort and funds expended for each integration task before and after process improvement

Process Task Outcomes
1. A product integration strategy is defined.
2. Unavoidable constraints of integration that influence requirements are defined.
3. A product capable of being verified against the specified requirements from architectural design is assembled and integrated.
4. Nonconformances due to integration actions are recorded.

Table 5.28   Product Verification Process (SEP-28)

Process purpose: The provider shall perform the verification process to confirm that the specified design requirements are fulfilled by the product. This process provides the information required to effect the remedial actions that correct nonconformances in the realized product or the processes that act on it.

Roles and agents:

- Project leader
- Sustainability engineer
- Integrated product team

Inputs:	**Outputs:**
• Verification plan, including the verification compliance requirement matrix • Sustainability engineering management plan and/or software development plan • Test and evaluation master plan • Independent verification and validation plan • Team work plan • Statement of objectives • Statement of work • System technical requirements • Preferred physical solution representation • Specified requirements	• Demonstration test readiness report • Design solution verification report • Design solution deficiency and discrepancy reports (hardware and software, if applicable)

Process diagram:

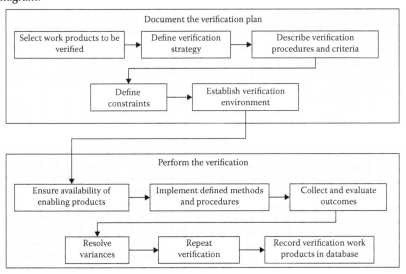

Figure 5.41   Product verification process.

(*Continued*)

Table 5.28   Product Verification Process (SEP-28) (*Continued*)

Step	Action
1	**Action statement:** Document the verification plan. **Notes and guidance:** The plans account for the sequence of configurations defined in the integration strategy and, where appropriate, take account of disassembly strategies for fault diagnosis. The schedule typically defines risk-managed verification steps that progressively build confidence in compliance of the fully configured product.
1.1	**Action statement:** Select the work products for verification based on requirements. **Notes and guidance:** The work products to be verified may include those associated with maintenance, training, and support services. The work product requirements for verification are included with the verification methods. The verification methods address the approach to work product verification and the specific approaches that will be used to verify that specific work products meet their requirements.
1.2	**Action statement:** Define the strategy and methods for verifying the product entities throughout the life cycle. **Notes and guidance:** This strategy applies to the product and to its descriptions (e.g., requirements, design definitions). It includes the context and purpose for each instance of verification action, e.g., verifying the design, ability to build the design correctly, ability to reproduce the product, ability to correct a fault arising, and ability to predict failures. Verification demonstrates, through assessment of the product, that the product is made "right" (i.e., fulfills the specified design against which the product was realized). During verification, wherever possible, the product includes its human operators. The nature and scope of the verification action (e.g., review, inspection, audit, comparison, static test, dynamic test, demonstration, or a combination of these) depend on whether a model, prototype, or actual product is being verified and on the perceived risks (e.g., safety, commercial criticality).
1.3	**Action statement:** Describe detailed verification procedures and criteria. **Notes and guidance:** Verification criteria are defined to ensure that the work products meet their requirements. Examples of sources for verification criteria include the following: 1. Product and product component requirements 2. Standards 3. Organizational policies 4. Test type 5. Test parameters 6. Parameters for trade-off between quality and cost of testing 7. Type of work products 8. Suppliers 9. Proposals and agreements

Table 5.28 Product Verification Process (SEP-28) (*Continued*)

Step	Action
1.4	**Action statement:** Define and communicate potential decision constraints.
	Notes and guidance: This includes practical limitations of accuracy, uncertainty, and repeatability that are imposed by the verification-enabling products; the associated measurement methods; the need for product integration; and the availability, accessibility, and interconnection with enabling products.
1.5	**Action statement:** Establish the verification environment to include verification-enabling products.
	Notes and guidance: An environment must be established to enable verification to take place. The verification environment can be acquired, developed, reused, modified, or a combination of these, depending on the needs of the project. The type of environment required will depend on the work products selected for verification and the verification methods used. A peer review may require little more than a package of materials, reviewers, and a room. A product test may require simulators, emulators, scenario generators, data reduction tools, environmental controls, and interfaces with other products.
2	**Action statement:** Perform the verification.
2.1	**Action statement:** Ensure that enabling resources are ready and available.
	Notes and guidance: Enabling resources include facilities, equipment, and trained operators who are prepared to conduct the verification.
2.2	**Action statement:** Implement defined verification methods and procedures to demonstrate compliance with specified design requirements.
	Notes and guidance: Noncompliance identifies the existence of random faults and/or design errors, and corrective actions are initiated as appropriate. Verification is undertaken in a manner, consistent with organizational constraints, such that uncertainty in the replication of verification actions, conditions, and outcomes is minimized. Approved records of verification actions and outcomes are made.
2.3	**Action statement:** Collect and evaluate verification outcomes.
	Notes and guidance: In accordance with agreement terms or organizational objectives, conduct verification to isolate that part of the product that is giving rise to a nonconformance. Fault diagnosis is conducted to a level of resolution consistent with cost-effective remedial action, including reverification following defect correction, and/or organizational quality improvement actions. Verification data is collected, classified, and collated according to criteria defined in the verification strategy. This categorizes nonconformances according to their source and corrective action and owner. The verification data is analyzed to detect essential features such as trends and patterns of failure, evidence of design errors, and emerging threats to services.
2.4	**Action statement:** Resolve variances as appropriate.
	Notes and guidance: resolve variances, as appropriate, and reverify to establish compliance when the cause of the variance was failure to properly complete the fully characterized design.

(*Continued*)

Table 5.28   Product Verification Process (SEP-28) *(Continued)*

Step	Action
	Any product requirements that are not controllable and observable shall be reported as an unverifiable requirement to process. Variances shall be documented in the design solution discrepancy reports and/or integrated enterprise data repository for evaluation and resolution.
2.5	**Action statement:** Repeat verification as appropriate. **Notes and guidance:** Reverify according to a redesign verification plan, test method, or procedure when variances were determined to be caused by poor verification or inadequate verification environmental preparation. The level of regression testing shall depend on the complexity of the design fix and the level necessary to ensure that the redesign has resolved the nonconformance and been readdressed in the test plan (refer to the testing process).
2.6	**Action statement:** Record verification work products and results in the established database, and make data available to stakeholders. **Notes and guidance:** Record verification results, including corrective actions taken; lessons learned; outcomes achieved; trade-off, effectiveness, and risk analyses completed with resulting key decisions; test activities completed; variances; and the verified design solution in the established enterprise data repository. Results should be included in the redesigned verification plan and shall be an output to the product analysis process (end product verification), so that the information can be included in the system verification process, and to the established enterprise data repository (configuration management process).

Measures:

 • Percent of verification schedules met

Process Task Outcomes
As a result of the successful implementation of the verification process: 1. A verification strategy is defined. 2. Verification constraints are provided as inputs to requirements. 3. Data providing information for corrective action is reported. 4. Objective evidence that the realized product satisfies the product requirements and the architectural design is provided.

Table 5.29   Product Analysis (SEP-29)

Process purpose: The supplier is to perform product analysis activities to ensure that decisions concerning the product design will meet enterprise, project, and other stakeholder requirements in terms of technical performance, total ownership cost, material readiness, reduced logistics footprint, and other defined stakeholder objectives.

Table 5.29 Product Analysis (SEP-29) (*Continued*)

Roles and agents:
• Project leader
• Sustainability engineer
• Integrated product team

Inputs:	Outputs:
• Available alternative evaluation methods, tool, and models	• Analysis plans
	• Validated system models, simulations, and prototypes
• Analysis of available methods, tool, and models	• Validated analysis methodologies
	• Analysis objectives and criteria
• Acquired evaluation methods, tools, and models	• Analysis approach description
	• System analysis records and reports
• Model requirements	• Analysis models
• Simulation requirements	• Production engineering assessment
• Prototype requirements	• Test and evaluation assessment
• Integrated stakeholder requirements	• Deployment and installation assessment
	• Operations assessment
• System-level functional requirements	• Support assessment
	• Training assessment
• System-level performance requirements	• Total ownership cost assessment
	• Environmental assessment
• Preferred product architecture	• Disposal assessment
	• Effectiveness analysis reports

Process diagram:

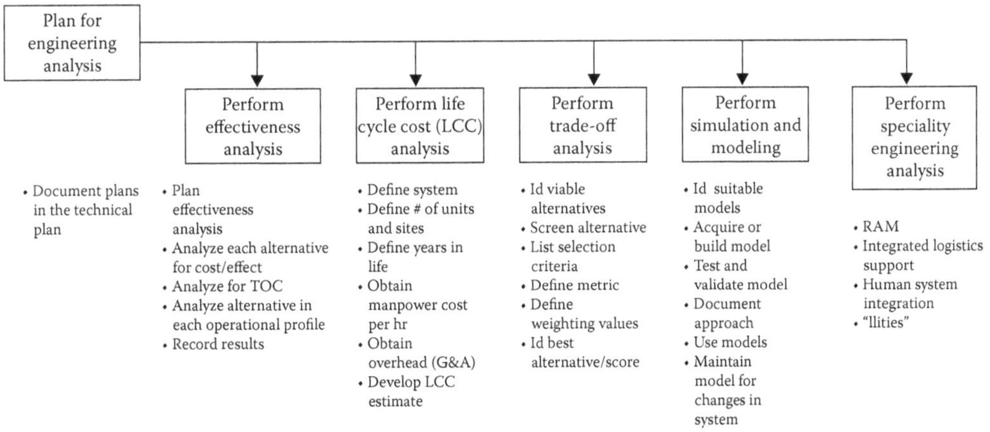

Figure 5.42 Product analysis process.

(*Continued*)

Table 5.29   Product Analysis (SEP-29) (*Continued*)

Step	Action
	Action statement: Plan for conducting sustainability engineering analysis.
1	**Action statement:** Assess available methodologies and tools for analyzing and evaluating technical alternatives, and describe their use in project technical plans. **Notes and guidance:** Describe how the project will measure progress, evaluate alternatives, select preferred alternatives, and document data and decisions used and generated. Systems analyses methods are to include trade-off studies, effectiveness analyses, modeling and simulation, and design analyses, including design for specialty engineering, to determine progress in satisfying technical requirements and program objectives and to provide a rigorous quantitative basis for performance, functional, and design requirements.
2	**Action statement:** Plan effectiveness analysis to include purpose, objectives, methodology, data collection, task schedule, resources needed, and expected outcomes. **Notes and guidance:** Effectiveness analyses are done to a. Measure the extent each alternative physical solution considered during design may be expected to achieve product requirements b. Assist in choosing the preferred physical solution for the end product being developed c. Aid in determining recommended courses of action and associated impacts for trade-off analyses Effectiveness analyses are also used during a. System technical requirements definition to support performance analyses to determine a "knee in the curve" or some other identifiable characteristic that provides an optimal set of requirements b. Progress against requirements assessments to determine how well the design solution is maturing toward meeting agreement requirements c. Technical reviews for providing the review decision makers with the maturity of the design solution The plans for doing effectiveness analyses should be done in conjunction with planning for product analysis and include definition of any special techniques, procedures, tools needed, and simulations and modeling. Effectiveness models should be created for specific characteristics of product functionality. These characteristics include, but are not limited to, operations (such as measures of effectiveness), supportability, reliability, maintainability, production, training, disposal, test/validation/verification, deployment/installation, environmental, and total ownership cost (including design to cost or cost as an independent variable [CAIV]). Effectiveness models should allow parameters to be varied so that relative, individual effect on total product performance and life cycle cost (LCC) can be determined. All effectiveness models must be validated to ensure valid analysis and simulation results.

Table 5.29   Product Analysis (SEP-29) (*Continued*)

Step	Action
3	**Action statement:** Analyze each alternative for product cost and effectiveness based on factors such as accuracy, availability, capacity, maintainability, reliability, responsiveness, operability, safety, security, spares, requirements, survivability, transportability, and vulnerability.
	Notes and guidance: Cost may be treated like a performance objective (design to cost) or as a CAIV. System and cost effectiveness analyses should include the following, as applicable:
	1. Production engineering analysis and assessment to determine what it will take to manufacture or produce, including assembly and integration, the resulting end product. This includes production-ability-related design factors; alternative manufacturing and production approaches; impacts of long-lead-time items; and material, capacity, tools, equipment, and people limitations.
	2. Test and evaluation analysis and assessment to determine what it will take to do necessary tests and evaluations on the resulting end products. This includes analyzing the various kinds of validations, verifications, demonstrations, qualification, acceptance and other testing that may be needed; testability-related design factors; and test and evaluation requirements such as testing sites, facilities, site/facility capacities and limitations, people, and life cycle testing consistency.
	3. Deployment and installation analysis and assessment to determine the requirements and constraints associated with deploying and/or installing the resulting end product. This includes factors for site/host selection, activation/installation, on-site assembly, and site-unique hazards; compatibility with existing infrastructures; environmental impact considerations; early deployment of training items and personnel; initial provisioning and spares; packaging, handling, storage, and transportation requirements and constraints; and site transition requirements.
	4. Operation analysis and assessment to determine what it will take to satisfy operational requirements for the resulting end product. This includes operation and support facility and equipment requirements; interoperability of interacting products required to execute operational functions in the intended use environments; required joint and combined operations including other services, contractors, and international partners; and planned and potential future operation uses.
	5. Supportability analysis and assessment to determine what it will take to support end products over the life cycle. This includes supportability-related design factors; all planned levels of maintenance; and support resources required such as people, parts, facilities, and materials.
	6. Training analysis and assessment to determine what it will take to train users of the resulting end product. This includes development of qualified personnel with appropriate skills, proficiencies, and capabilities; initial and follow-on training requirements; and training resources required such as people, facilities, training materials, and how often retraining will be required (perishability of previous training).

(Continued)

Table 5.29   Product Analysis (SEP-29) (*Continued*)

Step	Action
	Determine the sensitivity to constraints and uncertainties in input data and assumptions. When another product has comparable characteristics, it can be used as a baseline to support the determination, completeness, and achievability of effectiveness analysis requirements.
4	**Action statement:** Analyze each alternative for total ownership cost to the customer, enterprise, and users. The following costs are typically included in a total ownership cost analysis: development, production, test, deployment/installation, training, operations, support/maintenance, and retirement/disposal.
	Notes and guidance: Of interest is determining the economic consequences of each alternative in terms of costs to the enterprise and to the customer for each alternative physical solution representation, analysis of alternatives option, or proposed change. As a result of this analysis, design-to-cost targets (if applicable), current estimate of product total LCC, and known uncertainties in these costs should be established.
5	**Action statement:** Analyze the environmental impact of each alternative, including applicable environmental statutes and hazardous material lists, from an enterprise-based life cycle perspective.
	Notes and guidance: The system and its end products must operate within prescribed environmental definitions. The system/end products and the environment will interact in certain ways, and the goal is to minimize the adverse impact of the system/end products on its environment and the environment on the system/end products. Environmental impacts should include the natural environment (air, land, and water), organizational environment (enterprise and geopolitical), and social environment (people, animal, plant, cultures, and religions).
	It is important to understand the interfaces between the system/end products and the environment in terms of all materials and energies exchanged across the interface. Each interface is studied for ways of reducing environmental impact. Likewise, environmental laws and regulations must be studied for compliance. The supplier must adhere to all applicable statutes and agreements to designated hazardous material lists. Use of materials that present a known hazard will be avoided to the extent possible. Legal implications to the government should be identified and defined.
	An environmental impact analysis should include, as applicable, the following:
	1. Environmental analysis and assessment to determine the impact on and by each end product and enabling product alternative on factors such as noise pollution, quantities and types of hazardous materials used, hazardous waste disposal, and other defined environmental requirements applicable. This includes, from an enterprise-based life cycle perspective, the applicable federal, state, municipal, and international environmental statutes and applicable hazardous material lists affecting the project; endurance of compliance by each physical solution end product; and the effect on and by each end product and enabling product on the infrastructure, land and ocean, atmosphere, water sources, and animal, plant, and human life, as applicable.

Table 5.29   Product Analysis (SEP-29) (*Continued*)

Step	Action
	2. Disposal analysis and assessment to determine what it takes to dispose of end products and by-products. This includes disposability-related design factors; identifying environmental factors for process wastes and outputs as well as used end products and their subsystems; consideration of various disposal methods such as storage, dismantling, demilitarization, reusing, recycling, and destruction; and people, costs, sites, responsible agencies, handling and shipping, supporting items, and applicable federal, state, local, and host nation regulations.
6	**Action statement:** Analyze each alternative for each operational profile to provide an analytical confirmation that the alternative satisfies appropriate operational requirements. **Notes and guidance:** For analysis of alternative physical solution representations or of the preferred physical solution, satisfaction of the set of derived technical requirements should be confirmed. For analysis of alternative attributes (for requirement conflict resolution) or for evaluating logical solution representations, the impact on the ability to satisfy the defined product technical requirements within acceptable costs and risks should be considered.
7	**Action statement:** Record operational effectiveness analysis outcomes in the established project information database, including assumptions, details of the analysis, findings, lessons learned, models used, rationale for decisions made, and other pertinent information that affects the interpretation of the effectiveness analysis results. **Notes and guidance:** The results of the effectiveness analysis should be provided to the requesting source and recorded in the enterprise data repository (data management process). It is important for follow-on analyses that models, data files, and their documentation be maintained, updated, and modified as required. Each version of a model or data file that impacts requirements, design, or decisions should be entered into the enterprise data repository.
8	**Action statement:** Perform LCC analysis. **Notes and guidance:** LCC analyses are used in product cost/effectiveness assessments. The LCC is not necessarily the definitive cost proposal for a program. LCC estimates are often prepared early in a program's life cycle—during concept definition. At this stage, there is insufficient detail design information available to support preparation of a realistic, definitive cost analysis. These are much more detailed and prepared perhaps several years later than the earliest LCC estimates. Later in the program, LCC estimates can be updated with actual costs from early program phases and should be more definitive and accurate due to hands-on experience with the product.

(*Continued*)

Table 5.29   Product Analysis (SEP-29) (*Continued*)

Step	Action
	In addition to providing information for the LCC estimate, these studies also help to identify areas in which emphasis can be placed during the subsequent subphases to obtain the maximum cost reduction. Adequate documentation requires three basic elements:
	1. Data and sources of data on which the estimate is based
	2. Estimating methods applied to that data
	3. The results of the analysis
	The following are recommended activities to perform LCC analysis:
	1. Obtain a complete definition of the system, elements, and their subsystems.
	2. Determine the total number of units of each element, including operational units, prototypes, spares, and test units to be procured. If it is desired to develop parametric cost data as a function of the number of operational units, define the minimum and maximum number of operational units and how, if any, the number of spares and test units will vary with operational unit size.
	3. Obtain the life cycle program schedule, including spans for research and development (R&D), production and deployment, and operations and support (O&S) phases. The production and deployment phase length will vary with the number of operational units.
	4. Obtain manpower estimates for each phase of the entire program and, if possible, for each element and subsystem. These are especially important for cost estimating during R&D and O&S.
	5. Obtain approximate/actual overhead. General and administrative (G&A) burden rates and fees that should be applied to hardware and manpower estimates. Usually, this is only necessary for effort within your own company; suppliers will have already included it in their cost estimates. These data are not required to the accuracy that our finance department would use in preparing a formal cost proposal.
	6. Develop cost estimates for each subsystem of each product element for each phase of the program. This is, of course, the critical step. Generally, it should be done as accurately as time and resources allow.
9	**Action statement:** Perform trade studies.
	Notes and guidance: Trade studies provide an objective foundation for the selection of one of two or more alternative approaches to solution of an engineering problem. The trade study may address any of a range of problems from the selection of high-level system architecture to the selection of a specific commercial off-the-shelf processor.
	In developing a design, it is tempting to select a design solution without performing a formal trade study. The selection may seem obvious to the supplier—the other possible alternatives appear unattractive, particularly to

Table 5.29   Product Analysis (SEP-29) (*Continued*)

Step	Action
	other team members (e.g. design, manufacturing, quality, and other engineering disciplines). However, it will be far easier to justify the selected solution in a proposal or at a formal design review if we have followed certain procedures in making the selection. Use of a formal trade study procedure will provide discipline in our decision process and may prevent some ill-advised decisions. It is important, also, to recognize when a formal trade study is not needed in order to reduce project costs. Whenever a decision is made, a trade-off process is carried out, implicitly, if not explicitly. It is useful to consider trade studies in three levels of formality: • Formal: These trades use a standardized methodology, are formally documented, and are reviewed with the customer or internally at a design review. • Informal: These trade studies follow the same kind of methodology but are only recorded in the engineer's notebook and are not formally reviewed. • Mental: When a selection of any alternative is made, a mental trade study is implicitly performed. The trade study is performed with less rigor and formality than documented trades. These types of trade studies are made continuously in our everyday lives. These are appropriate when the consequences of the selection are not too important; when one alternative clearly outweighs all others; or when time does not permit a more extensive trade. However, if the rationale is not documented, it is soon forgotten and unavailable to those who may follow. One chooses the level of trade study depending on the consequences to the project, the complexity of the issue, and the resources available. The resources to perform trades are allocated based on the overall LCC differences (with provision for risk coverage) in alternative selection for the potential trades. Those with the largest overall LCC deltas are performed first. Since more informal trades can be performed with fewer resources than formal trades, the number and selection of trades and their formality need to be decided with the customer and with the necessary team members who might find some design solutions favorably or unfavorably impacting manufacturability, production ability, reliability, testability, maintainability, and so on. Remember, it takes minimal effort to document the rationale for informal and "mental" trade-off conclusions. Recommended activities for performing trade studies: There are multiple techniques for performing trade studies. These include multi-attribute utility analysis, decision trees, and maximum expected utility. There is no need to standardize on any one. One might be better for one trade study another better in another situation. The key components of a formal trade study are the following: a. A list of viable alternative solutions to be evaluated. b. A list of screening criteria. Any alternative that fails one of these criteria is ruled out of further consideration.

(*Continued*)

Table 5.29 Product Analysis (SEP-29) (*Continued*)

Step	Action
	c. A list of selection criteria; that is, a set of factors that characterize what makes a specific alternative desirable. This should include cost, risk, and performance factors.
	d. For each of the selection criteria, a metric characterizing how well the various solutions satisfy that criteria.
	e. Weighting values assigned to each of the selection criteria, reflecting their relative importance in the selection process.
	With these components, an objective measure of the suitability of each alternative as a solution to the problem is obtained. If this process is performed correctly and objectively, then the alternative with the best score is the best overall solution.
10	**Action statement:** Perform modeling, simulation, and prototyping. **Notes and guidance:** N/A
11	**Action statement:** Determine the appropriate methodologies and tools needed to support the product analysis effort. **Notes and guidance:** N/A
12	**Action statement:** Acquire needed methodology and tools. **Notes and guidance:** N/A
13	**Action statement:** Identify the requirements for and the approach for developing models, simulations, and prototypes. **Notes and guidance:** N/A
14	**Action statement:** Select or develop the required models, simulations, and prototypes. **Notes and guidance:** Care is needed in the design of the model to ensure that the general criteria are met. Usually, this requires some degree of fundamental analysis of the product: a. Identify the relevant product characteristics that are to be evaluated through use of the model. b. Determine the relevant measurable parameters that define those characteristics, and separate them from irrelevant parameters. c. Define the scope and content of data needed to support the decision economically and accurately. It is particularly important that the model be economical in use of time and resources and that the output data be compact and readily understandable to support efficient decisions. The Taguchi design of experiments process (identifying the sensitivity of the results to variation of key parameters and adjusting the spacing of sampling so that the total range of results is spanned with the minimum number of test points) can be very effective in determining the bounds and the limits of the model. This data can be used to estimate the value of the information gained by producing the model. The model itself can be considered as a product to which the requirements analysis, functional analysis, and system synthesis steps of the sustainability engineering process are applied to determine the requirements for the model and define

Table 5.29   Product Analysis (SEP-29) (*Continued*)

Step	Action
	the approach. This analysis provides an overall description of the modeling approach. Following its review and approval, the detailed definition of the model can be created according to usual practice for the type of model selected.
15	**Action statement:** Test and validate models, simulations, and prototypes prior to use.
	Notes and guidance: It is crucial to prove that the model is trustworthy and suitably represents reality, particularly in cases where a feel for product behavior is absent, or when serious consequences can result from inaccuracy. Models can be validated by the following:
	a. Experience with application of similar models in similar circumstances
	b. Analysis showing that the elements of the model are of necessity correct and are correctly integrated
	c. Comparison with test cases in the form of independent models of proven validity or actual test data
	The modeling schema itself can be validated by using small-scale models.
16	**Action statement:** Develop a detailed implementation approach for using the acquired tools and the newly developed models, simulations, and prototypes in support of the product analysis effort.
	Notes and guidance: Document the approach/methodology in the analysis report.
17	**Action statement:** Maintain the validated models, simulations, and prototypes in accordance with product changes.
	Notes and guidance: N/A
18	**Action statement:** Use appropriate models, simulations, and prototypes, and record results.
	Notes and guidance: Obtain needed input data to set the model's parameters to represent the actual product and its operating environment. In some situations, defining and acquiring the basis model data can be a very large effort, so care in design of the model is needed to minimize this problem. Perform as many runs as are needed to span the range of the product parameters and operating conditions to be studied, and in the case of statistical models, to develop the needed level of statistical validity.
19	**Action statement:** Perform design analysis to consider appropriate specialty engineering disciplines.
	Notes and guidance: The objective of specialty engineering analysis is to give enough information to sustainability engineers to appreciate the significance of various engineering specialty areas, even if they are not an expert in the subject. It is recommended that subject matter experts are consulted and assigned as appropriate to conduct specialty engineering analysis. With a few exceptions, the forms of analysis presented herein are similar to those associated with sustainability engineering. Most analysis methods are based on the construction and exploration of models that address specialized engineering areas, such as electromagnetic compatibility, reliability, safety, and security. Not every kind of analysis and associated model will be applicable to every application domain.

(*Continued*)

Table 5.29 Product Analysis (SEP-29) (*Continued*)

Measures:

- Number of documented SE analysis reports per project defined measurement period
- Percent of actual project budget expended on SE analysis vs. amount planned for this activity

Process Task Outcomes

Effectiveness analysis reports are provided to the requestor of the effectiveness analysis and captured in the enterprise data repository. Each report will document the results of the effectiveness analysis in accordance with the agreement and effectiveness analysis plan to include the following: outcomes from each analysis and assessment made, and who approved the results; input data used and who approved the data; models used; and related data files, assumptions, and lessons learned. Some examples of types of reports/analyses may include mission area analysis, measures of effectiveness, mission analysis, analysis of alternatives, and product concept analysis. For effectiveness analyses that support the stakeholder requirements definition process, customer requirement and other stakeholder requirements are analyzed to determine warfighter deficiencies and to analyze technology opportunities for increased systems effectiveness and/or cost reductions. For effectiveness analyses that support the system technical requirements, outcome data includes the following:

- The effectiveness of various mixes of requirements without regard to the means of implementation (except for legacy systems for which changes of performance are being considered)
- Effectiveness to help come up with a "knee in the curve" or some other identifiable characteristic that provides an optimal set of requirements

For effectiveness analyses that support the logical architecture design process, the outcome data are very similar to those for the physical architecture design process in that effectiveness of various logical representations are considered without regard to the means of implementation (except for legacy systems).

For effectiveness evaluations to support trade-off analyses of alternative physical solution representations or an evaluation of the preferred physical solution (the logical architecture design process), the outcome data provides a quantitative assessment of the value of a point design solution. The objective of these evaluations is to measure how well the point design meets its set of derived requirements. For systems effectiveness assessments that support the data management process, outcome data includes, as applicable, the following:

- Overall system or system product effectiveness for each operational profile with respect to satisfying customer requirements within acceptable risks
- Impact on enabling product requirements with respect to each associated process (development and integration, production/manufacturing, test, deployment, training, operations, support, and disposal)
- System cost effectiveness with respect to attributes such as capability (accuracy), dependability (availability, reliability, operability, survivability, and vulnerability), and suitability (capacity, maintainability, responsiveness, safety, security, spare requirements, and transportability)
- Total ownership costs to the enterprise, customer, and/or user, including the known uncertainties (risks) in these costs
- Compliance impacts of applicable federal, state, municipal, and international environmental statutes and applicable hazardous material lists, as well as legal liabilities
- Environmental impacts on the land and ocean; atmosphere; water sources; and animal, plant, and human life

Table 5.30   Testing Process (SEP-30)

Process purpose: The provider shall perform test and evaluation (T&E) to identify levels of product performance and to assist in correcting deficiencies.	

Roles and agents:	
• Project leader	
• Sustainability engineer	
• Integrated product team	

Inputs:	**Outputs:**
• Needs documents	• Updated needs documents
• Capability development documents	• Updated capability development documents
• Capability production documents	• Updated capability production document
• System design documents	• Updated system design documents
• System specifications	• Updated system specifications
• Prior test reports	• Test plans
	• Test procedures
	• Test case documents
	• Test reports

Process diagram:

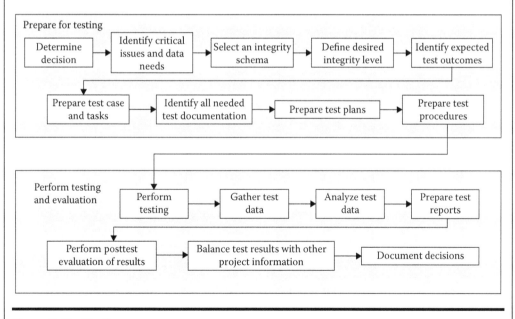

Figure 5.43   Testing process.

(*Continued*)

Table 5.30   Testing Process (SEP-30) (*Continued*)

Step	Action
1	**Action statement:** Determine the issue and decision to be made.
	Notes and guidance: The testing process is an iterative process that provides answers to critical T&E questions for decision makers at various times during a product acquisition. The T&E process begins during the formative stages of the program with the T&E coordination function, in which the information needs of the various decision makers are formulated in conjunction with the development of the program requirements, acquisition strategy, and analysis of alternatives.
2	**Action statement:** Identify critical issues and data requirements.
	Notes and guidance: This step is the identification of T&E information required by the decision maker. The required information usually centers on the current product under test, which may be in the form of concepts, prototypes, Electronic Document Management Systems (EDMs), or production representative/ production systems, depending on the acquisition phase. The required information consists of performance evaluations of effectiveness and suitability, providing insights into how well the product meets the user's needs at a point in time.
3	**Action statement:** Select an integrity level schema.
	Notes and guidance: There shall be a documented definition of the integrity levels or the decision to not use an integrity level scheme. The integrity level (or the decision to not use an integrity level scheme) shall be assigned as a result of agreements among all specified parties (or their designated representative(s)), such as the customer, supplier, developer, and independent assurance authorities (e.g., a regulatory body or responsible agency).
4	**Action statement:** Identify the desired integrity level for the product or software.
	Notes and guidance: This guide uses integrity levels to determine the testing tasks to be performed. Integrity levels may be applied to requirements, functions, groups of functions, components, and subsystems. Some of these may not require the assignment of an integrity level because their failure would impart no negative consequence on the intended product operation. The integrity scheme may be based on functionality, performance, security, or some other product or software characteristic. Whether an integrity level scheme is mandatory is dependent on the needs of the stakeholders for the product. The user may follow the four-level schema as defined in Appendix A as an example in this guide or may use a different schema. However, if a different schema is used, a mapping should be made between the user's schema and the example. Some software elements and components may not require the assignment of an integrity level (i.e., not applicable) because the failure would impart no consequences on the intended product operations. The user may want to add a Level 0 to Table 1. Level 0 would cover failures that would cause no consequences or are not applicable.

Table 5.30   Testing Process (SEP-30) (*Continued*)

Step	Action
5	**Action statement:** Prepare pretest and evaluation plan to include expected outcomes. **Notes and guidance:** In this step, the pretest analysis of the evaluation objectives from step 2 determine the types and quantities of data needed, the results expected or anticipated from the tests, and the analytical tools needed to conduct the tests and evaluations. The use of validated models and simulation systems during pretest analysis can aid in determining: how to design test scenarios; how to set up the test environment; how to properly instrument the test; how to staff and control test resources; how best to sequence the test trials; and how to estimate outcomes.
6	**Action statement:** Prepare test case plan and test tasks needed. **Notes and guidance:** In this step, test activity and data management is the actual detailed test activity planning. Tests are conducted, and data management for data requirements is identified in Step 5. T&E managers determine what valid data exist in historical files that can be applied and what new data must be developed through testing. The necessary tests are planned and executed to accumulate sufficient data to support analysis. Data are screened for completeness, accuracy, and validity before being used for Step 4.
7	**Action statement:** Identify all test documentation needed based on test tasks, to include document contents. **Notes and guidance:** Commonly used test documents are as follows: • Master test plan (MTP) • Level test plan (LTP) • Level test design (LTD) • Level test case (LTC) • Level test procedure (LTPr) • Level test log (LTL) • Anomaly report (AR) • Level interim test status report (LITSR) • Level test report (LTR) • Master test report (MTR)
8	**Action statement:** Perform detailed test planning. **Notes and guidance:** Two levels of detailed test planning need to be documented: the test and evaluation master plan (TEMP) and the LTP. *TEMP*: The purpose of the TEMP is to provide an overall test planning and test management document for multiple levels of test (either within one project or across multiple projects). In view of the software requirements and the project's (umbrella) quality assurance planning, master test planning as an activity comprises selecting the constituent parts of the project's test effort; setting the objectives for each part; setting the division of labor (time, resources) and the interrelationships between the parts; identifying the risks, assumptions, and standards of workmanship

(Continued)

Table 5.30   Testing Process (SEP-30) (*Continued*)

Step	Action
	to be considered and accounted for by the parts; defining the test effort's controls; and confirming the applicable objectives set by quality assurance planning. It identifies the integrity level schema, the integrity level selected, the number of levels of test, the overall tasks to be performed, and the documentation requirements.
	LTP: Specify for each LTP the scope, approach, resources, and schedule of the testing activities for its specified level of testing. Identify the items being tested, the features to be tested, the testing tasks to be performed, the personnel responsible for each task, and the associated risk(s). In the title of the plan, the word "level" is replaced by the organization's name for the particular level being documented by the plan (e.g., component test plan, component integration test plan, system test plan, and acceptance test plan).
	In most projects, there are different test levels requiring different resources, methods, and environments. As a result, each level is best described in a separate plan. Different LTPs may require different usage of the documentation content topics listed below. Some examples of test levels for the development activity to undertake are as follows:
	a. Each software unit and database.
	b. Integrated units and components.
	c. Tests for each software requirement.
	d. Software qualification testing for all requirements.
	e. Systems integration: aggregates of other software configuration items, hardware, manual operations, and other systems. It is not unusual for large systems to have multiple levels of integration testing.
	f. System qualification testing for system requirements.
9	**Action statement:** Perform test design for test cases and test procedures.
	Notes and guidance: The purpose of the LTD document is to specify any refinements of the test approach and to identify the features to be tested by this design and its associated tests. Example content topics are as follows:
	LTD outline (full example)
	1. Introduction
	1.1. Document identifier
	1.2. Scope
	1.3. References
	2. Details of the LTD
	2.1. Features to be tested
	2.2. Approach refinements
	2.3. Test identification
	2.4. Feature pass/fail criteria
	2.5. Test deliverables

Table 5.30 Testing Process (SEP-30) (*Continued*)

Step	Action
	3. General
	3.1. Glossary
	3.2. Document change procedures and history
	The purpose of the LTC is to define (to an appropriate level of detail) the information needed as it pertains to inputs and to outputs from the software or software-based system being tested. The LTC includes all test case(s) identified by the associated segment of the LTD (if there is one). Example content topics are as follows:
	LTC outline (full example)
	1. Introduction (once per document)
	1.1. Document identifier
	1.2. Scope
	1.3. References
	1.4. Context
	1.5. Notation for description
	2. Details (once per test case)
	2.1. Test case identifier
	2.2. Objective
	2.3. Inputs
	2.4. Outcome(s)
	2.5. Environmental needs
	2.6. Special procedural requirements
	2.7. Intercase dependencies
	3. Global (once per document)
	3.1. Glossary
	3.2. Document change procedures and history
	The purpose of an LTPr is to specify the steps for executing a set of test cases or, more generally, the steps used to exercise a software product or software-based system item in order to evaluate a set of features. Example content topics are as follows:
	LTPr outline (full example)
	1. Introduction
	1.1. Document identifier
	1.2. Scope
	1.3. References
	1.4. Relationship to other procedures

(*Continued*)

Table 5.30 Testing Process (SEP-30) (*Continued*)

Step	Action
	2. Details
	2.1. Inputs, outputs, and special requirements
	2.2. Ordered description of the steps to be taken to execute the test cases
	3. General
	3.1. Glossary
	3.2. Document change procedures and history
10	**Action statement:** Conduct tests.
	Notes and guidance: Perform tests designed to support the test requirements such as those listed below:

Test Type	Purpose
Development test (DT)	Support of design and technical reviews. DT&E is T&E conducted throughout the acquisition process to assist in engineering design and development and to verify that technical performance specifications have been met. The DT&E is planned and monitored by the developing agency and is normally conducted by the contractor. However, the development agency may perform technical compliance tests before OT&E. It includes the T&E of components, subsystems, preplanned product improvement (P3I) changes, hardware/software integration, and production qualification testing.
Operational test (OT)	The field test, under realistic combat conditions, of any item of (or key component of) weapons, equipment, or munitions for the purposes of determining the effectiveness and suitability of the weapons, equipment, or munitions for use in combat by typical military users and the evaluation of the results of such test.
Production qualification test (PQT)	Qualification testing is a form of development testing that verifies the design and manufacturing process. PQTs are formal contractual tests that confirm the integrity of the product design over the operational and environmental range in the specification. These tests usually use production hardware fabricated to the proposed production design specifications and drawings.
Integration test	As subsystems and components are assembled they are tested to insure that the integrated system performs satisfactorily to satisfy specified requirements.

Table 5.30   Testing Process (SEP-30) (*Continued*)

Step	Action
11	**Action statement:** Gather test data. **Notes and guidance:** N/A
12	**Action statement:** Analyze test data. **Notes and guidance:** The DT&E results are evaluated to ensure that design risks have been minimized, and the product will meet specifications. The results are also used to estimate the product's utility when it is introduced into service.
13	**Action statement:** Prepare test reports. **Notes and guidance:** The purpose of the LTL is to provide a chronological record of relevant details about the execution of tests. An automated tool may capture all or part of this information. An example of an LTL outline is as follows: 1. Introduction 1.1. Document identifier 1.2. Scope 1.3. References 2. Details 2.1. Description 2.2. Activity and event entries 3. General 3.1. Glossary The purpose of the AR is to document any event that occurs during the testing process that requires investigation. This may be called a problem, test incident, defect, trouble, issue, anomaly, or error report. An example of an AR outline is as follows: 1. Introduction 1.1. Document identifier 1.2. Scope 1.3. References 2. Details 2.1. Summary 2.2. Date anomaly discovered 2.3. Context 2.4. Description of anomaly 2.5. Impact 2.6. Originator's assessment of urgency (see IEEE 1044-1993 [B13]) 2.7. Description of the corrective action 2.8. Status of the anomaly 2.9. Conclusions and recommendations

(*Continued*)

Table 5.30   Testing Process (SEP-30) (*Continued*)

Step	Action
	3. General
	3.1. Document change procedures and history
	The purpose of the LTR is to summarize the results of the designated testing activities and to provide evaluations and recommendations based on these results. It is customary to replace the word "Level" in the title of the document with the organization's name for the particular test level (e.g., acceptance test report). There is one LTR for each test level defined by the organization or project. Small projects may merge reports for multiple levels. They may vary greatly in the level of detail of documentation (e.g., a unit test report may simply be a statement that it passed or failed, whereas an acceptance test report may be much more detailed). An example of an LTR outline is as follows:
	1. Introduction
	1.1. Document identifier
	1.2. Scope
	1.3. References
	2. Details
	2.1. Overview of test results
	2.2. Detailed test results
	2.3. Rationale for decisions
	2.4. Conclusions and recommendations
	3. General
	3.1. Glossary
	3.2. Document change procedures and history
	The purpose of the MTR is to summarize the results of the levels of the designated testing activities and to provide evaluations based on these results. This report may be used by any organization using the MTP. Whenever an MTP is generated and implemented, there needs to be a corresponding MTR that describes the results of the MTP implementation. An example of an MTR outline is as follows:
	1. Introduction
	1.1. Document identifier
	1.2. Scope
	1.3. References
	2. Details of the MTR
	2.1. Overview of all aggregate test results
	2.2. Rationale for decisions
	2.3. Conclusions and recommendations

Table 5.30 Testing Process (SEP-30) (*Continued*)

Step	Action
	3. General 3.1. Glossary 3.2. Document change procedures and history
14	**Action statement:** Perform posttest evaluation of results against expected outcomes. **Notes and guidance:** In this step, posttest synthesis and evaluation is the comparison of the measured outcomes (test data) with the expected outcomes, tempered with technical and operational judgment. This is where data are synthesized into information. When the measured outcomes differ from the expected outcomes, the test conditions and procedures must be reexamined to determine if the performance deviations are real or were the result of test conditions, such as lack of fidelity in computer simulation, insufficient or incorrect test support assets, instrumentation error, or faulty test processes. The assumptions of tactics, operational environment, systems performance parameters, and logistics support must have been carefully chosen, fully described, and documented prior to test. Modeling and simulation may normally be used during the data analysis to extend the evaluation of performance effectiveness and suitability.
15	**Action statement:** Balance test results with other project information. **Notes and guidance:** In this step, the decision maker weighs the T&E information against other programmatic information to decide a proper course of action. This process may identify additional requirements for test data and iterate the T&E process again.
16	**Action statement:** Document decisions. **Notes and guidance:** Documents that are inputs to the test and evaluation process are requirements documents, product specifications, and other design- and support-related documents. The decisions that result from test reports are documented in the form of updates to these input documents, and these updated system documents are then used as inputs in the next iteration of the recursive performance of design and testing down the levels of the system's work breakdown structure. System performance parameters are also updated in top-level requirements documents used to support project decision gate and technical reviews.

Table 5.31 Product Validation Process (SEP-31)

Process purpose: The provider shall perform the validation process to confirm that the specified design requirements are fulfilled by the product in the operational environment.

This process provides the information required to effect the remedial actions that correct nonconformances in the realized product or the processes that act on it.

Roles and agents:

- Project leader
- Sustainability engineer
- Integrated product team

(*Continued*)

Table 5.31   Product Validation Process (SEP-31) (*Continued*)

Inputs:	Outputs:
• Validation plan • Sustainability engineering management plan and/or software development plan • Test and evaluation master plan (TEMP) • Independent verification and validation plan • Team work plan • Statement of objectives • Statement of work • System technical requirements • Preferred physical solution representation • Specified requirements	• Demonstration test readiness report • Design solution validation report • Design solution deficiency and discrepancy reports (hardware and software, if applicable)

Process diagram:

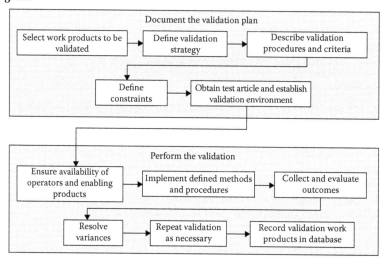

Figure 5.44   Product validation process.

Step	Action
1	**Action statement:** Document the Validation Plan
1.1	**Notes and guidance:** The plans account for the sequence of configurations defined in the integration strategy and, where appropriate, take account of disassembly strategies for fault diagnosis. The schedule typically defines risk-managed validation steps that progressively build confidence in compliance of the fully configured product.

Table 5.31 Product Validation Process (SEP-31) (*Continued*)

Step	Action
1.2	**Action statement:** Select the work products for validation based on requirements. **Notes and guidance:** The work products to be verified may include those associated with maintenance, training, and support services. The work product requirements for validation are included with the validation methods. The validation methods address the approach to work product validation and the specific approaches that will be used to verify that specific work products meet their requirements.
1.3	**Action statement:** Define the strategy and methods for verifying the product entities throughout the life cycle. **Notes and guidance:** This strategy applies to the product and to its descriptions: For example, requirements and design definitions. It includes the context and purpose for each instance of validation action (e.g., verifying the design, ability to build the design correctly, ability to reproduce the product, ability to correct a fault arising, and ability to predict failures). Validation demonstrates, through assessment of the product, that the product is made "right" (i.e., fulfills the specified design against which the product was realized). During validation, wherever possible, the product includes its human operators. The nature and scope of the validation action (e.g., review, inspection, audit, comparison, static test, dynamic test, demonstration, or a combination of these) depend on whether a model, prototype, or actual product is being verified and on the perceived risks (e.g. safety, commercial criticality).
1.4	**Action statement:** Describe detailed validation procedures and criteria. **Notes and guidance:** Validation criteria are defined to ensure that the work products meet their requirements. Examples of sources for validation criteria include the following: 1. Product and product component requirements 2. Standards 3. Organizational policies 4. Test type 5. Test parameters 6. Parameters for trade-off between quality and cost of testing 7. Type of work products 8. Suppliers 9. Proposals and agreements
1.5	**Action statement:** Define and communicate potential decision constraints. **Notes and guidance:** This includes practical limitations of accuracy, and uncertainty, repeatability that are imposed by the validation-enabling products; the associated measurement methods; the need for product integration; and the availability, accessibility, and interconnection with enabling products.
1.6	**Action statement:** Obtain the test articles or aggregation of test articles, and establish the validation environment to include validation-enabling products.

(*Continued*)

Table 5.31 Product Validation Process (SEP-31) (*Continued*)

Step	Action
	Notes and guidance: Acquire the test article, or aggregation of end products, for the validation as appropriate to the project life cycle phase and level of system structure. Test articles for operational test (OT) must be representative of production and are usually procured as part of a low-rate initial production contract. In early phases where an OT is being conducted, the test article may be a prototype or even a model. The number of test articles and their configuration need to be planned in conjunction with the test and evaluation master plan (TEMP). The "test article" should include any support equipment, trainers, or other items necessary to test the article under operationally representative conditions.
	NOTE: The test article is typically the product, or an aggregation of products, that is to be delivered or that has been delivered and that has already been verified. In early enterprise-based life cycle developments, the product or aggregation of products undergoing validation can be a virtual prototype or model. Thus, a detailed simulation, operated so that customer perceptions can be evaluated, is a possible means of validation.
	An environment must be established to enable validation to take place. The validation environment can be acquired, developed, reused, modified, or a combination of these, depending on the needs of the project.
	The type of environment required will depend on the work products selected for validation and the validation methods used. A peer review may require little more than a package of materials, reviewers, and a room. A product test may require simulators, emulators, scenario generators, data reduction tools, environmental controls, and interfaces with other systems.
2	Perform the Validation
2.1	**Action statement:** Ensure that enabling resources are ready and available.
	Notes and guidance: Enabling resources include facilities, equipment, and trained operators who are prepared to conduct the validation.
2.2	**Action statement:** Implement defined validation methods and procedures to demonstrate compliance with specified design requirements.
	Notes and guidance: Noncompliance identifies the existence of random faults and/or design errors, and corrective actions are initiated as appropriate. Validation is undertaken in a manner, consistent with organizational constraints, such that uncertainty in the replication of validation actions, conditions, and outcomes is minimized. Approved records of validation actions and outcomes are made.
2.3	**Action statement:** Collect and evaluate validation outcomes.
	Notes and guidance: In accordance with agreement terms or organizational objectives, conduct validation to isolate that part of the product that is giving rise to a nonconformance. Fault diagnosis is conducted to a level of resolution consistent with cost-effective remedial action, including revalidation following defect correction, and/or organizational quality improvement actions. Validation data is collected, classified, and collated according to criteria defined in the validation strategy. This

Table 5.31   Product Validation Process (SEP-31) (*Continued*)

Step	Action
	categorizes nonconformances according to their source and corrective action and owner. The validation data is analyzed to detect essential features such as trends and patterns of failure, evidence of design errors, and emerging threats to services.
2.4	**Action statement:** Resolve variances as appropriate. **Notes and guidance:** Resolve variances, as appropriate, and reverify to establish compliance when the cause of the variance was failure to properly complete the fully characterized design. Any product requirements that are not controllable and observable shall be reported as an unverifiable requirement to the interface management process via the logical architecture design process but should be confirmed as part of the standards task in task 1 above as well. Variances shall be documented in the design solution discrepancy reports and/or integrated enterprise data repository for evaluation and resolution.
2.5	**Action statement:** Repeat validation as appropriate. **Notes and guidance:** Reverify according to a redesign validation plan, test method, or procedure when variances were determined to be caused by poor validation or inadequate validation environmental preparation. The level of regression testing shall depend on the complexity of the design fix and the level necessary to ensure that the redesign has resolved the nonconformance and been readdressed in the test plan (refer to the testing process).
2.6	**Action statement:** Record validation work products and results in the established database, and make data available to stakeholders. **Notes and guidance:** Record validation results, including corrective actions taken; lessons learned; outcomes achieved; trade-off, effectiveness, and risk analyses completed with resulting key decisions; test activities completed; variances; and the verified design solution in the established enterprise data repository. Results should be included in the redesigned validation plan and shall be an output to the product analysis process (end product validation), so that the information can be included in the product validation process, and to the established enterprise data repository (configuration management process).

Measures:

- Percent of verification schedules met

Process Task Outcomes	
Representative Tasks	*Process Task Outcomes*
Determine validation exit criteria	The type of validation required and the requirements to be used are determined. The types include the following: 1. Validation against customer requirements in the anticipated usage environment, with test conditions that span the expected range of actual operating conditions, to the extent practical, and in conjunction with stakeholders, as appropriate

(*Continued*)

Table 5.31   Product Validation Process (SEP-31) (*Continued*)

	2. Certification tests against established certification requirements
	3. Acceptance tests using operational processes and personnel in operational environments
	4. As specified in the agreement
	NOTES:
	• Validation tests are conducted during the test and evaluation phase of the engineering life cycle, after the end products have been verified against specified requirements, from the lowest level of the system structure to the end products that will be delivered to the marketplace to satisfy validated customer requirements.
	• Validations of Types 1 through 3 are satisfied with the same tests, when appropriate.
	• Validation can be for a single end product or an aggregation of end products for the same building block.
Acquire appropriate test article	The test article, or test articles, used for the validation is determined to be appropriate to the enterprise-based life cycle phase and the level of system structure.
	NOTE: End products validation consists of one or more tests using a version of the product (or products) as nearly like the final version as is practical and necessary, taking into account the enterprise-based life cycle phase and the nature of the product. If the nature of either product, its operating conditions, or the enterprise-based life cycle phase of development precludes use of actual products or prototypes, then breadboards, hardware-in-the-loop simulations, virtual-reality simulations, or other models and simulations are applicable for end products validation.
Conduct validation	• Validation is completed in accordance with the validation plan, as required in the agreement.
	• Validation outcomes are compiled, analyzed, and compared to the validation exit criteria; variations and anomalies have been identified; and corrective actions are defined.
	• When outcome variances from exit criteria were not caused by improper test performance of validation procedures or by improper data collection: Replanning, redefinition of the design solution, and the implementation process, as appropriate, are reaccomplished.

Table 5.31   Product Validation Process (SEP-31) (*Continued*)

Representative Tasks	Process Task Outcomes
	NOTE: Care is to be taken to ensure that the requirements derive to remove variances do not conflict with customer or other stakeholder requirements or other validated technical requirements without coordinating such change with the appropriate stakeholders.
Perform revalidation	If variances were caused by poor test conduct, retesting, using improved or correct test equipment and procedures, is performed.
Record validation results	Validation procedures, compliance data, outcomes, assumptions, corrective actions, lessons learned, and so on, are recorded in the established project information database.

Table 5.32   Enabling Product Readiness Determination (SEP-32)

Process purpose: The supplier shall determine readiness of enabling products for development, production, test, deployment/installation, training, support/maintenance, and retirement or disposal. This process determines the readiness of enabling products furnished by the supplier to support each life cycle phase of the product.

Roles and agents:

- Project leader
- Sustainability engineer
- Integrated product team

Inputs:	**Outputs:** All outputs should be archived
• Enabling products • List of methods and tools, facilities equipment, and training • Specified requirements • Enabling products development projects	• Enabling products readiness assessment plan • Enabling products readiness determination

Process diagram:

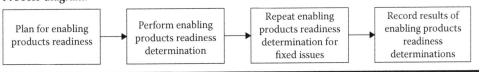

Figure 5.45   Product readiness process.

Step	Action
1	**Action statement:** Plan enabling product readiness determination and associated process proofing in accordance with the appropriate plan, maturity of related end products, agreement, applicable enterprise-based life cycle phase, and level in the system structure.

(*Continued*)

Table 5.32   Enabling Product Readiness Determination (SEP-32) (*Continued*)

Step	Action
	Notes and guidance: Include the following: 1. Selection and definition of the appropriate method for the enabling product readiness determination and for proofing for each applicable associated process 2. Readiness determination procedures to be followed for the method selected, the purpose and objective of each procedure, pretest and posttest actions, and the criteria for determining the success or failure of the procedure 3. Establishment and checkout (for example, adequacy and completeness) of the environment (for example, climatic conditions, equipment, facilities, and measuring devices, etc.) in which the readiness determination method and procedures will be implemented 4. Assurance that required information regarding the status and maturity of enabling product development or requirements definition is available and that nondevelopmental enabling products are available and, if appropriate, integrated with the environment according to appropriate plans and schedules A comprehensive plan to conduct the readiness review should be developed and agreed-to by the contractor and government. Plan should include resources needed to conduct review, method of establishing contractor's readiness, environment or facilities necessary for the assessment, metrics to ensure mitigation of supplier's risk, and follow-up/corrective action plans.
2	**Action statement:** Perform the planned enabling product readiness determination and associated process proofing, using the selected methods and procedures within the established environment: 1. Collect and evaluate readiness determination outcomes to either show compliance or identify variances (untraceable requirements and constraints, anomalies, variations, voids, and conflicts) 2. For variances not caused by poor readiness determination, or process proofing conduct or conditions, complete appropriate tasks of the planning process, control process, requirements definition process, and solution definition process to resolve variances, and then repeat the readiness determination or proofing **Notes and guidance:** Readiness reviews should be conducted to assess risk of enabling products supporting each life cycle phase of the product. Actions (with milestones) to mitigate risk should be identified in readiness reports to stabilize product configuration and minimize change activity in later phases. Examples of readiness review reports include the integrated training plan, production readiness review report, initial operating supportability capability review report, and logistics support analysis. Any design, test, manufacturing, logistics, and disposal issue should be identified in the readiness reviews for an effective product development.

Table 5.32   Enabling Product Readiness Determination (SEP-32) (*Continued*)

3	**Action statement:** Reaccomplish readiness determination. Reaccomplish readiness determination according to redesigned plans, test method, or procedure when variances were determined to be caused by poor readiness or proofing conduct or by inadequate environmental preparation. A follow-up or another readiness review can be conducted if the risk was considered excessive in the original readiness review.
	Notes and guidance: Supplier must provide evidence that risk has been effectively mitigated to ensure a smooth transition into the next planned life cycle phase. After the exit criteria have been met and risk has been lowered, the supplier is ready to enter the next planned life cycle phase.
	Action statement: Record results. Record readiness determination and process proofing results, including corrective actions taken; lessons learned; outcomes achieved; trade-off, effectiveness, and risk analyses completed, with resulting key decisions; test activities completed; variances; and the verified enabling products and proofing of associated processes in the established enterprise data repository.
	Notes and guidance: N/A

Measures:

- Percent of enabling products in place at time of need
- Enabling product readiness determination execution time and cost

Process Task Outcomes

The following types of enabling products will be provided:

1. Fleet assets—fleet-owned assets being modified (e.g., mission computer, radar system, flight control system), operational assets (support aircraft, ship assets, drones, weapon targets, satellites), and so on.

2. Development—Computer-Aided Engineering (CAE) tools, prototypes, life cycle analysis, laboratories/facilities, requirements management and system architecture database, software development facility, and so on.

3. Production—tooling and facilities, manpower, and so on.

4. Test—test equipment and software, verification plans and procedures, test ranges, government-furnished equipment, and so on.

5. Deployment—staging facilities, warehouses, shipping containers, and so on.

6. Training—class rooms, flight simulator, instructors, and so on.

7. Support—repair facilities, diagnostic equipment, shipping services, staffing, and so on.

8. Disposal—disposal site, refurbishment facilities, removal tools, safety bulletins, and so on.

Table 5.33   Product Deployment and Transition (SEP-33)

Process purpose: The supplier is to transition verified products to the customer or user of the products in accordance with the agreement.

Roles and agents:

- Project leader
- Sustainability engineer
- Integrated product team

Inputs:	Outputs:
• Verified and validated integrated end product or enabling product report • Manufacturing process and personnel system • Integrated master schedule • Specified requirements (for packaging and handling) • Enabling products readiness determination • Integrated logistics support certification	• Operational system products

Process diagram:

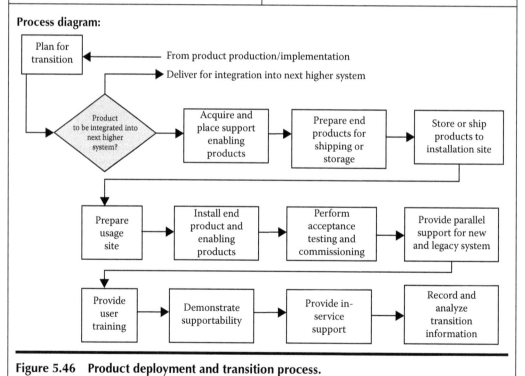

Figure 5.46   Product deployment and transition process.

Table 5.33 Product Deployment and Transition (SEP-33) (*Continued*)

Step	Action
	Plan for Deployment and Transition
1	**Action statement:** Determine if product is to be integrated into a next higher system of systems. **Notes and guidance:** The transition strategy includes installation and commissioning of the system in accordance with agreements. Wherever possible, this includes human operators.
	Perform the Deployment and Transition
2	**Action statement:** Acquire and put in place appropriate enabling products to carry out relevant transition to use requirements. **Notes and guidance:** Enabling products specifically looked for are as follows: 1. Delivery addresses 2. Fleet release message 3. Installation procedures 4. Training 5. Operation and maintenance manuals 6. Spares and repair parts, along with package handling, storage and transportation 7. In-service support equipment
3	**Action statement:** Prepare end products for shipping and storage, as required by the agreement. **Notes and guidance:** N/A
4	**Action statement:** Store end products awaiting shipping and ship or transport to the customer at the intended usage sites, in accordance with the agreement. **Notes and guidance:** N/A
5	**Action statement:** Prepare sites where end products will be stored, installed, used or maintained, or serviced, as required by the agreement. **Notes and guidance:** Site preparation is conducted in accordance with applicable health, safety, security, and environmental regulations.
6	**Action statement:** Install end products at the appropriate sites, as required by the agreement. **Notes and guidance:** The product is configured with required operational data.
7	**Action statement:** Perform commissioning, as required by the agreement, to bring delivered or installed end products to operational readiness with appropriate acceptance and certification tests completed in accordance with the product validation process. **Notes and guidance:** N/A

(Continued)

Table 5.33   Product Deployment and Transition (SEP-33) (*Continued*)

Step	Action
8	**Action statement:** Provide, if required by the agreement, a parallel operation (ghosting) of the new and the legacy end products so that service is continuous during the transition period.
	Notes and guidance: Acceptance tests, as specified in agreements, can define the criteria that demonstrate that the product entity possesses the capability to deliver the required services when installed in its operational location and staffed by operators.
9	**Action statement:** Provide training for users, maintenance, and other personnel, in accordance with the agreement.
	Notes and guidance: N/A
10	**Action statement:** Provide in-service support in accordance with the agreement.
	Notes and guidance: N/A
11	**Action statement:** Deliver all planned support elements.
	Notes and guidance: N/A
12	**Action statement:** Analyze, record, and report the transition information, including results of transition actions, nonconformances, and corrective actions taken.
	Notes and guidance: Postimplementation reporting includes flaws in the product requirements as well as technical features. When inconsistencies exist at the interface between the product, its specified operational environment, and any products that enable the utilization stage, the deviations lead to corrective actions and/or requirement changes. Lessons learned should also be recorded.

Measures:

- Percentage of on-time delivery

Process Task Outcomes	
Representative Tasks	*Process Task Outcomes*
Plan effectiveness analyses	A plan is prepared to include the purpose, objectives, execution and data collection requirements, schedule of tasks, availability of required resources, expected outcomes, and the general approach for required effectiveness analyses.
Analyze product cost effectiveness	For each alternative physical solution representation, as well as for the design solution, the product cost effectiveness is determined with respect to the following attributes, as applicable: accuracy, availability, capacity, maintainability, reliability, responsiveness, operability, safety, security, survivability, spare requirements, transportability, vulnerability, and so on.

Table 5.33   Product Deployment and Transition (SEP-33) (*Continued*)

Process Task Outcomes	
Representative Tasks	*Process Task Outcomes*
Analyze total ownership cost	Costs to the enterprise and to the customer for alternative physical solution representations, for analysis of alternatives options, or for proposed changes and the known uncertainties (risks) in these costs are determined. NOTE: The following costs are typically included in a total ownership cost analysis: development, production, test, deployment/installation, training, operations, support/maintenance, and retirement/disposal.
Analyze environmental impacts	Applicable federal, state, municipal, and international environmental statutes and applicable hazardous material lists affecting the project and endurance of compliance by each physical solution are determined; the effect on and by each end product and enabling product on the infrastructure, land and ocean, atmosphere, water sources, and animal, plant and human life, as applicable, has been determined, from an enterprise-based life cycle perspective.
Analyze product effectiveness	For each operational profile, each alternative physical solution representation and the design solution are assessed by analytic confirmation to satisfy appropriate requirements.
Record outcomes of effectiveness analyses	Effectiveness analysis outcomes, as well as the details of the analyses performed, including rationale, assumptions, and lessons learned, are captured and recorded in the established information database.

5.6 Product Utilization Processes

The product utilization processes are used to (1) operate the product and monitor its service and performance; (2) prepare a maintenance strategy based on the stakeholder's preferred way of maintaining the product; and (3) at the end of the product's life cycle, perform a disposal or closeout process to deactivate, disassemble, and remove the product or service and its waste products, consigning them to a final condition and returning the environment to its original or an acceptable condition.

There are three processes associated with product utilization, as shown in Figure 5.47.

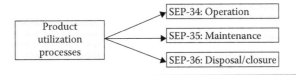

Figure 5.47   Product utilization processes.

Table 5.34 Operation (SEP-34)

Process purpose: The provider shall perform the operation process to use the product in order to deliver its services and to assess needs for product improvement. This process assigns personnel to operate the product, monitor its service, and monitor operator-product performance. It identifies and analyzes operational problems in relation to stakeholder requirements and organizational constraints.

Scope: The sustainability engineering process uses a requirements loop and a design loop in a progressive, iterative analytical approach to make operational requirements and design decisions at successively lower levels. As this process iterates, requirements are planned, documented, developed, identified, controlled, tracked, and verified within the configuration management (CM) process. CM provides the common approach necessary to minimize variation and improve information integrity. Many projects will need to perform contract closure at the end of several life cycle phases.

Roles and agents:

- Project leader
- Product support
- Integrated product team

Inputs:	**Outputs:**
• Operational product	• Product operations reports
• Trained operators	• Engineering change proposals (ECPs)
• Consumable materials	

Process diagram:

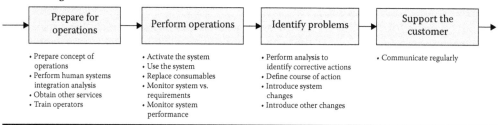

Figure 5.48 Operation process.

Step	Action
1	**Action statement:** Prepare for operations.
1.1	**Action statement:** Prepare a concept of operations.
	Notes and guidance:
	1. The availability of services as they are introduced, routinely operated, and withdrawn from service. Where appropriate, it includes coordination with preexisting, concurrent, or continuing services delivered by other products that provide identical or similar services.
	2. The staffing strategy and schedules for operators.

Table 5.34   Operation (SEP-34) (*Continued*)

Step	Action
	3. Where appropriate, the release and reacceptance criteria and schedules of the product to permit modifications that sustain existing or enhanced services.
1.2	**Action statement:** Perform human systems integration (HSI) analysis. Perform HSI analysis to identify personnel skills needs, numbers of skilled personnel needed, and which functions are best performed by the operator of the product. **Notes and guidance:** The HSI analysis is performed during the early product development life cycle.
1.3	**Action statement:** Obtain other services. Obtain other services related to operation of the product. **Notes and guidance:** N/A
1.4	**Action statement:** Train operators. Assign trained, qualified personnel to be operators. **Notes and guidance:** This may include awareness of the product in its operational environment and a defined program of familiarization, with appropriate failure detection and isolation instruction. Operator knowledge, skill, and experience requirements guide the personnel selection criteria, and where relevant, their authorization to operate is confirmed. Selection and training of instructors to perform training that employs the operational product may be an aspect of staffing. A training mode of the operational product may impact service availability.
2	**Action statement:** Perform operational activation and checkout.
2.1	**Action statement:** Activate the product. Activate the product in its intended operational situation to deliver instances of service or continuous service according to its intended purpose. **Notes and guidance:** Where agreed, maintain continuous service capacity and quality when the product replaces an existing product that is being retired. During a specified period of changeover or concurrent operation, manage the transfer of services so that continuing conformance to persistent stakeholder needs is achieved.
2.2	**Action statement:** Use the product. Use the product to perform the operational mission. **Notes and guidance:** N/A
2.3	**Action statement:** Use consumables. Consume materials, as required, to sustain the services. **Notes and guidance:** This includes energy sources for hardware and provisions for operators.
2.4	**Action statement:** Monitor operations versus evolving requirements. Monitor the product to ensure that it continues to satisfy the users original and evolving requirements. **Notes and guidance:** N/A
2.5	**Action statement:** Monitor product performance. Monitor the product operation to confirm that service performance is within acceptable parameters. **Notes and guidance:** The product may exhibit unacceptable performance when elements implemented in the hardware have exceeded their useful life or the product's operational environment affects the operating and maintenance personnel (including staff turnover, operator stress, and fatigue).

(Continued)

Table 5.34   Operation (SEP-34) (*Continued*)

Step	Action
3	**Action statement:** Perform problem identification.
3.1	**Action statement:** Perform analysis. Perform failure identification actions when noncompliance has occurred in the delivered services. **Notes and guidance:** N/A
3.2	**Action statement:** Identify corrective actions. Determine the appropriate course of action when corrective action is required to remedy failings due to changed need. **Notes and guidance:** The appropriate course of action may include, but not be limited to, introducing minor hardware or software adaptations or modified operator action, changes to the stakeholder requirements, changes to the design and/or implementation of the product, or tolerating diminished services.
3.3	**Action statement:** Introduce nonproduct changes. Introduce remedial changes to operating procedures, the operational environment, human–machine interfaces, and operator training as appropriate when human error contributed to failure. **Notes and guidance:** N/A
3.4	**Action statement:** Introduce product changes. **Notes and guidance:** This includes processing ECPs. The design and implementation of hardware or software changes requires reentering the engineering life cycle and using appropriate processes of this guide.
4	**Action statement:** Support the customer.
4.1	**Action statement:** Communicate with users. Continuously or routinely communicate with users to determine the degree to which delivered services satisfy their needs. **Notes and guidance:** The results are analyzed, and required action to restore or amend services in order to provide continued stakeholder satisfaction is identified. Wherever possible, the benefit of such action is agreed with stakeholders or their representatives.

Process institutionalization: The project management plan(s) will describe how policy, planning, resources, responsibility assignment, stakeholder involvement, monitoring, control, status reviews, feedback, quality assurance, evaluations, and audits are to be used to ensure institutionalization of this process.

Process tailoring:

- The following process elements may *not* be tailored: process owner and process objective.
- The following process elements may only be tailored by obtaining a process waiver: process activities.
- The following process elements may be tailored: guidance and notes, process roles, process implementation assets, and process-related measures.

Table 5.34 Operation (SEP-34) (*Continued*)

Measures:
• Effort and funds expended for process tasks (planned vs. actual)
• Timeliness:
• Process cycle time: The total elapsed time to move a proposed change from the beginning to the end of any configuration control process that either establishes or changes a baseline.
• Touch time: Time spent actually processing the proposed change.
• Wait time: Time spent by the proposed change in a queue.
Process Task Outcomes
As a result of the successful implementation of the operation process:
a. An operation strategy is defined.
b. Services that meet stakeholder requirements are delivered.
c. Approved corrective action requests are satisfactorily completed.
d. Stakeholder satisfaction is maintained.

Table 5.35 Maintenance Support (SEP-35)

Process purpose: The provider shall prepare a product-level maintenance strategy based on the user community preferred way of maintaining the product and to prepare a maintenance support plan based on that strategy.	
Roles and agents: • Project leader • Product support • Integrated product team	
Inputs: • Maintenance enabling resources • Maintenance concepts/plans • Technical manuals • Spares and repair parts • Testing equipment • Trained maintainers	**Outputs:** • Product maintenance data collection (MDC) reports • Engineering change proposals (ECPs)

(*Continued*)

Table 5.35   Maintenance Support (SEP-35) (*Continued*)

Process diagram:

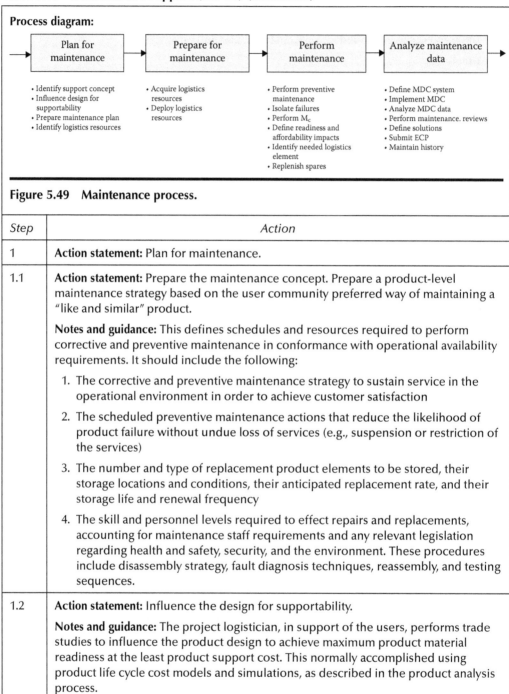

Figure 5.49   Maintenance process.

Step	Action
1	**Action statement:** Plan for maintenance.
1.1	**Action statement:** Prepare the maintenance concept. Prepare a product-level maintenance strategy based on the user community preferred way of maintaining a "like and similar" product.
	Notes and guidance: This defines schedules and resources required to perform corrective and preventive maintenance in conformance with operational availability requirements. It should include the following:
	1. The corrective and preventive maintenance strategy to sustain service in the operational environment in order to achieve customer satisfaction
	2. The scheduled preventive maintenance actions that reduce the likelihood of product failure without undue loss of services (e.g., suspension or restriction of the services)
	3. The number and type of replacement product elements to be stored, their storage locations and conditions, their anticipated replacement rate, and their storage life and renewal frequency
	4. The skill and personnel levels required to effect repairs and replacements, accounting for maintenance staff requirements and any relevant legislation regarding health and safety, security, and the environment. These procedures include disassembly strategy, fault diagnosis techniques, reassembly, and testing sequences.
1.2	**Action statement:** Influence the design for supportability.
	Notes and guidance: The project logistician, in support of the users, performs trade studies to influence the product design to achieve maximum product material readiness at the least product support cost. This normally accomplished using product life cycle cost models and simulations, as described in the product analysis process.

Table 5.35 Maintenance Support (SEP-35) (*Continued*)

Step	Action
1.3	**Action statement:** Prepare the maintenance plan. **Notes and guidance:** The maintenance plan defines requirements for product support based on known product configuration and on results of logistics support analysis (LSA) or equivalent analysis. The maintenance plan includes spares planning factors for each product replacement item and repairable candidate.
1.4	**Action statement:** Identify enabling products for maintenance. **Notes and guidance:** Based on the concept of operations and the product deployment locations, the LSA identifies the needed product support elements (spares, support equipment, technical manuals, and trained personnel) for each operating and maintenance site.
2	**Action statement:** Prepare for maintenance.
2.1	**Action statement:** Acquire logistics resources. Obtain the enabling systems, product elements, and services to be used during maintenance of the product. **Notes and guidance:** N/A
2.2	**Action statement:** Deploy logistics resources. This involves deploying of support equipment, spare parts, technical manuals, and trained maintenance personnel to the operational sites. **Notes and guidance:** N/A
3	**Action statement:** Perform maintenance.
3.1	**Action statement:** Perform preventive maintenance. Perform preventive maintenance by replacing or servicing product elements prior to failure, according to planned schedules and maintenance procedures. **Notes and guidance:** Perform preventive maintenance by inspecting, servicing, or replacing product elements, prior to failure, according the planned maintenance schedules and maintenance procedures, as identified in reliability-centered maintenance analysis.
3.2	**Action statement:** Isolate product failures. Perform failure identification actions when a noncompliance has occurred in the product. **Notes and guidance:** Use built-in-test and manual fault isolation techniques to isolate the failure to the single failed product replaceable unit.
3.4	**Action statement:** Identify readiness and/or affordability impacts. Identify corrective actions to correct system problems associated with readiness or affordability. **Notes and guidance:** Refer to the note in step 3.5 below. This specific action is to identify which specific product components are driving readiness or affordability problems. These components may be candidates for a product improvement ECP.

(*Continued*)

Table 5.35 Maintenance Support (SEP-35) (*Continued*)

Step	Action
3.5	**Action statement:** Identify needed logistics element. Identify corrective actions associated with elements of logistics. **Notes and guidance:** In most all cases, having enough of the right spare parts available at the operational site can support readiness requirements. However, to provide the most cost-effective levels of support, improvements in fault isolation procedures, improved support equipment, and support of additional training of operators and maintainers is required. The action is to identify which element of logistics can provide the needed "affordable-readiness" solution.
3.6	**Action statement:** Replenish spares. Confirm that logistics actions satisfy the required replenishment levels so that stored product elements meet repair rates and planned schedules. **Notes and guidance:** Monitor the quality and availability of spares, their transportation, and their continued integrity during storage. Acquire, train, and accredit, as necessary, personnel to maintain operator numbers and skills.
4	**Action statement:** Perform MDC, analysis, and reporting.
4.1	**Action statement:** Implement MDC. Implement maintenance data reporting and other problem reporting information systems. **Notes and guidance:** Maintenance and material management (3M) is a standard MDC system for the military. 3M data is routinely collected on a wide variety, but not every type, of system. It may be necessary to make a special request to collect needed MDC data or to make separate arrangements with the user community for sample data collection for specific analysis purposes.
4.2	**Action statement:** Analyze MDC data. Analyze maintenance data to identify product problems associated with maintaining product readiness and affordability requirements. **Notes and guidance:** The assessment of the performance of a product requires the availability of operational and maintenance histories of the various product elements. Performance and effectiveness parameters are established early in the product life cycle with the development of operational requirements and the maintenance concept, as described in step 1.1 above. These parameters describe the characteristics of the product that are considered paramount in fulfilling the need objectives. Now with the product deployed and in full operational status, the following questions arise: 1. What is the true performance and effectiveness of the product? 2. What is the true effectiveness of the logistics support product? 3. Are the initially specified requirements being met? 4. Is the product "cost effective"? 5. Are all customer expectations being fulfilled?

Table 5.35 Maintenance Support (SEP-35) (*Continued*)

Step	Action
4.3	**Action statement:** Perform maintenance reviews. The user preferred maintenance support concept is to be part of the product technical reviews. **Notes and guidance:** During the design phases of the life cycle, supportability and maintainability are reviewed as part of the following reviews: a. Systems requirements review b. Systems functional review c. Preliminary design review d. Technology readiness assessment e. Integrated baseline review After product deployment, two important reviews are conducted to assess supportability: a. In-service review b. Postdeployment review These reviews for supportability are to a. Assess product operations in field conditions b. Assess support of fielded product
4.4	**Action statement:** Define solutions. Corrective actions may be needed in response to a product/equipment deficiency (i.e., the product fails to meet the specified requirements) or may be accomplished to improve product performance, effectiveness, or logistics supportability. If corrective action is to be accomplished, the necessary planning and implementation steps are a prerequisite to ensure complete compatibility of all elements of the product throughout the change process. **Notes and guidance:** Changes are documented in ECPs, and the ECP is analyzed to evaluate the logistics support impact.
4.5	**Action statement:** Maintain history. Maintain a history of problem reports, corrective actions, and trends to inform operations and maintenance personnel and other projects that are creating or utilizing similar product entities. **Notes and guidance:** N/A

Process institutionalization: The project management plan(s) will describe how policy, planning, resources, responsibility assignment, stakeholder involvement, monitoring, control, status reviews, feedback, quality assurance, evaluations, and audits are to be used to ensure institutionalization of this process.

Process tailoring:
- The following process elements may *not* be tailored: process owner and process objective.

(*Continued*)

Table 5.35   Maintenance Support (SEP-35) (*Continued*)

• The following process elements may only be tailored by obtaining a process waiver: process activities.   • The following process elements may be tailored: guidance and notes, process roles, process implementation assets, and process-related measures.

Measures:
 • Effort and funds expended for process tasks (planned vs. actual)
 • Timeliness:
   • Process cycle time: The total elapsed time to move a proposed change from the beginning to the end of any configuration control process that either establishes or changes a baseline.
   • Touch time: Time spent actually processing the proposed change.
 • Wait time: Time spent by the proposed change in a queue.

Process Task Outcomes
As a result of the successful implementation of the maintenance support process: a. A maintenance strategy is defined.   b. Maintenance constraints are provided as inputs to requirements.   c. Services that meet stakeholder requirements are sustained.   d. Replacement product elements are made available.   e. Approved corrective action requests are satisfactorily completed.   f. Product life cycle and failure data are recorded and maintained.

Table 5.36   Disposal/Closure (SEP-36)

Process purpose: The provider shall perform the disposal/closure process to end the existence of a product. This process deactivates, disassembles, and removes the product, as well as any waste products, consigning them to a final condition and returning the environment to its original or an acceptable condition. The process destroys, stores, or reclaims products and waste products in an environmentally sound manner, in accordance with legislation, agreements, organizational constraints, and stakeholder requirements. Where required, it maintains records in order that the health of operators and users, and the safety of the environment, can be monitored.
Roles and agents: • Project leader   • Product support   • Integrated product team

Table 5.36   Disposal/Closure (SEP-36) (*Continued*)

Inputs:	Outputs:
• Project artifact files • Project life cycle documentation including disposal plan • Project sponsor direction	• Practical constraints of disposal that influence design are defined. • The product elements are stored, reclaimed, recycled, or destroyed. • The environment is returned to its original or an agreed-to state. • Knowledge gained from product creation and operation is retained. • Records allowing analysis of lessons learned are available.

Process diagram:

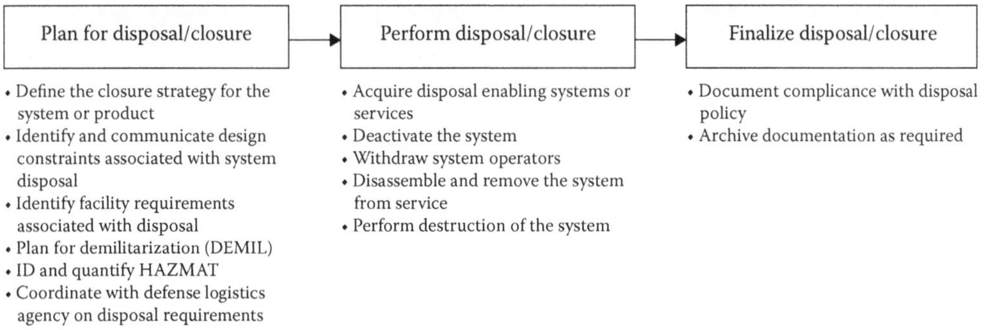

Figure 5.50   Disposal process.

Step	Action
	Plan for Disposal/Closure
1	**Action statement:** Define a closure strategy for the product to include each product element and any resulting waste products. **Notes and guidance:** This defines schedules, actions, and resources that 1. Permanently terminate the product's delivery of services 2. Transform the product into, or retain it in, a socially and physically acceptable state, thereby avoiding subsequent adverse effects on stakeholders, society, and the environment 3. Take account of the health, safety, security, and privacy applicable to disposal actions and to the long-term condition of resulting physical material and information
2	**Action statement:** Unavoidable constraints on the product design arising from the disposal strategy are communicated. **Notes and guidance:** This includes issues of disassembly, including their associated enabling products, access to and availability of storage locations, and available skill levels.

(*Continued*)

Table 5.36 Disposal/Closure (SEP-36) (*Continued*)

Step	Action
3	**Action statement:** Specify containment facilities, storage locations, inspection criteria, and storage periods if the product is to be stored. **Notes and guidance:** N/A
4	Plan for Material "Demilitarization" (DEMIL)
4.1	**Action statement:** Identify and quantify all hazardous material used on the product, and map these to the location on the product. **Notes and guidance:** N/A
4.2	**Action statement:** Coordinate with the defense logistics agency and component logistics and explosive safety agencies to identify applicable DEMIL, reuse, and hazardous material disposal requirements. **Notes and guidance:** N/A
5	Perform Disposal/Closure
5.1	**Action statement:** Acquire the enabling products or services to be used during disposal/closure of a product. **Notes and guidance:** N/A
5.2	**Action statement:** Deactivate the product to prepare it for removal from operation. **Notes and guidance:** Interfaces to other products are considered; for example, power, fuel, and are disconnected in accordance with disassembly instructions and relevant health, safety, security, and privacy legislation.
5.3	**Action statement:** Withdraw operating staff from the product, and record relevant operating knowledge. **Notes and guidance:** This is conducted in accordance with relevant safety, security, privacy, and environmental standards, directives, and laws.
5.4	**Action statement:** Disassemble the product into manageable elements to facilitate its removal for reuse, recycling, reconditioning, overhaul, archiving, or destruction. **Notes and guidance:** N/A
5.5	**Action statement:** Remove the product from the operational environment for reuse, recycling, reconditioning, overhaul, or destruction. **Notes and guidance:** This is conducted in accordance with relevant safety, security, privacy, and environmental standards, directives, and laws. Elements of the product that have useful life remaining, either in their current condition or following overhaul, are transferred to other products of interest or organizations. Where appropriate, recondition product elements to extend their useful life. Reallocate, redeploy, or retire operators.

Table 5.36   Disposal/Closure (SEP-36) (*Continued*)

Step	Action
5.6	**Action statement:** Conduct destruction of the product, as necessary, to reduce the amount of waste treatment or to make the waste easier to handle.
	Notes and guidance: This activity includes obtaining the destruction services required in order to melt, crush, incinerate, or demolish the product or its elements as necessary. For military items such as guns, rifles, and cannons, this includes "demilitarization" by cutting the barrels off by use of a cutting torch. Act to safeguard and secure knowledge and skills possessed by operators.
6	Finalize the Disposal/Closure
6.1	**Action statement:** Confirm that no detrimental health, safety, security, and environmental factors exist following disposal.
	Notes and guidance: N/A
6.2	**Action statement:** Archive information gathered through the lifetime of the product to permit audits and reviews in the event of long-term hazards to health, safety, security, and the environment and to permit future product creators and users to build a knowledge base from past experiences.
	Notes and guidance: N/A

Measures:

- Percent completion of all disposal activities, as assessed by an independent agency
- Amount of resources utilized to perform the disposal process

Process Task Outcomes

As a result of the successful implementation of the disposal process:

a. A product disposal strategy is defined.

b. Disposal constraints are provided as inputs to requirements.

c. The product elements or waste products are destroyed, stored, reclaimed, or recycled.

d. The environment is returned to its original or an agreed state.

e. Records allowing knowledge retention of disposal actions and the analysis of long-term hazards are available.

Chapter 6

Tailoring Sustainability Engineering Processes

6.1 Sustainability Engineering Tasking Document

Prudent planning suggests the establishment of a planning and control system for the sustainability effort. The sustainability engineering tasking (SET) document describes how a project will be implemented and controlled. The capability to define and execute sustainability engineering responsibilities can impact the level of commitment necessary to achieve acceptable risk in project execution.

Commitments to the effort are in two basic forms: The first is through planned outcomes of sustainability engineering processes (i.e., process products) from its program-specific implementation. Some of these outcomes will go to contract. These include sustainability engineering process products essential to the program's success (e.g., specifications) as well as process products confirming that key characteristics of the product have been achieved (e.g., that requirements are traceable and verifications confirm achievement of requirements). Other outcomes will be used internal to the project with varying degrees of tasking activity. Examples of such outcomes are as follows:

1. Development and use of lower-level specifications of the product, such as the following:
   a. Development and product design/fabrication requirements
   b. Verification criteria for the requirements, including incremental demonstrations, confirmations, and acceptance criteria necessary to achieve product success
2. Risk avoidance events to achieve success, and the criteria used to adjudicate successful achievement
3. Process metrics (cost, schedule, performance, quality, etc.)
4. Definition of the numbers and qualifications of people needed to execute responsibilities
5. Budgets and schedules

The second form of commitment deals with SEM implementation. Three options are considered in obtaining this commitment: The first is to allow the project leadership to have complete control over the SET document. The project leadership retains SET ownership but is required to inform each tasking activity of changes to the SET. The second option is to not cite the SET in the contract. In this case, the project leadership retains ownership and is not obliged to submit revisions to the tasking activity even for the purpose of information. The third option is to cite the SET in the contract and the tasking activity would have control and not the project leadership. The type of commitment expected depends on the criticality of the process to the program and the project leadership's capability to execute the process. Although the third option is not desirable from a performance-based perspective, it may be needed to achieve a minimally acceptable level of process excellence. Performing activities that have not earned/demonstrated a sufficiently capable level of self-governance may need additional oversight to ensure that the program succeeds.

The bottom line is thus: Regardless of the specifics of the implementation, four basic factors are balanced to arrive at the level of commitment to a process expected from the project leadership. These four factors are as follows: (1) The insight the tasking activity needs to execute risk management responsibilities, (2) the risk of the program, (3) the criticality of the process to the program, and (4) the demonstrated process excellence of project leadership.

A consistent implementation of this planning and control system will have the project leadership and engineer develop a SET to support planning requirements. The engineer may elect to provide the SET, in part or total, to the project leadership for use in proposal preparation. The project leadership's SET would provide detailed information for the next acquisition phase or engineering effort or both to identify specific events, accomplishments, and criteria necessary to satisfy planned and required technical exit criteria.

This chapter outlines the components of the SET. The project leadership is responsible for defining them. The engineer determines how the SET will be conducted to satisfy the program's objectives. Each task needs to be traceable to the product's definition and requirements. The SET should provide a summary, with reference to the detailed plans, for all the required technical activities.

6.1.1 IMP

The first important component of the SET is an IMP, which is developed by the engineer. The IMP provides top-level events, proposed accomplishments, and accomplishment criteria throughout the entire program. The IMP should contain the following:

1. A general life cycle road map of the key engineering activities to be accomplished and specification of who will be responsible for the accomplishment
2. The makeup of multidisciplinary IPT for tasking activities and their specific responsibilities
3. Plans and criteria for transitioning critical product and process technologies
4. Identification of trade-off studies, the scope and depth of product effectiveness assessments, current measures of effectiveness hierarchy, technical risk management plans, critical technical parameters, and tracking requirements for those parameters
5. Identification of existing simulations

The IMP is a programmatic technical events plan. It is not calendar driven. The events in an IMP are key demonstration, progress assessment, and decision points. These events should include reviews, for example, the PDR; major verification efforts; or other technical events where necessary to measure and demonstrate progress before proceeding with follow-on technical efforts. For each event in the IMP, one or more accomplishments are identified. The IMP reflects both entry accomplishments (what must be done to initiate a review or demonstration milestone) and exit accomplishments (what must be done to know that the event has been successfully completed) for an event. Accomplishments are critical tasks/activities that must be done prior to entering (or exiting) an event. The criteria used to measure the successful completion of each accomplishment must be defined. An accomplishment is complete when all the accomplishment criteria are satisfied and can be demonstrated. The event can be initiated (or completed) when all identified accomplishments are done.

The IMP events should be identified in the format of entry events (initiating the PDR) and exit events (completing the PDR). Entry and exit accomplishments should also be defined for each event. The IMP should reflect integration of the efforts necessary to achieve the required accomplishments along with the success criteria. All IMP accomplishments should be event related and not time driven. The IMP accomplishments should have one or more of the following characteristics:

1. They define a desired result at a specified event that indicates design maturity or progress directly related to each product and process.
2. They define the completion of a discrete step in the progress of the planned development.
3. They define activities that provide product and process functionality.

IMP accomplishment criteria should be measurable. The criteria need to provide a definitive measure or indicator that the required level of maturity or progress has been achieved.

As the key control element, IMP should be implemented in a manner that requires a change proposal to modify it. In cases where demonstration milestones are used to determine whether or not to exercise an option or move to the next phase of a program, a demonstration milestone is normally established. Demonstration milestones should be incorporated into the program.

6.1.1.1 Use of IMP

The IMP is the top-level process control and progress measurement tool for an effort. This plan provides a concise mechanism for in-process verification. The IMP accomplishments need to reflect an integration of all technical factors. The IMP accomplishments with associated criteria establish confirmation requirements for completion of critical work tasks and for product and process maturity demonstrations. In this manner, the IMP provides the link between acquisition strategy and phase exit criteria by demonstrating accomplishments at events. Phase exit criteria are included in the IMP as required accomplishments for one or more events. Program baseline parameters are incorporated into the IMP through the inclusion of accomplishment criteria related to progress in achieving requirements (i.e., TPMs, discussed in Section 6.1.3). Thus, the events provide a mechanism to demonstrate and confirm progress and to assess risk.

The IMP is developed to address specific program needs and the technical effort to be accomplished. In addressing the features of the technical effort, the IMP identifies the accomplishments needed to ensure that required technical progress is being made, risks are being controlled, and the system is maturing as needed to satisfy requirements. Thus, every IMP reflects the unique aspects of a particular program by documenting the demonstrations and risk mitigations necessary for overall program success.

6.1.1.2 IMP Format

The IMP has four critical elements: (1) Events, (2) required accomplishments for each event, (3) criteria for each accomplishment, and (4) event sequencing (or timeline). A numbering scheme is useful to simplify tracking of events, accomplishments, and accomplishment criteria to the statement of work (SOW), WBS, and the IMS (discussed in Section 6.1.2). Additionally, an IMP typically includes definitions for key terms used throughout the document.

6.1.1.2.1 IMP Events

Events in an IMP can include demonstration milestones, reviews, and other key progress demonstration points. Normally, the project leadership selects the dates for most milestones. The IMP is not date driven; however, a limited number of events may be date-related demonstration milestones. Technical reviews (discussed in Section 6.2) are commonly used IMP events. The performing activity is normally allowed to select milestones pertinent to the proposed approach. The IMP events need to be organized to reflect the necessary sequence of milestone events. Other types of events are possible, such as initiation of significant demonstrations. These types of events may be identified by either the tasking activity or the project leadership. When building the proposed IMP, the project leadership is generally responsible for proper event sequencing.

6.1.1.2.2 IMP Accomplishments

The IMP accomplishments delineate interim and final steps to defining, developing, and producing a system. They need to be properly sequenced and organized around the product and event they support. Examples are "structural definition complete" and "software documented." Additionally, product and process maturation accomplishments are included to reflect progress toward achieving required performance and required verifications. Defining the scope of accomplishments is a factor critical to the success of the IMP. If the scope of an accomplishment is too narrow, the flexibility in executing the technical effort can be curtailed. For example, defining accomplishments that represent only a couple of months of effort is probably too detailed an approach for most program applications. Top-level accomplishments go in the IMP. The supporting detail, representing smaller pieces of an effort, is incorporated into the IMS. Accessibility to the IMS gives the tasking activity the capability of seeing the performing activity's technical progress. Alternatively, accomplishments that are too broad (e.g., if they span multiple events) need to be broken up into manageable parts.

6.1.1.2.3 IMP Accomplishment Criteria

An accomplishment is not complete unless all its associated criteria have been satisfied. It is essential that all accomplishment criteria are measurable and useful. For example, noting "test plan complete" is measurable and useful, whereas "test plan 85% complete" is neither measurable nor useful because the last 15% may include the hard-to-do elements that could require more effort than the first 85%. Likewise, the annotation "100 pages of test plan complete" is measurable but it is not useful because document length provides no meaningful measure on whether the planning is near or far from completion. Accomplishment criteria can include completed work efforts such as "product analysis complete" and "programmer's manual done."

Progress in achieving TPMs (discussed in Section 6.1.3) should also be included. At successive events, technical progress can be checked by using accomplishment criteria that confirm that the value of a technical parameter is within an allowed tolerance band and that the current estimate of the technical parameter satisfies the threshold value. Thus, the data associated with the accomplishment criteria provide quantitative inputs to program decision points.

6.1.1.2.4 IMP Timeline

The IMP is an event-based schedule; it is not time based. The logical sequence of events, however, does need to be evident. Many activities can occur in parallel, but when one effort logically depends on the completion of another this relationship must be evident in the IMP. A simple example is that an entry accomplishment for product design review is the completion of system PDR. The logical sequencing, or critical accomplishment path, is critical to overall process control. This sequencing may or may not be evident in the event-numbering scheme used. The natural tendency is to require that event number three always follows event number two. This order is not obvious because the initial time-based scheduling provides the planned time sequence of events if all goes well. However, significant problems have been experienced with time (calendar) sequences used to structure successive event dependencies. For example, perhaps event number three does not rely on the completion of event number two. For this reason, event dependencies in the IMP need to be based on a critical accomplishment path. This structure enables truly independent events and accomplishments to proceed if difficulties arise in successfully reaching closure on a separate event.

6.1.1.3 IMP Structure

The IMP needs to be structured in a manner that effectively controls progressive development to ensure that all the tasks that need to be accomplished are completed prior to starting or continuing technical efforts that depend on those accomplishments. For example, one needs to consider the progression between CDR and SVR. The following suggested outline illustrates how an IMP can be used to control the effort to progressively verify a particular CI in the product:

1. Event: Accomplishment (entry or exit) criteria
2. The CDR: Product design released for building qualification unit (exit)
   2.1. Product design complete
   2.2. Product design confirmed by analysis to satisfy performance and functional requirements in the CI development specification
   2.3. Verification plan complete
3. Product test readiness review (TRR)
   3.1. Product CDR complete (entry)
   3.2. Product CDR exit accomplishments met
   3.3. Product verification procedures complete (entry)
   3.4. Procedures defined to verify all critical points in the operational envelope
   3.5. Product qualification unit ready for verification (entry)
   3.6. As-built unit conforms to design
   3.7. Verification resources defined and ready (entry)
   3.8. Trained personnel available
   3.9. Facilities equipped and available

3.10. Product qualification unit released for verification (exit)

3.11. All entry accomplishments demonstrated

4. Product FCA

4.1. Product TRR complete (entry)

4.2. Product TRR exit accomplishments met

4.3. Verification testing complete (entry)

4.4. Verifications conducted in accordance with established procedures

4.5. All verifications complete

4.6. Product satisfies CI development specification (entry)

4.7. All part 3 requirements verified in accordance with part 4 verification criteria

4.8. Product verified to meet established requirements (exit)

4.9. All entry accomplishments demonstrated

5. Product verification review

5.1. Product test readiness demonstrated (entry)

5.2. Product TRR completed

5.3. Product meets established requirements (entry)

5.4. Product FCA completed

6.1.1.4 Implementation of an IMP

The IMP should be realistic and illustrative of what the tasking activity requires. Experience indicates that engineers bias their development approach in an attempt to be responsive to a perceived requirement. The IMP needs to be compatible with the overall program schedule, and it needs to include tasking to require that progress be tracked and demonstrated or confirmed at specific events.

6.1.2 IMS

The IMS outlines the calendar dates of IMP events, showing when individual tasks will be done as well as how and when each of the accomplishment criteria will be met.

6.1.3 TPMs

In this component of the SET, the engineer identifies the parameters and metrics that track the performance of a product. Achievement assessments should be planned to support cost reporting, such as the cost performance report and cost and schedule status report. The technical parameters selected for tracking need to be critical indicators of technical progress and achievement, and they include either product parameters or confidence interval parameters, or both. Parameter descriptions should include identification of the related risk. The relationships between the selected parameters and any lower-level component parameters that must also be measured are determined.

The purpose of a TPM is to provide the project leadership with accurate data to monitor program execution. The TPM assesses technical characteristics of the product, and it identifies problems through engineering analyses or tests indicating that performance is being achieved in comparison with the values that were specified in contractual documents. Cost and schedule performance measures assess the effort from the point of view of the schedule for the phases of work and the cost of accomplishing those phases. By comparing the value of accomplished work with the planned values and the actual cost of the work performed, variances are identified that quantify the effects of problems. In addition to the problems resulting from unrealistic cost and schedule planning, cost and schedule performance measures may show up in technical inadequacies. The TPM and cost and schedule performance measures are complementary functions.

The TPM can also provide a basis to assess the adequacy of remaining cost performance budgets and to revise cost and schedule projections as necessary. The TPM assessment points should be planned to coincide with cost reporting as well as the planned completion of significant design and development tasks or an aggregation of tasks. Assessments facilitate the verification of achieved results in the completed task in terms of technical requirements and the verification that technical work still to be accomplished is within established budgets. Thus, TPM and cost and schedule performance measurement needs to be integrated in both task planning and contract reviews to achieve the ultimate goal of effective cost, schedule, and technical performance management.

6.1.4 Technical Integration Plans

Technical integration plans define and implement product functionality. How multidisciplinary teamwork must be implemented is defined in terms of how the organizational structure of an enterprise supports the time-phased needs of the technical effort. The project leadership describes the organizational responsibilities and authority for the effort, including the control of subcontracted technical efforts. The planning of technical tasks provides the foundation for cost and schedule planning. Technical tasks form the basis of allocating resources, scheduling task elements, assigning authority and responsibility, and integrating all aspects of the technical program. Technical planning is carried out to meet contractual requirements and is integrated with the cost and schedule control system at the appropriate level. The allocated resources form the performance measurement baseline for integrated cost, schedule, and technical management. This relationship pertains to both initial program definition and redefinitions occurring as part of the decision and control process.

6.1.5 Technical Transition Plans

The project leadership or engineer should establish, implement, and control a technology transition approach that identifies and applies relevant available and emerging technologies to program-specific efforts. The activities and the criteria for assessing, validating, and transitioning new technologies from development and demonstration programs, including commercially developed technologies, should be included in these plans. The plans should include the methods used to identify technology alternatives and the selection criteria used to determine when and which alternatives will be incorporated into product and process solutions. Performance requirements for technologies critical to system success are monitored using TPMs. When technologies cannot be effectively transitioned or when requirements can be defined only generally, opportunities for preplanned product improvements or evolutionary alternatives should be identified and documented in the transition plan. The criteria for validation include maturity in performance, affordability, and life-cycle processes.

6.1.6 System Configuration

The system configuration document in SET is an overview description of the integrated set of people, processes, and products that comprise the system effort. The products in this system comprise a group of items (parts or components and other subsystems) that satisfy a logical group of functions within the system. The idea of the system configuration document is to maintain

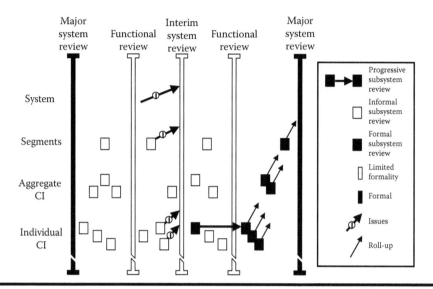

Figure 6.1 Example system configuration partitioning.

configuration control on the system and its subsystems. A subsystem identified and designated for configuration control is called a CI. A CI may include equipment, computers, material, software, and facilities. A CI that consists of software is a computer software CI (CSCI).

As CIs and their constituents are identified and developed, the interfaces between them are also identified and developed. Figure 6.1 illustrates, for example, how a system may be partitioned into CIs and components and how particular types of CIs would appear in this partitioning (a top-down process). Partitioning is organizing requirements (performance and function) into logical groupings. The characterization of the subsystem or CI in terms of a solution description is accomplished with a bottom-up approach to define the best combination of constituents that satisfies the partitioned requirements. At each node in the configuration, the definition of the subsystem/CI represents integrated performance and functionality of all lower-level items. Partitioning and developing solution descriptions are typically conducted iteratively. As cost, schedule, performance, and risk assessments of solutions are developed, alternative partitioning of requirements is accomplished to resolve problems (e.g., unacceptable risk) and to best provide overall integrated cost, schedule, and performance for the system.

6.2 Technical Reviews

Technical reviews are used to demonstrate progress, ensure that issues are resolved, verify the expected maturity of solutions, and confirm that risks are acceptable. Reviews require thorough up-front planning because they are event driven and are not held to a certain date. Therefore, scheduling must account for uncertainty in completion of all entry criteria as well as potential conflicts across events with common needs. Planned, formalized incremental reviews should be conducted to evaluate progress and resolve technical issues without reliance on technical interchange meetings, organizationally oriented reviews, working groups or committees, and program management reviews. Meetings of any kind should have well-defined objectives, a multidisciplinary approach, and a formalized format that documents proceedings (decisions, accomplishments, closures, etc.).

This format is necessary to ensure that any requirement/design change recommendations that result from these meetings are examined across the entire functionality of the people, products, and processes being impacted.

Technical reviews demonstrate and confirm the satisfactory completion of required accomplishments in accordance with the criteria stated in the IMP. Reviews that are required by the engineer, as well as all other reviews that are necessary to demonstrate, confirm, and coordinate progress, should be incorporated into an overall review plan. Reviews include structured meetings to assess progress toward satisfying IMP events. The project leadership defines the tasks associated with conducting each review, including the responsibilities of the personnel involved, review sites, necessary procedures (e.g., action item closeouts), entry criteria, and all required accomplishments. The engineer reviews the success criteria for the tasks and describes any additional success criteria necessary for review accomplishments, how compliance with requirements and criteria are determined, and how identified discrepancies are handled. Overall review planning, as well as the plan to conduct each review, is included in the SET document.

Major reviews should not be viewed as a single-point-in-time event. Major reviews culminate in a period of careful, disciplined technical effort, marking progress toward the end objective. Each major review has accomplishments that are integrated across the system (i.e., the integrated set of people, products, and processes), all technical areas, and the full spectrum of system functionality and that have measurable criteria to determine if each accomplishment is complete. The accomplishments and criteria are contained in the IMP. The event-based IMP provides the focus and objectives for the in-between smaller reviews and meetings. Formal reviews focus on demonstrating and confirming progress. A review is not complete until all IMP accomplishments are complete.

6.2.1 Structuring the Reviews

Technical reviews are structured within the total system context to confirm that required progress in satisfying requirements has been attained, functional and physical integration pertinent to the stage of product maturity has been achieved, risks are managed and in control, and continued development is warranted and executable or alternatives have been defined. Technical reviews are an integral control mechanism of the sustainability engineering process. Technical reviews provide opportunities to identify and resolve issues at the earliest time and lowest level; ensure that requirements are integrated and reflected in designs; provide progressive and in-process confirmation of product definition, design, integration, and verification. To provide the requisite level of control, four categories of technical reviews have been defined and are presented here: Major reviews are systemwide formal events. Subsystem, functional, and interim product reviews are components of an incremental and integrated confirmation process that leads to a major review.

Subsystem reviews focus on a single portion of the product and its interfaces. For any given system, a number of subsystem reviews may be used to reflect the partitioning of the product (e.g., a subsystem review of aggregate CIs and subsystem reviews of individual CIs). Functional reviews are systemwide events to confirm definition, traceability, integration, and implementation of an aspect of product functionality. Interim product reviews are systemwide events that occur between focusing on system progress in achieving objectives and issue resolution. Technical interchange meetings; independent, organizationally oriented reviews; working groups; committees; and program management reviews are replaced by subsystem review, functional review, and interim product review. The intent is not to turn every review into a fully formal review but to incorporate planning for all reviews (formal and informal) into the overall process.

Every review has some degree of formality, even if it is nothing more than recording the activities and agreements made among participants. Reviews are formal only to the extent necessary to document agreements and issues and to demonstrate (or confirm) IMP accomplishment criteria. For example, informal subsystem reviews may focus on work activities leading to demonstration that required accomplishments in the IMP for an upcoming formal review are done. Multidisciplinary teamwork is expected at reviews and meetings. Reviews are selected and held incrementally based on the complexity of the program and phase of development (or modification). An incremental review process provides a smaller forum for efficient exchange of information. One-on-one or few-on-few technical interchange meetings may be necessary to discuss a specific problem. However, changes to requirements or designs in terms of either the approach to be taken or the agreed-to numbers must be made in a multidisciplinary environment. Typically, participants in a review include the tasking and performing activity personnel responsible for the item or area being reviewed, key representatives for lower-level items, and other personnel who have a stake in the specific objectives of the review.

The review process is shown conceptually in Figure 6.2. The boxes in the figure identify formal reviews, such as major system reviews.

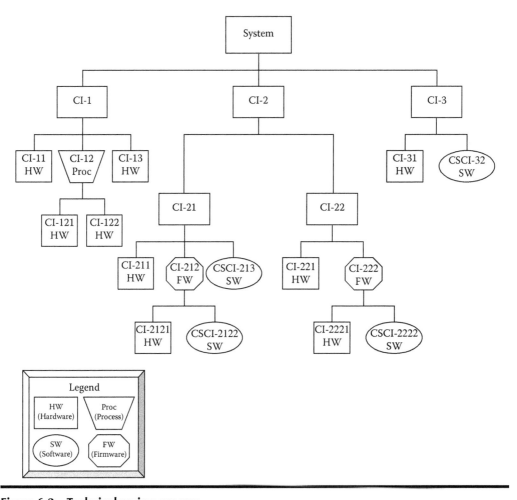

Figure 6.2   Technical review process.

These reviews can be viewed as working meetings. For a subsystem, the meeting represents all the disciplines (people) functioning as an IPT to define and design the subsystem. The term subsystem has a broad span, from a single component or CI to aggregate CIs to all the CIs satisfying a segmented portion of system requirements. In preparation for a major system review, a roll-up of subsystems reviews is done. Key concerns (such as risk and outstanding issues), findings, summaries, and so on are presented at the next higher level in the system hierarchy. Each such review demonstrates or confirms that required progress has been achieved. The traditional review work (i.e., evaluation of data) has already been done at the appropriate level by the people actually working on the subsystem. Demonstration and confirmation results are integrated at progressively higher levels. Finally, when there is joint agreement (i.e., the IMP) on what must be done to achieve success, consensus is reached on both when it is time and when it is not time to conduct a review. In addition to subsystem reviews, Figure 6.2 also depicts functional and interim system reviews. These reviews need not be held separate from other programwide reviews.

6.2.2 Sustainability Engineer's Role in Reviews

The responsibilities of the engineer include the following:

1. Ensure that the review process confirms that the requirements set (tasking activity responsibility) is balanced against the current design (performing activity responsibility). These objectives need to be well defined, reflected in engineering plans, and communicated to the performing activity.
2. Determine the degree to which the objectives of a review have been satisfied. This judgment can be made during the course of the review when demonstrating (or confirming) that the exit accomplishments of the review have been satisfactorily completed as measured by defined criteria in the IMP. At the end of the formal review, the engineer summarizes the results by notification of approval (review is satisfactorily completed and all accomplishments are demonstrated as measured by defined criteria in the IMP), contingent approval (review is not considered complete until the satisfactory completion of resulting action items to satisfy IMP exit accomplishments for the review), or disapproval (the review was inadequately executed or planned accomplishments were not complete).
3. Ensure that action plans are established to satisfy all accomplishments and criteria not demonstrated or confirmed so that the review can be completed when all action plans have been successfully executed. Action plans may be generated against both program activity and performing activity responsibilities.
4. Ensure the formation of a properly prepared program activity review team. This activity includes ensuring that team members understand roles, responsibilities, and specific objectives of the review. Training and prebriefings may be necessary. Formal reviews are focused on issues and confirmation of required accomplishments in the IMP building on an in-depth review of analyses performed, requirements needed, and solutions found before the formal review started.
5. Serve as cochair for formal reviews.
6. Provide formal acknowledgment to the performing activity of review accomplishment when proceedings are published.
7. Provide administrative information to the host, such as the name, organization, security clearance, and so forth for each tasking activity participant, prior to the review.
8. Ensure that all significant inputs have been included in the proceedings during or at the conclusion of the review.

During each review, the engineer needs to perform the following:

1. Host the review at an appropriate site (or sites).
2. Provide administrative support (e.g., resources, materials, meeting rooms, security, clerical personnel).
3. Provide other necessary information and items, including agendas and plans.
4. Ensure appropriate participation including that of subcontractors, vendors, and suppliers.
5. Provide information and items necessary to demonstrate and confirm that the IMP accomplishments associated with the review event have been satisfied.
6. Be able to substantiate trade-off decisions with technical details and associated rationale.
7. Document the proceedings, including noting key points, decisions, and issues with associated rationale as well as recording open and unresolved items with closure requirements and responsibilities.

6.2.3 Major Reviews

At a major review, subsystem leaders with key support staff demonstrate that the accomplishments and criteria for that review have been met. They will confirm that issues and concerns addressed during previous reviews have been satisfactorily resolved. Major reviews demonstrate that risk levels are acceptable and provide an opportunity to modify program emphasis for the next phase or effort. Proper integration and management of interim, subsystem, and functional reviews (as delineated in the SET) should mean that a detailed, total evaluation during the major review is unnecessary. Only those areas requiring close scrutiny should be addressed in detail. Such areas will be designated based on the results of the most recent interim system review. Prior to the production phase, the focus of reviews is on process, system concepts, system requirements, and interface requirements. Throughout the remaining system life cycle, major reviews are conducted to demonstrate system maturity during modification efforts. Major reviews are typically culminating events for one or more incremental reviews to confirm resolution of issues and demonstrate progress. Prior to a major review, the last subsystem review for each CI is normally formal. The major review is a formal system review to demonstrate system readiness to proceed with follow-on technical efforts. Sections 6.2.3.1 to 6.2.3.8 describe major reviews and outline when they typically occur. Each major review must demonstrate that there is an audit trail from the exit conditions of the previous review to the current conditions with changes substantiated as appropriate.

Major reviews include ASR, SRR, SFR, PDR, CDR, SVR, FCA, and PCA. These reviews reflect major system development milestones, each traceable to the preceding event or review. Major reviews have well-defined entry and exit criteria. Although it is called an audit, system PCA uses the same concept of identifying accomplishments and success criteria. Incremental reviews facilitate the conduct of major reviews as system demonstration and confirmation events.

6.2.3.1 ASR

The ASR normally occurs during the concept exploration and definition phase for new developments. It is used as a phase exit review to ensure that all necessary efforts have converged on the information necessary to support a decision. The ASR focuses on the confirmation that a preferred system concept with cost, schedule, and performance objectives has been defined; the preferred system concept can provide a cost-effective, operationally effective, and suitable solution to needs; and an executable development and risk management approach (including technology transitions, verifications, prototyping, and risk reduction efforts) has been defined to reduce the risk of the preferred system concept to the point where commitment to a system specification is appropriate. Review increments may include an interim system review to assess the scope of the alternatives considered and to reveal interface and interoperability issues. Functional reviews may be considered to raise issues and support system planning. Modification programs should consider subsystem reviews to assess impacts of technology application. The ASR should demonstrate the following:

1. The system concept is traceable to and can satisfy mission needs and other identified customer requirements.
2. Life cycle resource requirements, significant potential environmental consequences, timing to need, and other factors designated by the program have been identified.
3. The system concept is documented and defines cost, schedule, and performance objectives and thresholds.
4. Pertinent technologies (product and process) have been identified and the approach to validation, including prototyping and simulation, and transition is defined.
5. Risks and risk drivers have been identified, quantified, and prioritized, and an effective risk management approach is defined.
6. The critical accomplishments, success criteria, and metrics have been defined for the next acquisition phase or continued technical effort including technical exit criteria.
7. A draft specification tree and planned WBS for the next phase of technical effort are defined and traceable to the physical architecture.

6.2.3.2 SRR

Normally, an SRR is held early in the product definition and risk reduction phase for new developments. This review is also held in the production, fielding/deployment, and operational support phase for modifications, upgrades, and product and process improvements. This review serves to ensure that a balance is struck between requirements and solution approach risk (i.e., convergence has occurred on a system solution that has acceptable risk and the system requirements satisfy customer requirements).

An SRR is conducted to demonstrate progress in converging on viable, traceable system requirements that are balanced with cost, schedule, and risk by confirming the following:

1. Customer requirements (including environments, usage modes, and other pertinent factors) were analyzed and translated into system-specific functional and performance requirements.
2. Technology validation and demonstration plans are complete and closure plans on technical demonstrations and maturations are achieving required progress.
3. Critical technologies for people, product, and process solutions have been identified and assessed.
4. Risks are identified and quantified, and risk mitigation actions are achieving required progress.
5. The total system approach to satisfying requirements (including interfaces) for primary system functions has been identified (draft system and initial development specifications).

6.2.3.3 SFR

The SFR refocuses on system design review. Many activities associated with a system design review are still pertinent; however, the fundamental objective here is to establish and verify an appropriate set of functional and performance requirements for the system. Particular attention should be paid to the results of trade-off studies conducted to define requirements in areas that have moderate-to-high risk. The SFR is normally held during

1. Product definition and risk reduction phase for new developments requiring technology validation prior to establishing a functional baseline
2. System demonstration phase for systems using sufficiently mature technologies that do not require a demonstration and validation phase
3. Production, fielding/deployment, and operational support phase for modifications, upgrades, and product and process improvements

An SFR is conducted to demonstrate convergence on and achievability of system requirements and readiness to initiate preliminary design by confirming the following:

1. System functional and performance requirements have converged and characterize a design approach that satisfies established customer needs and requirements.
2. The physical architecture and draft allocated configuration documentation establish the adequacy, completeness, and achievability of functional and performance requirements (sufficient design and systems analyses including assessment and quantification of cost, schedule, and risk).
3. Critical technologies for people, product, and process solutions are verified for availability, achievability, needed performance, and readiness for transition.
4. The process completely defines functional and performance requirements including that
   a. Solutions for people, products, and processes satisfy all primary system functions.
   b. An audit trail from SRR is established with changes substantiated.
   c. Risks are mitigated and remaining risks are deemed acceptable.
   d. The system functional baseline can be established.
5. The draft specification tree has been assessed (based on the physical architecture) for the next phase of engineering effort to include any effect on the planned or approved WBS.
6. Planned contractor WBSs (CWBSs) for the next phase or technical effort have been assessed based on planned or approved WBS.
7. The risk-handling approach has been defined for the next phase or technical effort.
8. Preplanned product and process improvement and evolutionary acquisition requirements and plans have been defined.
9. Implementation requirements for technology transition have been defined.
10. The critical accomplishments, success criteria, and metrics have been defined for the next acquisition phase or continued technical effort.

6.2.3.4 PDR

The PDRs are also held in production, fielding/deployment, and operational support phase for modifications, upgrades, and product and process improvements. A PDR is held for each CI or aggregate CIs in the specification tree. Individual CI PDRs should ensure that a preliminary CI architecture is complete, a CI development specification is complete or development specification is approved, and a preliminary allocated baseline is complete or allocated baseline is approved. A system PDR is held after completion of all CI and aggregate CIs PDRs.

A PDR is conducted to confirm that the total system detailed design approach (as an integrated composite of people, product, and process solutions) satisfies the functional baseline, risks are mitigated with closure plans for remaining risks demonstrating required progress, and the total system is ready for detailed design. A PDR confirms that the following areas have been covered:

1. The process completely defines system requirements for design including that
   a. The design approach is balanced across cost, schedule, performance, and risk for the life cycle.
   b. The system physical architecture is an integrated detailed design approach for people, products, and processes to satisfy requirements, including interoperability and interfaces.
   c. An audit trail from SFR is established with changes substantiated.
   d. The system design approach is consistent with available DT&E results.
   e. Risks are mitigated and remaining risks are deemed acceptable.
   f. The allocated baselines for system CIs are defined.
2. Issues for the system, CIs, functional areas, and subsystems are resolved.
3. Sufficient detailed design has been accomplished to verify the completeness and achievability of defined requirements.
4. The risk-handling approach is refined for the next phase or technical effort.
5. Preplanned product and process improvement and evolutionary acquisition requirements and plans have been refined.
6. Critical accomplishments, success criteria, and metrics are valid for continued technical effort.

6.2.3.5 CDR

A series of CDRs is normally held for new developments and can be held in product definition and risk reduction phases for major prototyping activities. The CDRs are also held in production, fielding/deployment, and operational support phase for modifications, upgrades, and product and process improvements. A CDR is held for each CI as well as aggregate CIs in the specification tree. A system CDR is held after completion of all CI or aggregate CI CDRs. Even when the government elects not to bring the allocated baseline under configuration control by the time of this review, an assessment of the flow down of requirements from the functional baseline to the lowest-level CI for each item in the specification tree should be included in the review. Any changes in draft configuration documentation since performing the PDR are reviewed by the tasking activity and impact on the functional baseline is assessed and validated.

A CDR is conducted to demonstrate that the total system detailed design (as an integrated composite of people, product, and process solutions) is complete and meets requirements and the total system is ready for manufacturing and coding. A CDR confirms the following:

1. Issues for the system, functional areas, and subsystems are resolved.
2. The process completely defines system design requirements including
   a. The design is balanced across cost, schedule, performance, and risk for the life cycle.
   b. The system physical architecture is an integrated detailed design for people, products, and processes to satisfy requirements, including interoperability and interfaces.
   c. An audit trail from a PDR is established with changes substantiated.
   d. Allocated baselines for system CIs are refined.
3. The system design compatibility with external interfaces (people, products, and processes) has been established.
4. System design and interface requirements and design constraints are consistent with DT&E results.

5. The DT&E results support critical system design and interface requirements and design constraints.
6. The risk-handling approach is refined for the next phase or technical effort.
7. Preplanned product and process improvement and evolutionary acquisition requirements and plans have been refined.
8. Critical accomplishments, success criteria, and metrics are valid for continued technical effort.

6.2.3.6 SVR

This review represents the culmination of incremental reviews to support a decision that the total system (as represented by all its people, products, and processes) is ready to enter the production, fielding/deployment, and operational support phase. The accomplishments formally confirmed at the culmination of SVR need to be incrementally demonstrated as part of system verification throughout the period between CDR and the conclusion of SVR. These accomplishments should be defined for subsystem or interim system reviews (events) to ensure logical, progressive, and comprehensive verification of system performance and function as well as to ensure total system readiness to enter the production, fielding/deployment, and operational support phase. The SVR is normally held during new developments, but it is also held in production, fielding/deployment, and operational support phase for modifications, upgrades, and product and process improvements. If a system FCA (Section 6.2.3.7) is planned, it may be held in conjunction with the SVR.

Special attention is critical for products that have a degree of concurrency in development and production. Products with development continued into the production phase may need to plan and conduct SVR as two separate reviews: (1) The first review addresses all verification and readiness requirements to satisfy the phase exit criteria supporting a milestone decision (or similar decision for smaller programs). (2) The second review addresses the completion of all activities necessary for full production release in the production, fielding/deployment, and operational support phase.

An SVR is conducted to demonstrate that the total system (people, products, and processes) was verified to satisfy requirements in the functional and allocated configuration documentation and to confirm readiness for production, support, training, deployment, operations, continuing verifications, continuing development (if any), and disposal. An SVR confirms the completion of all incremental accomplishments for system verification (e.g., TRRs, CI, and system FCAs) and validates that the following areas are complete:

1. Issues for the system, functional areas, and subsystems are resolved.
2. System and CI verification procedures are complete and accurate (including verification by test and demonstration of critical parameters as well as key assumptions and methods used in verifications by analytic models and simulations).
3. The system and CIs are confirmed ready for verification.
4. Verifications were conducted in accordance with established procedures; they were completed for people, products, and processes; and system processes are current, executable, and meet the need.
5. An audit trail from a CDR is established with changes substantiated and the system and CIs verified.
6. The risk-handling approach is refined for the next phase or technical effort.
7. Preplanned product and process improvement and evolutionary acquisition requirements and plans have been refined.

8. Planning is complete and procedures; resources; and other requisite people, products, and processes are available (or programmed to be available) to initiate operations, support, training, production, deployment, disposal, and continuing verifications and development (if any).

9. Critical accomplishments, success criteria, and metrics have been refined and validated for the next acquisition phase or continued technical effort.

6.2.3.7 FCA

A series of FCAs is normally held during the phase for each CI in new development and can also be held in production, fielding/deployment, and operational support phase for modifications, upgrades, and product and process improvements. The entry and exit accomplishments for this review and any other pertinent critical accomplishment and associated success criteria are to be included in the IMP. The FCAs are an incremental part of the SVR.

6.2.3.8 PCA

A PCA is conducted to confirm that all CIs have been satisfactorily completed; the current state of the decision database is valid and represents the system; items (including processes) that can be baselined only at the system level have been baselined; required changes to previously completed baselines have been implemented (e.g., deficiencies discovered during testing have been resolved and implemented); and system processes are current, can be executed, and meet the need. A system PCA may be conducted after a full set of production-representative CIs has been baselined. This review is conducted in accordance with contractually established CM procedures.

The series of PCAs is normally held in the production, fielding/deployment, and operational support phase for each CI in a new development and can also be held in operations and support phases for modifications, upgrades, and product and process improvements. The entry and exit accomplishments for this review and any other pertinent critical accomplishment and associated success criteria are to be included in the IMP. For this major demonstration/confirmation event, additional program-specific accomplishments should be incorporated into the IMP. These specific accomplishments may include accomplishments related to resolving design issues uncovered during continuing verification efforts (including operational test and evaluation [OT&E]).

6.2.4 Subsystem Reviews

Subsystem reviews are multidisciplinary formal and informal reviews to assess progress in defining and satisfying subsystem requirements. Initially, subsystem reviews focus on process and requirements in examining alternative solutions (e.g., design approaches) for performance and functional requirements to establish requirements feasibility and risk in the solution. As the system matures, emphasis shifts to solution approach, design, design implementation and, finally, confirmation that the solution satisfies requirements.

A subsystem review requires participation from all functional areas and technical disciplines needed to address life cycle requirements and actions to satisfy these requirements and all elements of a subsystem. User, supplier, performing activity, and subcontractor organizations need to participate to confirm and demonstrate progress. The main thrust of any given subsystem review is considering pertinent subsystem accomplishments and criteria defined in the IMP and risk associated with the development of the subsystem. The formality of these reviews depends on content. Most subsystem reviews can be held as working meetings. Some, such as the PDR of a CI or the review for a CSCI, are formal demonstrations and confirmations of the accomplishments and criteria identified in the IMP. Subsystem reviews should replace some of the numerous, informal, technical interchange meetings and other working group meetings that are normally held during a program. Subsystem reviews below the CI level on contract may not require tasking activity oversight.

Subsystem reviews are conducted to ensure that the requirements (including interface requirements) for the subsystem have been identified, balanced across segments and interfaces, documented, and met. These reviews address issues and assess progress of a subsystem and ensure that the subsystem is developed in a life-cycle context (development through disposal). Each review focuses on required accomplishments for the IMP event that the review supports as well as upcoming system reviews. The subsystem review addresses impacts on and by interfaces with other subsystems and systems; documentation; risk; and, to the extent that they apply, designs, verification readiness, and documentation. Generally, a subsystem review confirms that the specifications required for a CI, its materials, and its processes are defined adequately to ensure the following:

1. At subsystem requirements reviews, the requirements allocated to the CI are complete and incorporated into the specification and that pertinent interface control documentation is established.
2. At subsystem design reviews, the requirements allocated to the CI are viable and the necessary process and material specifications have been developed.

6.2.4.1 Software Specification Review

Software specification reviews are conducted to demonstrate convergence on CSCI requirements as an integrated part of system and CI requirements and readiness to initiate preliminary design for the CSCI by confirming that the following areas are complete:

1. Subsystem and functional issues have been resolved.
2. The system physical architecture has converged on and characterizes a software design approach that includes design allocation of functional and performance requirements, interface requirements, and constraints to the CSCI as well as derived requirements for the CSCI.
3. The CSCI requirements are traceable to higher-level requirements and the set of requirements incorporates the functionality that must be implemented in the CSCI.
4. The relationship between CSCI and associated computer hardware requirements has been identified and the design compatibility between hardware and software has been established.
5. The CSCI requirements needed to ensure that CSCI performance and system compatibility satisfy higher-level and interfacing requirements have been captured in a completed software requirements specification and, if applicable, completed interface requirements specification.
6. Cost, schedule, and performance risks have been identified, quantified, and prioritized.
7. Risks are acceptable and risk management planning for the CSCI has been incorporated into overall technical risk management.
8. The CSCI life cycle resource requirements are compatible with, and incorporated into, system life cycle resource requirements.

6.2.4.2 Subsystem PDR

Subsystem PDRs are conducted to confirm that the CI detailed design approach (as an integrated composite of applicable people, product, and process solutions) provides required functionality; risks are mitigated with closure plans for remaining risks demonstrating required progress; and the CI is ready for detailed design. A subsystem PDR confirms the following items:

1. Subsystem and functional issues have been resolved.
2. The process completely defines CI requirements for design including the following areas:
    a. The design approach is balanced across cost, schedule, performance, and risk for the life cycle.
    b. The CI physical architecture is an integrated detailed design approach for applicable people, products, and processes to satisfy requirements, including interoperability and interfaces.
    c. An audit trail from SFR is established with changes substantiated.
    d. The CI design approach is consistent with available DT&E results.
    e. Risks are mitigated and remaining risks are deemed acceptable.
    f. The allocated baseline for the CI is defined.
3. Sufficient detailed design has been accomplished to verify the completeness and achievability of defined requirements.
4. The risk-handling approach is refined for the next phase or technical effort.
5. Applicable preplanned product and process improvement and evolutionary acquisition requirements and plans have been refined.
6. Critical accomplishments, success criteria, and metrics are valid for continued technical effort.

6.2.4.3 Subsystem CDR

Subsystem CDRs are conducted to demonstrate that the CI detailed design (as an integrated composite of applicable people, product and process solutions) is complete and meets requirements and the CI is ready for fabrication, coding, assembly, and integration of qualification units. A subsystem CDR confirms the completion of the following items:

1. Subsystem and functional issues have been resolved.
2. The process completely defines CI design requirements:
    a. The design is balanced across cost, schedule, performance, and risk for the life cycle.
    b. The CI physical architecture is an integrated detailed design for applicable people, products, and processes to satisfy requirements, including interoperability and interfaces.
    c. An audit trail from CI PDR is established with changes substantiated.
    d. Allocated baselines for the CI are refined.
3. The CI design compatibility with external interfaces (people, products, and processes) has been established.
4. The CI design and interface requirements and design constraints are consistent with DT&E results.
5. The risk-handling approach is refined for the next phase or technical effort.
6. Preplanned product and process improvement and evolutionary acquisition requirements and plans have been refined.
7. Critical accomplishments, success criteria, and metrics are valid for continued technical effort.

6.2.4.4 The TRR

The TRRs are conducted, as needed, for each CI to confirm completeness of test procedures, ensure that the CI is ready for testing, and ensure that the performing activity is prepared for formal testing. A TRR confirms that the following subsystem test requirements are complete:

1. Test procedures comply with test plans and descriptions, demonstrate adequacy to accomplish test requirements, and satisfy CI specification requirements for verifications.
2. Pretest predictions and informal test results (if any) indicate that testing confirms necessary performance.
3. New or modified test support equipment, facilities, and procedure manuals are required to accomplish planned DT&E and OT&E are available and satisfy requirements.
4. The requisite operational and support documents are complete and accurate.

6.2.4.5 Subsystem FCA

Subsystem FCAs are conducted to verify that CIs have achieved the requirements delineated in the functional, if any, and allocated configuration documentation. These audits are conducted in accordance with contractually established CM procedures.

6.2.4.6 Subsystem PCA

Subsystem PCAs are conducted on the as-built version of a CI to assess CI physical attributes against CI design documentation. The PCAs are normally conducted when production-representative articles are available and establish or verify the product baseline for the CI. They are conducted in accordance with contractually established CM procedures.

6.2.5 Functional Reviews

Functional reviews are conducted across a system by representatives from all involved disciplines to address progress and issues associated with one functional area or a similar group of functional areas. These reviews are structured to assess and confirm that the functionality of the system, within the area being reviewed, is traceable throughout the architecture from top-level customer needs and requirements to solutions. Functional reviews examine system functionality. Although the review does not have to start at the top of the system architecture, it does address functionality from the starting point all the way to the bottom. In a sustainability engineering environment, solutions are affected by integrated consideration of all areas of system functionality, and not just a single area.

Integrated, multidisciplinary functional reviews are conducted across the system to demonstrate the following:

1. Progress in defining requirements for system functionality
2. Vertical traceability of the functionality from needs/customer requirements to solutions
3. Integration and balance of the functionality across interfaces
4. Progress in converging on design solutions that provide the required functionality

With this perspective, these reviews are conducted to assist in identifying and resolving issues, support the identification of alternatives to satisfy higher-level requirements, and support technical plan development to ensure that product plans are functionally integrated systemwide.

Functional reviews focus on problem identification, confirmation of functionality implementation, and alternative identification. By examining the functionality throughout the architecture, progress and issues with other vertical (needs/requirements to solutions) and horizontal (integration of the functionality across all interfacing items) traceability issues can be determined. Functional reviews focus on progress in satisfying the accomplishments and criteria associated with a functional area for the next formal review. Within this overall perspective, functional reviews also support the functional integration of product planning across the system.

Functional reviews should be implemented with a well-defined charter. In concept, these reviews can provide a leveraging mechanism to promote functional integration across the system (and especially across all interfaces) as well as requirements traceability. In practice, without disciplined definition, structure, and implementation, a review could degenerate to advocacy and confrontational events. When problems/issues are identified alternatives and opportunities for resolution should be defined, not new requirements, specific solutions, or implementations. Staying focused on requirements traceability, functionality integration (especially across interfaces), issue identification, and approaches or opportunities for resolution should serve to minimize the opportunities for dissension. In an integrated multidisciplinary environment, a functional review should be viewed as a facilitation tool. Decisions to implement solutions need to be made in an environment broader than the scope of a functional team. Subsections 6.2.5.1 through 6.2.5.6 identify some of the possible functional reviews.

6.2.5.1 Support Review

Support reviews (SRs) are conducted to evaluate support requirements integration. These reviews assess support status and issues related to meeting system requirements with functional alternatives and solution alternatives such as new or modified support equipment, facilities, integrated diagnostics, technical orders/manuals, and skill levels. In addition, interface issues and the consistency and validity of support concepts for the system are addressed.

6.2.5.2 Training Review

Training reviews (TRs) are conducted to evaluate training requirements integration, interface issues, and the consistency and validity of training concepts for a system. These reviews assess training status and issues related to meeting system requirements with functional alternatives and solution alternatives such as new or modified training equipment, facilities, training manuals and materials, and training of the trainers.

6.2.5.3 Development Review

Development reviews (DRs) are conducted to evaluate requirements integration. These reviews assess status of and issues related to meeting system requirements that are not addressed in other functional reviews but are critical to satisfying system requirements such as interoperability, interface management, systems integration, system security, system safety, and computer resources. These reviews specifically track critical system parameters including high-impact parameters (e.g., survivability) and cost drivers (e.g., reliability).

6.2.5.4 Verification Review

Verification reviews (VRs) are conducted to evaluate verification requirements integration. These reviews assess verification status and issues related to meeting system requirements with functional alternatives and solution alternatives such as test and evaluation plans, procedures, equipment, personnel, ranges, special facilities, limitations, and posttest analysis support. In addition, interface issues and the consistency and validity of verification concepts for the system are addressed.

6.2.5.5 Manufacturing Review

Manufacturing reviews (MRs) are conducted to evaluate manufacturing requirements integration. These reviews are used to assess manufacturing status and issues related to meeting system requirements with functional alternatives and solution alternatives such as high-risk/low-yield manufacturing processes or materials, manufacturing developments, and planned use of existing manufacturing elements and processes. As product design matures, the reviews focus on production planning (identification of special manufacturing processes, process controls, special tooling requirements, layout, inventory management, and material-handling requirements), facility needs, product changes, fabrication of tools/test equipment/special machines, software requirements, skill levels required, training needs, procedure manuals, and long-lead item acquisition. In addition, interface issues and the consistency and viability of manufacturing concepts for the system are addressed.

6.2.5.6 Disposal Review

Disposal reviews are conducted to evaluate disposal requirements integration in terms of meeting system requirements and constraints with functional alternatives and solution alternatives such as disposal of system elements including demilitarization, destruction, mothballing, hazardous material containment and substitutions, and recycling for reuse and recovery. Interface issues and the consistency and viability of disposal concepts for the system are addressed. The topic of integrating requirements that result from defined hazardous material and environmental impact requirements and constraints is also addressed.

6.2.6 Interim System Reviews

Interim system reviews are "across-the-system" reviews held between major reviews. Interim system reviews can be informal or formal depending on the purpose of the review. These reviews may be held concurrently with program management reviews that occur during a program. Participants of interim reviews should be limited to principal program personnel. Interim reviews provide avenues to address issues and demonstrate required systemwide progress and maturity. As needed, interim system reviews are conducted across the system between major reviews to provide the following information:

1. For coordinated senior management action, present surface issues and concerns that are not resolvable at the subsystem level. These issues could include identification of incompatibilities between subsystems or between subsystem and functional approaches that must be resolved to satisfy accomplishments and criteria in the IMP.
2. Provide system status on progress toward meeting IMP accomplishments and criteria for the next major review. An interim system review can serve as a stimulator/facilitator to emphasize critical problem areas to help ensure a successful major review.
3. Address progress toward achieving IMP accomplishments and criteria that can only be demonstrated through a total-system look.
4. Integrate progress in contractual efforts to assess status in meeting government-to-government obligations.

5. Provide in-process assessment of system maturity/risk reduction efforts.
6. Verify that action plan closure documentation is complete.
7. Ensure that closure plans are defined and implemented for subsystem and functional issues.
8. Confirm incremental progress toward meeting system-level IMP accomplishments.
9. Confirm that system maturity, including risk mitigation, is achieving needed progress.

Interim system reviews should be planned in advance. Additional interim system reviews should be held when one or more issues arise that threaten the successful completion of an upcoming major review. In summary, these reviews occur between major reviews when necessary to address critical risk areas and ensure that IMP accomplishments are being met.

6.3 Functional Tasks

The tasks presented in Sections 6.3.1 through 6.3.9 reflect important areas in product development. For each task, determining factors for the degree of performance required are satisfying total cost, schedule, and performance requirements and objectives at an acceptable level of risk. The engineer incorporates these tasks into the SEM areas of requirements analysis, functional analysis/allocation, synthesis, and systems analysis and control, and any impacts are included in product LCC estimates. The tasks are not intended to preclude or supersede requirements applied from other required standardization documents, such as the SET.

6.3.1 Human Factors

The engineer should identify and define the functional, performance, and solution-dependent requirements to ensure that human factors are integrated into product and process designs. Objectives include balancing system performance and cost of ownership by ensuring that item designs are compatible with the capabilities and limitations of the personnel who will operate, maintain, transport, supply, control, and dispose of items. Requirements and designs must be defined to minimize characteristics that require extensive cognitive, physical, or sensory skills; the performance of unnecessarily complex tasks; tasks that unacceptably impact manpower or training resources or result in frequent, repetitive, or critical errors.

6.3.2 System Safety and Health Hazards

The engineer should identify and define the requirements to put into effect the safe use of system items and to control hazards associated with these items. The total system of people, products, and processes, including verification, manufacture, support, and disposal activities, is analyzed to identify potential hazards for the life cycle. Identified hazards associated with the use of end items are documented to establish criteria for mitigating or defining and categorizing high and serious risks. Materials categorized as having high and serious risks are characterized in terms of risks related to producing, deploying, operating, supporting, training with, and disposing of system end items that incorporate such materials. Use of materials that present a known hazard is avoided to the extent practical. If use of hazardous materials is an essential element of the solution, a containment program, including procedures for safe use and disposal of such materials, is developed and implemented. This program includes eventual substitution for hazardous materials except for those explicitly accepted by the engineer for a specific application. Handling and disposal of hazardous material are included in LCC estimates.

6.3.3 System Security

The engineer should identify and define the requirements to eliminate or contain vulnerabilities to known or postulated security threats. Item, information, and database susceptibility to damage, compromise, or destruction should be reduced. Control of compromising emanations is explicitly addressed early in the acquisition of items that have a potential to emanate sensitive information. All items and processes, including system information flows, are evaluated for known or potential vulnerabilities for the entire life cycle. The engineer will establish the level to which the vulnerability is controlled.

6.3.4 Producibility

The engineer should identify and define the requirements for producibility, that is, the capability of an item or service to be produced. Multidisciplinary IPTs help ensure that items are producible and generate simple designs and stable manufacturing processes that reduce risk, manufacturing cost, lead time, and cycle time and that minimize use of strategic and critical materials. As part of system design, manufacturing methods, processes, and process controls are defined, evaluated, and selected based on total system cost, schedule, performance, and risk.

Prior to full-rate production, the engineer should ensure that the product design has stabilized; ensure that manufacturing processes and process controls have been proved; and rate production facilities, equipment, capability, and capacity are in place (or are about to be put in place) to support the approved schedule. Value-engineering concepts should be employed to assist in the identification of requirements that add cost to the system but add little or no value to users.

6.3.5 Integrated Logistics Support

The engineer should identify and define requirements to ensure that items can be supported. Suggestions to help ensure item support include the following:

1. Ensure that requirements are related consistently to readiness objectives, design, and each other.
2. Integrate support factors into item and system element design interactively with the design of support products and processes.
3. Identify cost-effective approaches to supporting an item when that item is deployed and installed.
4. Identify and define requirements for support structure elements so that the item is both supportable and supported when deployed or installed.
5. Plan for postproduction support to ensure continued, economic logistics support.

6.3.6 Test and Evaluation

The engineer should identify and define requirements that all item characteristics are verifiable. Verification of the acceptability and compatibility of human performance requirements, personnel selection, training, and human–machine interfaces of system procedural data is integrated into the system test program. The objectives, scope, and type of system test and evaluation need to reflect an integrated approach for functionality verification to conserve resources. Test and evaluation planning addresses performance, functional, and design requirements with appropriate quantitative criteria, test events or scenario descriptions, resource requirements (e.g., test range, special test facilities), and test limitations. Wherever practicable, tests for different objectives are combined. Test and evaluation efforts are structured to

1. Provide information to assess technical risks and aid in decision making
2. Generate information to determine whether items have met technical performance requirements, specifications, and objectives
3. Verify that items are operationally effective and suitable for intended use
4. Verify the critical assumptions, data, and methods used to derive critical item requirements (e.g., safety and survivability)
5. Verify the critical assumptions, data, and methods used in the verification of item performance.

6.3.7 Integrated Diagnostics

The engineer should identify and define requirements to incorporate diagnostics to provide unambiguous detection and isolation of faults that occur when system end items are in use. Factors to be considered in developing these requirements include embedded testability; built-in tests; automatic, semiautomatic, and manual testing; common test data; technical information; consistent detection and isolation; and training.

6.3.8 Transportability

The engineer should identify and define requirements to ensure that items are transportable. The performing activity identifies the limiting characteristics of transportation systems that apply to item requirements, designs, and development. The performing activity uses these data to derive and refine item requirements; designs; and impact-associated packaging, handling, storage, and transportation solutions. The performing activity addresses transportability with the development of new, modified, and NDIs and when developing integrated logistics support for items.

6.3.9 Infrastructure Support

The engineer should identify and define requirements for a compatible interface with the infrastructure supporting the system to identify unique infrastructure support requirements and to ensure timely planning to provide needed infrastructure support. The performing activity assesses each item for interaction with and integration into the command, control, communications, and intelligence structure. The engineer identifies the support that the system will require from other support agencies and commands.

6.4 System-Cost-Effectiveness Analysis Tasks

System-cost-effectiveness analysis and assessment tasks should be integrated into sustainability engineering processes to support the development of life-cycle-balanced products and processes. Critical requirements and verifications identified by analyzing each primary system function serve as constraints on other items and areas of impact. These constraints are included in requirements documentation and specifications for impacted items and areas. System-cost-effectiveness analyses and assessments help to serve the following functions:

1. Support the identification of mission and performance objectives and requirements.
2. Support the allocation of performance to functions.
3. Provide criteria for the selection of solution alternatives.
4. Provide analytic confirmation that designs satisfy customer requirements.
5. Support verification of people, product, and process solutions.

The engineer should identify the parameters that drive solutions and establish sensitivity variables for uncertainties in input data and assumptions. When another system has comparable characteristics, it is included as a baseline system to support the determination, completeness, and achievability of requirements. Subsections 6.4.1 through 6.4.9 identify a few such system-cost-effectiveness tasks.

6.4.1 Manufacturing Analysis and Assessment

Manufacturing analyses and assessments support the development of people, product, and process requirements and solutions necessary to produce end items. Manufacturing analyses include production analyses and manufacturing and production inputs to system effectiveness, trade-off studies, and LCC analyses. Alternative designs and capabilities of manufacturing are evaluated. Long-lead-time items, material source limitations, availability of materials and manufacturing resources, and production cost are identified, assessed, and documented. Manufacturing-critical characteristics of people, product, and process solutions would be identified and the risks are included in risk management efforts. Results from these activities and solution alternatives are assessed interactively with other system solution alternatives.

6.4.2 Verification Analysis and Assessment

Verification analyses and assessments support the development of people, product, and process solutions necessary to verify that system end items satisfy the necessary requirements. Verification analyses address verification requirements and criteria for solution alternatives; definition of verifications to demonstrate proof of concept; and development, qualification, acceptance, pertinent operational, and other testing. Life-cycle requirements for test consistency in and across the solution set are determined. These analyses address the requirements and procedures needed to verify critical verification methods and processes (such as key methods, assumptions, and data used in verifications by analysis). Verification-critical characteristics of people, product, and process solutions are identified and these risks are included in risk management efforts. Results of these activities and solution alternatives are assessed interactively with other system solution alternatives.

6.4.3 Deployment Analysis and Assessment

Deployment analyses and assessments support the development of people, product, and process solutions necessary to deploy end items. Deployment analyses and assessments address the following areas:

1. Factors for site/host selection, activation/installation, field assembly, and checkout requirements including identification of site-unique hazard classification and explosive ordnance disposal requirements
2. Operational and maintenance facilities and equipment requirements

3. Compatibility with existing infrastructure (e.g., computer/communication systems)
4. Determination of environmental impacts and constraints (environmental impacts on the system and system impacts on the environment) at deployment sites as defined by the environmental analysis and impact assessment task
5. Early deployment of training items and personnel
6. Initial provisioning and spares
7. Packaging, handling, storage, and transportation
8. Site transition requirements

Deployment-critical characteristics of people, product, and process solutions are identified and these risks are included in risk management efforts. Results of these activities and solution alternatives are assessed interactively with other system solution alternatives.

6.4.4 Operational Analysis and Assessment

Operational analyses and assessments support the development of people, product, and process solutions necessary to satisfy operational requirements for system end items. The following analyzes and assessments address the operational use of alternative solutions:

1. The way solutions will be used to accomplish required tasks in the intended environments
2. Interfacing systems required to execute operational functions in the intended environment
3. Required joint and combined operations
4. Identified modes of operational deployment and employment

Operations-critical characteristics of people, product, and process solutions are identified and the risks are included in risk management efforts. Results of these activities and solution alternatives are assessed interactively with other system solution alternatives.

6.4.5 Supportability Analysis and Assessment

Supportability analyses and assessments assist in the development of people, product, and process solutions to support system end items. Supportability analyses are used to assist in identifying the data and procedures required in specifications and other development documentation to ensure system life-cycle support (e.g., additional interface information and verification requirements to use "used" parts). Supportability analyses address the following:

1. All specified levels of operation, maintenance, and training for system end items
2. The planned life cycle to ensure that system end items satisfy intended use
3. Identification of supportability-related design factors
4. The development of an integrated support structure (people, products, and processes)
5. Support resource needs including parts, people, facilities, and materials

Supportability-critical characteristics of people, product, and process solutions are identified and the risks are included in risk management efforts. Results of these activities and solution alternatives are assessed interactively with other system solution alternatives.

6.4.6 Training Analysis and Assessment

Training analyses and assessments support development of people, product, and process solutions to train those personnel who use the system end items. Training analysis includes the development of personnel capabilities and proficiencies to accomplish tasks at any point in the system life cycle to the level they are tasked. These analyses address initial and follow-on training necessary to execute required tasks associated with system end item use. Training-critical characteristics of people, product, and process solutions are identified and risks are included in risk management efforts. Results of these activities and solution alternatives are assessed interactively with other system solution alternatives.

6.4.7 Disposal Analysis and Assessment

Disposal analyses and assessments support development of people, product, and process solutions to dispose of products and by-products. Environmental factors for process wastes and outputs as well as used products and components are included. Alternative disposal methods for system parts and materials are evaluated, and requirements for new or modified methods are determined. Methods addressed include storage, dismantling, demilitarization, reusing, recycling, and destruction. Costs; sites; responsible agencies; handling and shipping; supporting items; and applicable federal, state, local, and host nation regulations are factors in the analyses. Disposal-critical characteristics of people, product, and process solutions are identified and the risks are included in risk management efforts. Results of these activities and solution alternatives are assessed interactively with other system solution alternatives.

6.4.8 Environmental Analysis and Impact Assessment

The project must adhere to all applicable statutes and to designated hazardous material lists. Environmental analyses are performed to determine the impact on and by each system product and process alternative on factors such as noise pollution, quantities and types of hazardous materials used, hazardous waste disposal, and other defined environmental requirements as applicable. Methods to mitigate problems identified from this analysis are defined and an assessment of impacts is generated. Assessment results are factored into effectiveness analyses as well as system definition, design, and verifications. Analysis outputs are documented appropriate to the acquisition phase and used in conjunction with cost and performance analyses outputs to support acquisition phase exit criteria. The use of materials that present a known hazard to the environment are avoided to the extent practical. Environment-critical characteristics of people, product, and process solutions are identified and the risks are included in risk management efforts.

6.4.9 LCC Analysis and Assessment

The LCC analyses and estimates, including the cost of development, acquisition, ownership, and disposal, are conducted and updated as designated in the program to support decisions, assessments of system cost-effectiveness, and trade-off studies. This effort identifies the economic consequences of solution alternatives. These analyses develop the requisite cost information to support decisions on alternative people, product, and process solutions and risk assessments. These analyses include established design-to-cost targets, a current estimate of these costs, and known uncertainties in these costs.

6.5 Project Implementation Tasks

The following tasks should be performed as needed to satisfy program implementation requirements:

1. Conduct developmental test and evaluation to validate technologies for application to system solutions, acquire definition information to support synthesis, and acquire verification information to support assessments in systems analysis and control.
2. Implement engineering test models and other related items needed to conduct developmental test and evaluation. This requirement does not include items for those tests conducted on low-rate or full-rate production hardware.
3. Generate and reuse, as appropriate, software to satisfy sustainability engineering requirements.
4. Provide sustained engineering and problem solution support.

Project leadership should also verify people, product, and process solutions by design analysis, design simulation, inspection, demonstration, or test. Required performances of all critical characteristics are verified by demonstration and test. Design analysis and simulation are used to complement, not replace, demonstration and test. Tests include system effectiveness evaluations and manufacturing process proofing. Where total verification by test is not feasible, testing is used to verify key characteristics and assumptions used in the design analysis or simulation. Commensurate with these efforts, the following activities should also be covered:

1. Conduct verifications of the physical architecture (including interfaces) from the lowest level up to the total system to ensure that functional and performance requirements are satisfied.
2. Generate evidence necessary to confirm that the system and CIs meet requirements.
3. Validate technologies for use in people, product, and process solutions considering cost, schedule, performance, and risk using established criteria.
4. Verify that materials employed in system solutions can be disposed of in a safe, environmentally compliant manner.

6.6 Other Pervasive Considerations

The tasks discussed in Sections 6.6.1 through 6.6.6, as selected and tailored for a particular program application, should be integrated into sustainability engineering processes.

6.6.1 Computer Hardware

Computer resources that are needed for end items should be managed as an integral part of overall project development. Computer hardware resource decisions are not finalized until the software design demonstrates a maturity that minimizes the risk of inadequate processor throughput and memory. Similarly, software design decisions are not finalized until computer hardware resource designs demonstrate a maturity that minimizes the risk of incompatibility.

6.6.2 Use of Software

A project needs to consider the computer models and simulations that will be used when such use contributes to decision process. The models need to allow parameters to be varied so that relative and individual parameter effects on total system performance and LCC can be determined. Performance characteristics allocated to system functions will correlate to parameters in the models. The models, data files, and documentation are maintained, updated, and modified as required. Each version of a model or data file that impacts requirements, designs, or decisions is entered into a decision database. Requirements need to be addressed for software development tools and software development, integration, and test environments. The engineer evaluates the extent of the simulation application required to refine requirements and designs and to evaluate solutions for people, products, and processes by simulating interactions with the environment. Additionally, the simulation application is evaluated as an adjunct to prototyping. The performing activity employs simulation when it is cost-effective to do so.

6.6.3 Decision Database

The decision database is a repository of information generated and used during the lifetime of a project. The intent of the database is that when properly structured it provides access to technical information, decisions, and rationale that describe the current state of system development and evolution. Among other things, the database does the following:

1. Illustrates intraproduct, interproduct, and component interfaces
2. Permits traceability among components at various levels of product detail
3. Provides a means for complete and comprehensive change control
4. Includes techniques and procedural information for development, manufacture, verification, deployment, operation, support, training, and disposal of a product
5. Provides information to verify adequacy of product design
6. Provides data for trade-off studies and assessments of item capability to satisfy project objectives
7. Provides complete documentation of the design (products and processes) to support progressive system development and subsequent iterations of the sustainability engineering process.

The decision database should be on electronic media. The specific format and structure of the database may be defined by the tasking activity or left open for project leadership definition. Standardization of the format and structure should be an issue early in the acquisition, upgrade, or modification of a product to document and maintain necessary audit trail.

Off-the-shelf automated tools for database maintenance and data transfer are generally preferred over the development of new tools. Selection of tools may be predicated on the ability to satisfy prescribed security and data interchange requirements. Where multiple projects are supporting the same overall development effort, the project leadership should consider the requirement that each project use the same automated tools to facilitate the transition and translation of system data and provide a common format and interface among all participants.

The engineer evaluates the use of integrated computer-aided engineering, design, manufacturing, test, and support methods to support design integration through shared product and process models and databases. When it is cost-effective over the system life cycle, documentation of accomplishments and exchange of product and process information is implemented consistent with standard interchange formats such as continuous acquisition life cycle support (CALS).

6.6.4 Open System Architectures

Open system architectures (OSAs) should be identified and evaluated for use in system solutions. These architectures are evaluated for applications in systems that use preplanned product improvement or evolutionary acquisition strategies, when required solution functionality and mission application is expected to vary, and in circumstances where technologies change rapidly. Additionally, OSAs are evaluated for their application to system interoperability and use of solutions across multiple items. The OSAs are used where practical when they meet requirements and are cost-effective over the entire life cycle.

6.6.5 Prototyping

The engineer evaluates whether prototyping should be used to assist in identifying and reducing risks associated with integrating available and emerging technologies into item design to satisfy requirements. When employed, prototyping addresses all aspects of emerging technology that bear upon its successful application, including, for example, hardware, software, and manufacturing processes. Prototyping (experimental, rapid, or developmental) is used to provide timely assessment of item testability to identify the need for new or modified test capabilities. The project leadership would conduct the same type of evaluations, and for the same purpose, when supporting product improvements and modifications to fielded (operational) systems.

6.6.6 Materials, Processes, and Parts Control

A materials, processes, and parts control program should be established and controlled. Such a program focuses on standardization of parts, materials, and processes. The program addresses design, procurement, and availability of parts through the expected life of each item; considers the environment that the item is required to operate in; and accounts for life cycle support costs. The program emphasizes reducing the variety of parts, variability in processes, and associated documents used with items.

6.7 Tailoring Process

Tailoring is an activity that takes the form of deletion (removal of processes that are not applicable), alteration (modifying processes to more explicitly reflect application to a particular effort), or addition (adding processes to satisfy unique program requirements). Tailoring involves selecting the most appropriate or preferred processes and determines their precedence to ensure that the complete set of processes is integrated, harmonized, and balanced to best achieve optimum product requirements. By integrating sustainability processes, product requirements are addressed concurrently with performance requirements. In this way sustainability activities are integrated with all engineering and design activities, thereby avoiding duplicative effort and making the best use of activity outputs. An integrated approach to sustainability is essential because sustainability is related to other product characteristics. Harmonization includes provisions to resolve conflicts among process requirements and stakeholder positions or recommendations. These provisions can include procedures ranging from simple notification to notification with solution recommendations and justifications. Each selected sustainability engineering process needs to contribute, in a cost-benefit sense, to the final objective. Processes are not viewed as an end in themselves. They should be in balance with other processes. The systemic approach embodied in the concept of the SEM provides this balance.

Tailoring is performed to both breadth and depth of application based on the project and phase in its development. Breadth means dealing with the type and number of products impacted by the project. Tailoring for depth involves decisions concerning the level of detail needed to generate and substantiate the output required to satisfy program objectives. Specific processes require defining the depth of detail, level of effort, and the data expected. The depth of the engineering effort varies from program to program according to complexity, uncertainty, urgency, and the willingness to accept risk. The objectives of the program and the inputs to sustainability engineering processes demarcate the breadth and depth of application. To assist in defining the depth of application and level of effort, the following inputs should be considered:

1. The level of detail in a product's definition: For example, during conceptual investigations, a complete functional decomposition of the product for each alternative is not usually necessary. However, sufficient depth is necessary to provide confidence in cost, schedule, performance objectives, and related risk estimates. Different depths may be identified for areas in relation to the application of new technologies.
2. Program direction and limitations including the willingness to accept more risk.
3. Scenarios to be examined for each primary product function.
4. A set of MOEs organized hierarchically: The relative importance of each metric at the top level in the hierarchy is also identified.
5. Known constraints and requirements in areas where they are likely to exist, but quantitative data are not available.
6. The technology database including the identification of key technologies, performance, maturity, cost, risks, and any limiting criteria on the use of technology.
7. The factors essential to product success, including factors related to major risk areas (e.g., budget, resources, and threat).
8. Technical demonstration and confirmation events that must be conducted, including technical reviews.

The level of detail expected from end products of an effort needs to be identified as this determines the depth to which the sustainability engineering process is executed. For example, functional analysis and synthesis are conducted to a sufficiently detailed depth to identify areas of technical risk appropriate for consideration for the acquisition phase or effort. The term sufficiently detailed is determined based on the objectives of the program effort and can be characterized by the information content expected from the physical architecture. For example, during concept exploration the physical architecture describes the product concept. During program definition and risk reduction, the architecture describes the product in terms of its specifications (typically in draft form) and the concept of CIs that make up the product. The physical architecture provides the detailed design requirements for all product elements and the drawings for product CIs.

Throughout the development of a product, the level of detail may vary because the baseline product may be at one level of detail and product or process improvements or other modifications may be at a different level of detail. Note that level of detail needed from the technical effort to ensure adequacy of technical definition, design, and development is not synonymous with the level of detail expected for management control and reporting (e.g., cost performance reports). The basic sustainability engineering processes described in Chapter 5 can be applied to any development effort (including new developments, modifications, and products improvements), regardless of size or complexity. Attention to scope of the effort and level of output expected is, however, essential. For example, an unprecedented new product development in concept exploration phase is not likely to require CM audits or formal change control mechanisms. However, conceptual exploration investigation of modifications to an existing or foreign developed product may need this type of activity for a variety of reasons (e.g., to verify interface constraints).

The purpose of tailoring sustainability engineering processes for a specific project is to ensure that appropriate amount of effort is devoted to appropriate activities to reduce project risk to an acceptable level while making the most cost-effective use of engineering resources. It is often difficult to determine exactly how much sustainability engineering is "enough" on a given project (except in hindsight; e.g., when it is clear that more sustainability engineering could have prevented some disaster, which occurred in system integration). A general guideline, however, is that enough sustainability engineering should be performed on a project to ensure that the system, its requirements, its configuration, and its performance are well defined and verified; that all engineering risks have been identified and assessed; and that engineering resources in appropriate engineering disciplines (including systems engineering) are allocated throughout the program/project to deliver the required products and keep schedule, cost, and technical risks at an acceptable and cost-effective level.

6.7.1 Levels of Tailoring

There are two levels of tailoring of sustainability engineering processes discussed in Chapter 5:

1. Level one involves top-level tailoring by selection of specific processes that add value to a project. The selection is performed by executing the process implementation strategy (SEP-15) presented in Chapter 5.
2. Level-two tailoring involves a selection of internal tasks and activities from individual sustainability engineering processes. The selected processes should be described in the project SEMP, which was introduced in Chapter 1, along with perhaps the rationale for not selecting the other processes.

6.7.2 Tailoring Process

The steps involved in tailoring are outlined in Table 6.1. These steps take the form of deletion (removal of processes that not applicable), alteration (modifying processes to more explicitly reflect the application to a particular effort), or addition (adding processes to satisfy unique program requirements).

Table 6.1 Tailoring Process

Process purpose: This process is used to tailor the content and scope of sustainability engineering processes in Chapter 5 to fit the needs of a project. Tailoring consists of identifying specific process activities and process-related product and outputs products appropriate to the specific project being planned or replanned. The selection and mapping of these process activities to the project/product life cycle can vary substantially, and the time, energy, and effort devoted to each should reflect the economics and risks of the project being addressed. This process is a companion to process implementation strategy (SEP-15), which focuses on selecting the applicable processes.
Roles and agents: • Project systems engineer or technical lead • Project sponsor representative • Line management representative

(Continued)

Table 6.1   Tailoring Process (*Continued*)

Inputs:	Outputs:
• The goals and constraints of the project • Organizational and sponsor tasking requirements for the sustainability engineering process or products • The baseline SEM (Chapter 4) for the organization and any tailoring guidelines • Any cost targets and the acceptable level of risk for the project	• A documented description of the sustainability engineering activities planned for the project

Process diagram:

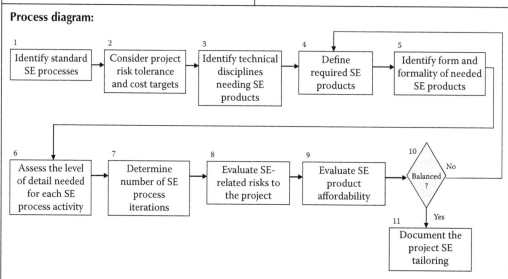

Figure 6.3   Tailoring process (SE in figure denotes sustainability engineering).

Step	Action
1	**Action statement:** Identify the process baseline. **Notes and guidance:** Identify the standard SEM from which tailoring is done.
2	**Action statement:** Determine the cost targets and risk tolerance of the project. **Notes and guidance:** If the project goals are unachievable at an acceptable cost and risk level, the acceptable combination of project goals, costs, and risk level must be negotiated until it is acceptable to management and seen as achievable by the engineers.
3	**Action statement:** Identify other technical and engineering disciplines needing sustainability products. **Notes and guidance:** Characterize what other engineering disciplines (designers, testers, etc.) on the project will need from sustainability engineering. This, together with the size of the total engineering team, will determine the type and content of the products that sustainability engineering needs to produce for the engineering effort to be a success.

Table 6.1 Tailoring Process (*Continued*)

Step	Action
4	**Action statement:** Identify required sustainability engineering deliverable products.
	Notes and guidance: Identify the deliverable documents for which sustainability engineering is responsible. Also identify any other products that are in the baseline methodology (from aforementioned step 1), which cannot be tailored per any tailoring guideline.
5	**Action statement:** Identify the form or format of needed sustainability products.
	Notes and guidance: For each product identified in steps 3 and 4, identify the form the product should take and the level of detail necessary to achieve the purpose of each product. This can often be done by citing examples of products from previous projects to give team members a common understanding of both the format and the level of detail planned. A data item description or template can provide the format of a document. The level of detail is typically best described by examples of the content for each section.
6	**Action statement:** Identify the level of detail needed for each process activity.
	Notes and guidance: One way to do this is to use the standard process that describes a set of process activities and identify which subparagraphs apply. Another approach is to write down the purpose of the activity and the risks to the project if it is not done adequately and then derive the level of detail needed to serve this purpose and avoid these risks. If this level of detail in the process activity is not affordable, determine in which areas risks can be allowed to rise.
7	**Action statement:** Determine number of iterations for the sustainability processes on the project.
	Notes and guidance: Identify the life cycle strategy planned for the project. This gives guidance on the number of iterations through related processes that should be planned. If the project is part of a larger program, this may also clarify which process activities may have been partially completed.
	Decisions about the number of iterations appropriate for the project depend on goals and constraints of the project. For a project that has a design-to-cost goal, you may choose to iterate through selected process activities several times to assure that all requirements that drive a design above cost targets are identified and modified.
8	**Action statement:** Evaluate product affordability.
	Notes and guidance: Assess whether any products, or their forms, or their level of detail (as determined in step 5) are unaffordable given the project goals, cost targets, and level of tolerable risk (as determined in step 2). In other words, look at what products are needed to enable the process to work well, given the circumstances (the project team, their familiarity with engineering processes involved in the project, their familiarity with the applications and product area technology, suspected staff turnover, etc.).
	In general, the less experienced the team or the more likely the personnel turnover the more explicit/formal the products for the process to be successful. The basic purpose of most of the products is communication between engineering project team members about what to do and how to do it.

(Continued)

Table 6.1   Tailoring Process (*Continued*)

Step	Action
9	**Action statement:** Evaluate sustainability-related risks. **Notes and guidance:** Identify and assess the sustainability-related risks on the project. For each risk that sustainability engineering can affect, determine cost-effective actions required to bring the risk levels after mitigation to acceptable levels.
10	**Action statement:** Determine if sustainability needs have been balanced with project risk and affordability. **Notes and guidance:** If adequate balance has been achieved, go on to step 11. If not, iterate back to step 4.
11	**Action statement:** Document project tailoring. **Notes and guidance:** Plans for tailoring should be included in project planning documents. The implementation of tailoring can use forms and spreadsheets. Document the tailoring planned to the baseline SEM and obtain approval. If no formal authorization is required, request an informal review of the proposed tailoring from an engineer who has experience with the same customer.

6.7.3 Process "Tailoring Traps"

Enterprise organizational units are to avoid the tailoring traps described in the INCOSE handbook (INCOSE 2007) and summarized as follows:

Reuse of a tailored baseline from another project without repeating the tailoring process: It is fallacious to assume that previously tailored baselines are appropriate for all projects. Prior successes are not a guarantee of future success. There is something unique in each project.

Using all processes and activities, "just to be safe": The trap is that each process carries an overhead cost. If this approach is taken, the quality of the product may actually degrade because of application of an inappropriate process. It cannot be called tailoring if there is no clear justification for the inclusion of every process in the plan.

Using a preestablished tailored baseline: Shortcuts to create templates of baselines that can be taken off the shelf and applied to work based on arbitrary categorizations such as high-, medium-, and low-risk projects can be counterproductive. They carry the same hazards as the aforementioned traps 1 and 2. Tailoring is important because emphasis is placed on the project and products and only processes that support attainment of the objective in terms of quality and performance should be retained.

Failure to include relevant stakeholders: The tailoring process itself can become a unifying activity that establishes shared visions and understanding of objectives/costs/risks. Suppliers, or other organizations, that are identified and not included in the process may feel disenfranchised with the result that they feel a lower level of commitment to the process baseline. When new parties are added, they should be made familiar with the baseline and asked to make constructive contributions.

Reference

1. International Council of Systems Engineering (INCOSE). 2007 (August). *Systems Engineering Handbook*, Version 3.1. San Diego, CA: International Council of Systems Engineering.

Chapter 7

Designing for Sustainability

7.1 Objectives

Sustainability should be viewed as a common sense ingredient of design. However, other performance requirements and budget concerns can frequently override design decisions made to improve sustainability. Now, the ownership costs and availability requirements of increasingly complex modern systems and equipment demand that designing for sustainability is as important as designing for other performance characteristics. The design engineer must ensure that considerations for sustainability are an integral element of every design trade study or design change activity.

The basic objectives of designing for sustainability are to meet the operational readiness requirements for the product and to reduce support costs. A design engineer committed to these objectives will continually challenge the design to uncover weaknesses and potential sustainability problems. The objective is to design-in sustainability. If this objective is not met and the product fails to meet sustainability objectives, corrective design changes will have to be made later in the product's life cycle at significant expense. The primary objective of a sustainability program is to identify and correct sustainability problems early in the design process when correction simply requires changing drawings.

7.1.1 Support Concepts

Support concepts are the methods by which the customer intends to sustain the product. They can be as varied as the design itself and range from discard at failure to a complete overhaul at failure. They may include periodic or scheduled maintenance or overhaul. They can include maintenance performed by the customer, the supplier, a third party, or some combination of the three. Within the military services, three levels of maintenance are normally defined: organizational (on-site), intermediate (local shops), and depot (an overhaul facility). Table 7.1 provides a brief description of each level of maintenance. No one definition of maintenance levels could be found for the commercial sector. However, perhaps defined somewhat differently or combined in some way, the

Table 7.1   Levels of Maintenance

Level of Maintenance	Where Normally Performed	Description
Organizational	On the product at the customer's operational or product site	Normally is limited to periodic performance checks, visual inspections, cleaning, limited servicing, adjustments, and removal and replacement of some components (i.e., constituent module, part, item, etc., of the product). Repair of removed components is normally not made at this level. Instead, the failed component is replaced with a spare. The removed component is then sent to the next level of maintenance (usually intermediate) for repair. Diagnostics, accessibility, and ease of removal and replacement are very important and should be key design considerations. At this level, the primary goals are keeping the product in a serviceable condition and rapidly restoring the product to an operable condition after failure using low-to-moderate skilled personnel.
Intermediate	On the product or a repairable component of the product at a customer's "shop location"	Products might be repaired by removal and replacement of parts or modules, or the parts or modules of a product might be repaired. The skill level of personnel is usually higher than at the organizational level of maintenance. Intermediate level of repair facilities may also be tasked with doing limited depot/overhaul level repairs. These types of repairs are typically based upon technical knowledge, facilities, and potential cost savings.
Depot	On the product or a repairable component of the product at a specialized repair facility operated by the customer or the original equipment manufacturer (OEM)	Facility may very well be structured like an assembly line. Maintenance includes rebuilding or overhauling a product and may be performed on a specific lot of failed equipment that has been screened for similarity in failure type. The most highly skilled and trained technical personnel are assigned to depots. Test equipment is very complex, technical publications are more detailed, and manufacturing source data are frequently available. One specific depot might be structured to support all forms of communication radios or all types of pumps.

same levels of maintenance are considered representative of those used by commercial industry. Maintenance performed at these levels keeps the product serviceable or restores it to an operational condition after a failure.

Maintenance can include two basic types of tasks. The first, called preventive maintenance (M_p), is usually performed at the organizational level. M_p retains a product in serviceable condition by inspections, servicing, and other preventive measures performed on a calendar, cyclical, or performance trending basis. The second is corrective maintenance (M_C). M_C is performed to return a product to operation after a failure and may be accomplished at the operational, intermediate, or depot level. The cost of maintenance, preventive or corrective, is directly determined by the maintenance of the design.

A support concept is more than simply identifying whether M_P and M_C are required and whether maintenance will be performed at one, two, or three levels of organization. It means deciding on a run-to-failure or on-condition maintenance approach. It also addresses who will provide support: the customer, the product manufacturer, or both. Often, the military services elect to plan for contractor support at the intermediate and depot levels until a product has been proven in actual use. Then responsibility for the maintenance may be transitioned to the military service. Such a strategy is called interim contractor support. Finally, a support concept can involve centralizing some organizational and intermediate level maintenance at one or two sites.

The approach to handling ambiguity groups is also a part of the support concept. Sometimes, factors make FI to a single replaceable unit (RU) or item impossible to achieve. These factors include the complexity that would be added by fault isolating to a single item, the total cost associated with fault isolating to a single item compared with the cost associated with fault isolating to two or more items, and the type of technology being used. Consequently, some failures will be detected by the integrated diagnostics and isolated to two or more items. To correct the failure, one of two basic approaches may be used. For relatively small ambiguity groups, the entire group will be replaced. For larger groups, items in the group will be iteratively replaced until the failure is corrected. The decision to use group or iterative replacement is primarily based on economics and the effect on predicted total DT.

A product may be a new development, a nondevelopmental item, or a COTS item. As shown in Table 7.2, the latitude that planners have in selecting a support concept is determined by the amount of new development involved.

For new development products, the support concept can and should greatly influence the design for maintenance. For example, ease of disassembly is not a concern for nonrepairable products that are discarded after failure. But if the product is a component or subsystem of a larger

Table 7.2 Latitude in Selecting a Support Concept

Type of Product	Degree of Latitude
New development	High—the designers can respond to the chosen concept as they design the product.
Nondevelopmental items	Less than for new development.
COTS	Little to none. It is unlikely that the engineering, design, and other detailed data needed to develop an organic repair capability will be available. Also, the supplier (OEM) will most likely sustain configuration control below the product level, not the customer. So, support for COTS will often consist only of removal and replacement at the operational level with depot and even intermediate maintenance usually performed by the OEM.

product, accessibility to facilitate removal and replacement is important. If a two-level maintenance (2LM) concept is desired, then reliable diagnostics are essential. If the diagnostics are not reliable, then items will be removed and shipped all the way to a depot, which may then determine that the item is good and place it back in the supply chain. Not only are many assets tied up in this situation but also it will take considerably longer to uncover the root cause of field problems. Finally, the design approach for a product can be very different depending on whether the customer or the contractor will be providing the support.

7.1.2 Operational and Support Environment

A supplier must understand the environment in which the customer will operate and sustain the product. Environmental factors, such as temperature and humidity, limit the way in which personnel can perform maintenance. For example, when maintaining products in very cold climates or under hazardous conditions (radioactive, biological, or chemical environments) personnel must wear heavy clothing and gloves. Such clothing restricts movement, requires more room for access, and reduces dexterity. In addition, materials can shrink or expand making it difficult to connect and disconnect mating parts. In hot climates with high humidity, perspiration can impair vision and affect a person's grip. If maintenance must be performed outside, the maintenance engineer must try to design access panels so that rain cannot penetrate into the interior of a product. Also, it might be necessary to perform maintenance during product operation. If so, the maintenance engineer's primary concern is to design the product and procedures to minimize the hazards involved with maintenance.

In addition to analytical techniques, the maintenance engineer has two excellent methods of characterizing the support environment. First, the customer's maintenance personnel can be brought in to participate in the design process at the earliest phase of product development. Second, maintenance and design engineers can visit the customer's operating sites to gain first-hand knowledge of the operational and support environment. Every product needs to be assessed for the environmental impact on maintenance.

7.1.3 Preventive versus Corrective Maintenance Requirements

M_P is usually self-imposed DT (although it may be possible to perform some M_P while the product is operating). M_P consists of actions intended to prolong the operational life of the equipment and keep the product safe to operate. Ideally, a product will require no servicing or other M_P and either the probability of failure is remote or redundancy makes failure acceptable (however, one often-required M_P task is to verify the operational status of redundant components prior to a mission). For such an ideal product, only M_C, if any, would be required. Most often, however, failure is not a remote possibility. Moreover, most products of any complexity require some servicing, even if that only consists of cleaning. Sometimes failures can actually be prevented by M_P. The goal, then, is to identify only the M_P that is absolutely necessary and cost-effective.

Figure 7.1 illustrates the two major categories of maintenance, M_P and M_C, and the tasks associated with each.

M_P is only applicable if the probability of failure is reduced by the M_P, or there is a quantitative indication of an impending failure. In the case of the former criterion, M_P has no benefit for items that have a random pattern of failure (i.e., constant failure rate [FR]). Consequently, we rarely, if ever, will use a M_P action for electronics since electronics exhibit a random pattern of failures. Mechanical items, on the other hand, usually have a limited useful period of life and then begin to wear out.

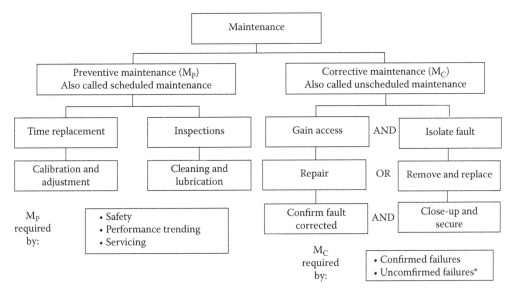

Figure 7.1   **Major categories of maintenance.**

The second criterion for determining if M_p is applicable is whether or not there is a quantitative indication of an impending failure (functional failure). If reduced resistance to failure can be detected (potential failure) and there is a consistent time between potential failure and functional failure, then M_p is applicable. Performance trending, as discussed in the Reliability chapter, has long been used to monitor operating parameters that have been shown to be dependable predictors of an impending failure. M_p may be applicable but it must also be effective. That is, it must

■ Reduce the FR to an acceptable level
■ Be less expensive than M_C and the cost of failure without M_p

7.1.4 Two-Level Maintenance

The objective of 2LM is to reduce manpower, equipment, facilities, and mobility footprint while still meeting the Air Force's Global Engagement mission objectives. The approach for meeting this objective is to modify or eliminate the intermediate (off-equipment) function when possible and consolidate that repair function at a depot or regional level.

Using state-of-the-art communications, item visibility, and fast transportation systems, unserviceable parts will be moved rapidly to and through the regional, depot, or contractor repair processes. 2LM will be performed at the appropriate organic Air Force regional, depot, or contractor repair activity. A regional repair center is a hybrid of three-level maintenance (3LM) and 2LM and combines the intermediate level maintenance function from multiple bases at one location. Therefore, from the perspective of the affected bases, the unserviceable assets are treated as 2LM and shipped to the regional repair center. The regional repair center performs the traditional intermediate level maintenance and the depot continues to perform the same type of repairs under the 3LM concept. This regional repair center concept should be applied where it makes good economic sense as it offers similar advantages to that of 2LM. Some of these advantages are

- Consolidates like-maintenance efforts
- Affords a mobility/surge option
- Provides a second source and/or multiple sources of repair
- Provides a source of experienced personnel
- Allows for increased repair flexibility while maintaining lower overall repair costs due to economies of scale

Good reliability performance is a desired attribute of a 2LM candidate. Good reliability indicates that a 2LM concept for the asset would not significantly increase transportation costs or depot workload. Good maintenance, specifically good diagnostics, is also important. Good diagnostics have a low CND rate. CND is an indicator of the frequency at which an asset appears to have failed on a weapon system or equipment end item when that failure cannot be duplicated in the repair shop. In such cases, the asset is returned to the supply system. A high CND rate would mean additional transportation costs and increased demand on the supply and repair systems in a pure two-level repair environment where all base level repair capability has been eliminated. Thus, a low CND rate is important in a 2LM environment.

7.2 Human Engineering

7.2.1 Human Factors

Human factors is an important design consideration. Often a specialist, a human engineering (HE) engineer, addresses the human factors element of design. The HE engineer has two roles. First, the HE engineer represents the potential user, operator, and sustainer and is concerned with ease of operation, safety, comfort, workloads, and so on. Second, the HE engineer evaluates people as "components" and their contribution to product effectiveness. The HE engineer is concerned with many design issues including:

- Safety of operators and sustainers
- Which functions to allocate to humans
- How best to present information to the user, operator, or sustainer
- Accessibility
- The design of tools and controls
- Anthropometry
- Required skill levels

A key focus of the HE engineer is on presenting information to operators and maintainers. Although information is usually presented through visual displays and auditory signals, other methods include touch, smell, the sense of balance (vestibular sense), or sensations of position and movement (kinesthesis). Each method has its own variables. Visual displays, for example, can be in color or black and white, use symbols or text, use moving scales with fixed indicators or fixed scales with moving indicators, and so on. Selecting the best method requires an understanding of how humans process, interpret, and store information; the detection and differential sensitivity of the human senses; and human psychology and physiology. Figure 7.2 illustrates some of the factors involved in human information processing.

The maintenance, HE, reliability, safety, and other design engineers must develop a product design that contributes to proper operator responses by creating perceivable and interpretable stimuli requiring reactions within the user's, operator's, or sustainer's capabilities. Feedback ought to be incorporated into the design to verify that operator responses are correct. In other words, product characteristics should serve as both input and feedback stimuli to the operator or sustainer. These interactions between the human and the product are depicted in Figure 7.3.

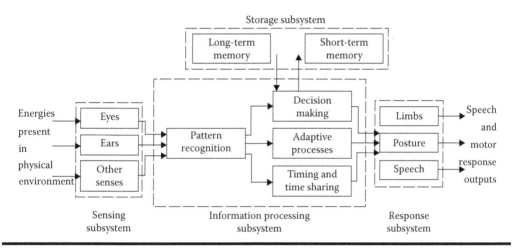

Figure 7.2   The human information processing system.

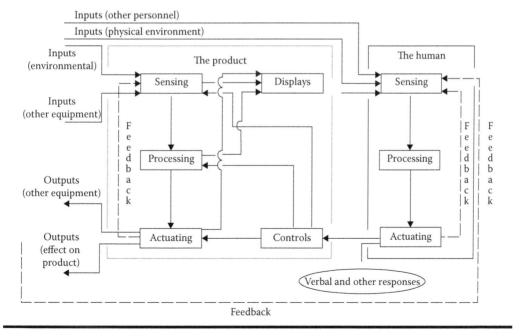

Figure 7.3   Interactions between human and product.

Humans control the functions of a product with switches, knobs, levers, wheels, and other devices. Collectively, such devices are called controls. In selecting the proper control for a specific function, the HE engineer must evaluate the function of the control, the requirements of the control task, the informational needs of the human, the requirements imposed by the work environment, and the consequences of inadvertent or accidental operation of the control.

Anthropometry is the science of measuring various human physical characteristics, primarily size, mobility, and strength. Using such measurements in designing a product, workplace, support equipment, and clothing, designers can enhance the efficiency, safety, and comfort of users, operators, and sustainers. People vary in size and strength within any group. This variance can be expressed statistically by taking appropriate measurements of the population and calculating the mean and standard deviation. Based on these statistics, percentiles can be calculated. For example, a 90th percentile height for American men means that only 10% of the males in the United States are taller than that height. Normally, the HE and maintenance engineer will design for people who are in the 95th or higher percentile for weight, stature, sitting height, and other anthropometric measurements. Anthropometric tables and charts are available in HE handbooks and military standards to help the engineer assess human physical interface factors. These tables and charts include information on percentile measurements of physical size; allowances for clothing; maximum strength (static forces and torque) of hands, fingers, and legs; and range of motion.

7.3 Tools and Support Equipment

Few products can be sustained without using some tools and support equipment. Maintenance of many consumer products requires only common hand tools, such as screwdrivers and pliers. Maintenance of other products can require test equipment, servicing stands, protective clothing, specialized tools, and so on. It is the sustainability engineer's responsibility to identify the tools and equipment needed by maintenance personnel to support the product.

To keep costs down and reduce the amount of specialized training required, the maintenance engineer should try to use tools and equipment already in use for other products. For example, airlines have a large investment in hand tools, support equipment, and other items with which aircraft are sustained. A commercial aircraft manufacturer who ignores this "in-place" inventory and designs an aircraft requiring all new tools and equipment will find it difficult to market a new aircraft, no matter how advanced it may be.

During design, the engineer must often deal with human factors considerations and this is especially important in the design of tools. Even the simplest of hand tools should be designed with a regard for the human form, manual dexterity, and other human factors. If they are not, they will be more difficult to use, will tire the user more quickly, and can result in injury or accidents. In the commercial marketplace, a variety of screwdrivers featuring new designs for the handle have been introduced. These new designs are intended to make gripping the screwdriver more natural with less chance of injury to the tendons and muscles of the hand and wrist.

Special tools, those not commonly found in the toolboxes of people who sustain a given product, should be avoided if at all possible. Using standard tools has been shown to minimize the possibility of damage or injury and the time associated with connecting and disconnecting items. In addition, special tools increase the cost of supporting the product during its useful life.

Support equipment includes diagnostic equipment (e.g., automatic test equipment [ATE]), dollies and lifting devices, stands and jacks, air conditioners and heaters, work lights, ladders and lifts, towing vehicles and tow bars, and other equipment needed to support the product. As is the case with tools, it is usually advisable to design the product to be supported with support equipment already used by those who will sustain the product. By using "standard" support equipment, costs and training requirements are reduced.

When newly designed tools and support equipment are needed, virtual reality (VR) techniques can be used to evaluate virtual "copies" of the support equipment or tool by "performing" maintenance activities with them. VR could allow technicians to view virtual information panels "superimposed" (using augmented reality techniques) on the actual equipment. Also, the support concept and any customer constraints or requirements regarding support equipment and tools must be understood and considered during all design trade-offs and analyses.

In designing new tools and support equipment, anthropometric measurements should be used to enhance the efficiency, safety, and comfort of users, operators, and sustainers. In addition, early estimates of maintenance time, labor hours, and other maintenance metrics can be used in making preliminary assessments of the support equipment and tools.

Finally, it is important to note that maintenance should be as much a consideration for the support equipment as it is for the product itself. Maintenance affects the availability and support costs of both the support equipment and the product. So in designing or selecting off-the-shelf support equipment, maintenance criteria should be part of requirements for the support equipment.

7.4 Maintenance Training

At the turn of the twentieth century, maintenance training was accomplished in large measure through apprenticeships, essentially on-the-job training (OJT). One learned how to do a task by watching and learning from an experienced person, a mentor. Most maintenance tasks were simple by today's standards and did not require a sophisticated or technical education. As products increased in complexity, incorporating ever more sophisticated technology, it became clear that OJT alone was insufficient. Classroom training and more stringent educational requirements quickly became commonplace in maintenance training, and in training in general. The training of maintenance personnel is now extremely important, expensive, and time-consuming. Many methods of training are now available and are often used in combination as part of a comprehensive training program. Table 7.3 lists some of the methods used to train maintenance personnel.

Traditionally, training focuses on

- Addressing performance deficiencies
- Individual skills and knowledge
- Improving the job–person fit

Selecting the most appropriate training method is critical. Trying to teach a manual skill through lecture, for example, will not produce very good results. At the other extreme, the theory of operation of a product does not lend itself to teaching by demonstration. Table 7.4 provides some prerequisites for selecting training methods for any training activity and some guidelines to follow in the selection process.

In designing the product, the engineer can and must consider the impact on training.

Table 7.3   Maintenance Training Methods

Method	Comments
• On-the-job	• Perhaps the oldest method of learning a trade or an activity requiring manual dexterity and skill • Time-consuming and requires skilled, experienced mentor
• Classroom	• Can include demonstration by the trainer with observation and imitation by the trainee, lectures • Best suited to theory and intellectual content
• Simulation	• Use of VR[a] and other simulation methods • Provides realistic "hands-on" experience in a controlled environment
• Computer aided	• Basically another form of classroom training in which the class size is one and the instructor's knowledge has been captured in software • A self-study approach • Allows self-paced learning
• Distance learning	• Can be used to deliver classroom training and self-study programs

[a]*Maintenance and manufacturing procedures, especially procedures that are seldom performed or are difficult to teach using conventional approaches, can be taught using VR. VR could also be used to train individuals in performing hazardous procedures, disposing of hazardous materials, or performing life-threatening procedures.*

Table 7.4   Selecting a Training Method

Prerequisites	Guidelines
• Set training objectives • Set training priorities • Design a curriculum	• Choose training methods suited to the training objective and material • Combine principles of andragogy (the art and science of helping adults learn) with a variety of teaching methods that allow for optimal training • Determine which training methods are most familiar to the trainees • Review the training activity objectives to make sure the training method is appropriate for achieving the objectives

Specific questions that can be asked of a design alternative are

- Does the design incorporate diagnostic resources at each level of maintenance that reduce the need for training?
- Does the design make use of tools and equipment already in use for other products thereby eliminating or reducing the amount of specialized training required?
- Does the design enhance standardization thereby helping to reduce training requirements both in number of personnel and the level of skill required?
- Can the planned maintenance personnel sustain the design using their existing skills?
- Is special training equipment req.uired?

7.5 Testability and Diagnostics

Testability is an inherent design characteristic, whereas diagnostics involves additional factors. Attention paid to both in design will reduce not only the cost of producing a product but certainly the cost of maintaining the fielded product. BIT is one typical approach to diagnostics and one that reduces maintenance labor and external test equipment. It is an important diagnostic tool and considering it at all levels of design is important for two reasons. First, densely packaged devices are increasingly used in the circuit card design. They decrease the accessibility required for and increase the risks of guided-probe testing. So including BIT in such designs is critical to effective diagnostics. Second, many integrated circuit vendors are incorporating some form of BIT into their designs. Designers must capitalize on this fact for higher-level designs (e.g., board or module) that use such devices by integrating lower-level BIT capabilities with higher-level BIT designs. Doing so increases the vertical testability of the system; that is, allows BIT tests performed at higher support levels (e.g., factory) to be used at lower levels (e.g., field). This characteristic will help sustain consistency across maintenance levels and may reduce the numbers of RTOKs or CNDs.

As is the case with many design approaches, the use of BIT is a matter of trade-offs. For example, as the coverage of BIT is increased, the chance for false alarms also increases, possibly increasing the demand for maintenance. The most important factor in BIT design is early planning. Without planning for BIT early in the life cycle, it will be harder to maximize any advantages offered by the use of BIT while minimizing any negative impacts such as increased design cost, higher hardware overhead, and increased FR. In one study, "Chip-To-System Testability," for Rome Laboratory (DoD 1997), Research Triangle Institute and Self-Test Services gave the following five axioms that allow designers to capitalize on BIT.

- Plan for BIT starting at the earliest stage of the program
- Design BIT in conjunction with the functional design, not as an afterthought
- Use the same high degree of engineering cleverness and rigor for BIT that is used for the functional design
- Use computer-aided design tools to design BIT whenever possible
- Incorporate the subject of BIT into peer, design, and program reviews

7.5.1 Testability

Testability is a subset of maintenance. However, because of the negative impact of poor testability on production and maintenance costs, testability is recognized as a separate design discipline in its own right and will continue to be treated as such, at least in the foreseeable future. Therefore, it is important to develop a testability program plan as an integral part of the systems engineering process and to elevate testability to the same level of importance accorded to other product assurance disciplines. Plans must include the requirement to analyze a design to ensure it provides for efficient and effective fault detection and isolation (FD&FI).

Ensuring that a product is testable requires adherence to some basic testability design principles. A list of the most common testability design principles, along with a brief description of each, is shown in Table 7.5. In addition to the principles shown in the table, checklists of testability design practices have been developed that are specific to technologies, such as analog, digital, mechanical, and so on. A detailed checklist is provided in Table 7.6.

Determining the amount of testability necessary in a design will be driven by the requirements for FD&FI. FD requirements are typically stated as the percentage of all possible faults that can be detected, using defined means (BIT, semiautomatic/automatic test, etc.). For instance, a system may have a requirement of 95% FD, indicating that 95% of all possible failures are to be detectable by the diagnostic capability of the system. FI requirements are typically stated as the percentage of time that FI is possible to a specified number of components. As an example, a system may have a requirement of 90% isolation to a single RU, 95% isolation to an ambiguity group of two or fewer RUs, and 100% isolation to an ambiguity group of three or fewer RUs. Table 7.7 lists some common measures of testability.

Table 7.5   Testability Design Principles

Principle	*Description*
Physical and functional partitioning	The ease or difficulty of FI depends to a large extent upon the size and complexity of the units that are replaceable. Partitioning the design such that components are grouped by function (i.e., each function is implemented on a single RU) or by technology (e.g., analog, digital) whenever possible will enhance the ability to isolate failures.
Electrical partitioning	Whenever possible, a block of circuitry being tested should be isolated from circuitry not being tested via blocking gates, tri-state devices, relays, and so on.
Initialization	The design should allow an item to be initialized to a known state so it will respond in a consistent manner for multiple testing of a given failure.
Controllability	The design should allow external control of internal component operation for the purpose of FD&FI. Special attention should be given to independent control of clock signals, the ability to control and breakup feedback loops, and tri-stating components for isolation.
Observability	Sufficient access to test points, data paths, and internal circuitry should be provided to allow the test system (machine or human) to gather sufficient signature data for FD&FI.
Test system compatibility	Each item to be tested should be designed to be electrically and mechanically compatible with selected or available test equipment to eliminate or reduce the need for a large number of ID designs.

Table 7.6   Inherent Testability Checklist

Mechanical design checklist (for electronic designs)		
• Is a standard grid layout used on boards to facilitate identification of components? • Is the number of I/O pins in an edge connector or cable connector compatible with the I/O capabilities of the selected test equipment? • Are connector pins arranged such that the shorting of physically adjacent pins will cause minimum damage? • Is each hardware component clearly labeled?	• Is spacing between components sufficient to allow for clips/test probes? • Has provision been made to incorporate a test header connector into the design to enhance ATE testing of surface-mounted devices? • Is defeatable keying used on each board so as to reduce the number of unique interface adapters required? • Is the design free of special setup requirements that would slow testing?	• Are all components oriented in the same direction (pin 1 always in same position)? • Does the board layout support guided-probe testing techniques? • When possible, are power and ground included in the I/O connector or test connector? • Have test and repair requirements impacted decisions on conformal coating? • Does the item warm up in a reasonable amount of time?
Partitioning checklist (for electronic functions)		
• Is each function to be tested placed wholly upon one board? • Within a function, is the size of each block of circuitry to be tested small enough for economical FD&FI? • Is the number of power supplies required compatible with the test equipment?	• If more than one function is place on a board, can each be tested independently? • If required, are pull-up resistors located on same board as the driving component? • Is the number and type of stimuli required compatible with the test equipment?	• Within a function, can complex digital and analog circuitry be tested independently? • Are analog circuits partitioned by frequency to ease tester compatibility? • Are elements included in an ambiguity group placed in the same package?
Test control checklist		
• Are connector pins not needed for operation used to provide test stimulus and control from the tester to internal nodes?	• Can circuitry be quickly and easily driven to a known initial state (master clear, less than N clocks for initialization sequence)?	• Are redundant elements in design capable of being independently tested?

(Continued)

Table 7.6   Inherent Testability Checklist (*Continued*)

Test control checklist		
• Is it possible to disable on-board oscillators and drive all logic using a tester clock? • Is circuitry provided to bypass any (unavoidable) one-shot circuitry? • In microprocessor-based systems, does the tester have access to the data bus, address bus, and important control lines? • Are input buffers provided for those control point signals with high drive capability requirements?	• Can long counter chains be broken into smaller segments in test mode with each segment under tester control? • Can feedback loops be broken under control of the tester? • Are active components, such as demultiplexers and shift registers, used to allow the tester to control necessary internal nodes using available input pins?	• Can the tester electrically partition the item into smaller independent, easy-to-test segments (placing tri-state element in a high impedance state)? • Have provisions been made to test the system bus as a stand-alone entity? • Are test control points included at those nodes which have high fan-in (test bottlenecks)?
Parts selection checklist		
• Is the number of different part types the minimum possible? • Is a single logic family being used? If not, is a common signal level used for interconnections?	• Have parts been selected that are well characterized in terms of failure modes?	• Are the parts independent of refresh requirements? If not, are dynamic devices supported by sufficient clocking during testing?
Test access		
• Are unused connector pins used to provide additional internal node data to the tester? • Are test access points placed at those nodes that have high fan-out? • Are active components, such as multiplexers and shift registers, used to make necessary internal node test data available to the tester over available output pins?	• Are signal lines and test points designed to drive the capacitive loading represented by the test equipment? • Are buffers employed when the test point is a latch and susceptible to reflections? • Are all high voltages scaled down within the item prior to providing test point access so as to be consistent with tester capabilities?	• Are test points provided such that the tester can monitor and synchronize to on-board clock circuits? • Are buffers or divider circuits employed to protect those test points that may be damaged by an inadvertent short circuit? • Is the measurement accuracy of the test equipment adequate compared to the tolerance requirement of the item being tested?

Table 7.6   Inherent Testability Checklist (*Continued*)

Analog design checklist		
• Is one test point per discrete active stage brought out to the connector? • Are circuits functionally complete without bias networks or loads on some other unit under test (UUT)? • Is a minimum number of complex modulation or unique timing patterns required? • Are response rise time or pulse width measurements compatible with test capabilities? • Are standard types of connectors used?	• Is each test point adequately buffered or isolated from the main signal path? • Is a minimum number of multiple phase-related or timing-related stimuli required? • Are stimulus frequencies compatible with tester capabilities? • Are stimulus amplitude requirements within the capability of the test equipment? • Does the design avoid external feedback loops?	• Are multiple, interactive adjustments prohibited for production items? • Is a minimum number of phase or timing measurements required? • Does the design allow testing without heat sinks? • Do response measurements involve frequencies compatible with tester capabilities? • Does the design avoid or compensate for temperature-sensitive components?
Radio frequency (RF) design checklist		
• Do transmitter outputs have directional couplers or similar signal sensing/attenuation techniques employed for BIT, off-line test monitoring purposes, or both? • Has provision been made in the off-line ATE to provide switching of all RF stimulus and response signals required to test the subject RF UUT? • Are the RF test I/O access ports of the UUT mechanically compatible with the off-line ATE I/O ports?	• If an RF transmitter is to be tested utilizing off-line ATE, has suitable test fixturing (anechoic chamber) been designed to safely test the subject item over its specified performance range of frequency and power? • Does the off-line ATE or BIT diagnostic software provide for compensation of UUT output power and adjustment of input power so that RF switching and cable errors are compensated for in the measurement data?	• Have suitable termination devices been employed in the off-line ATE or BIT circuitry to accurately emulate the loading requirements for all RF signals to be tested? • Does the RF UUT employ signal frequencies or power levels in excess of the core ATE stimulus/ measurement capability? If so, are signal converters employed within the ATE to render the ATE/UUT compatible?

(*Continued*)

Table 7.6 Inherent Testability Checklist (*Continued*)

Radio frequency (RF) design checklist		
• Have all RF testing parameters and quantitative requirements for these parameters been explicitly stated at the RF UUT interface for each RF stimulus/response signal to be tested?	• Have adequate testability (controllability/observability) provisions for calibrating the UUT been provided? • Has the UUT/ATE RF interface been designed so that the system operator can quickly and easily connect and disconnect the UUT without special tooling?	• Have RF compensation procedures and databases been established to provide calibration of all stimulus signals to be applied and all response signals to be measured by BIT or off-line ATE to the RF UUT interface? • Has the RF UUT been designed so that repair or replacement of any assembly or subassembly can be accomplished without major disassembly of the unit?
Electro-optical design checklist		
• Do all buses have a default value when unselected? • Have optical splitters/couplers been incorporated to provide signal accessibility without major disassembly? • Has temperature stability been incorporated into fixture/UUT design to assure consistent performance over a normal range of operating environments? • Can requirements for bore-sighting be automated or eliminated? • Do monitors possess sufficient sensitivity to accommodate a wide range of intensities?	• Does the design contain only synchronous logic? • Have optical systems been functionally allocated so that they and associated drive electronics can be independently tested? • Are the ATE system, light sources, and monitoring systems of sufficient wavelength to allow operation over a wide range of UUTs? • Has adequate filtering been incorporated to provide required light attenuation? • Can all modulation models be simulated, stimulated, and monitored? • Do optical elements possess sufficient range of motion to meet a variety of test applications?	• Do light sources provide enough dynamics over the operating range? • Does the test fixturing intended for the off-line test present the required mechanical stability? • Is there sufficient mechanical stability and controllability to obtain accurate optical registration? • Can targets be automatically controlled for focus and aperture presentation? • Are optical collimators adjustable over their range of motion via automation? • Are all memory elements clocked by a derivative of the master clock? (Avoid elements clocked by data from other elements.)

Table 7.6   Inherent Testability Checklist (*Continued*)

Electro-optical design checklist		
• Can optical elements be accessed without major disassembly or realignment? • Do test routines and internal memories test pixels for shades of gray?		• Are all clocks of differing phases and frequencies derived from a single master clock?
Digital design checklist		
• Does the design avoid resistance capacitance one-shots and dependence upon logic delays to generate timing pulses? • Is the design free of wired OR logic? • Will the selection of an unused address result in a well-defined error state? • Is the number of fan-outs for each internal circuit limited to a predetermined value? • Is the number of fan-outs for each board output limited to a predetermined value?	• Are latches provided at the inputs to a board in those cases where tester input skew could be a problem? • For multilayer boards, is the layout of each major bus such that current probes or other techniques may be used for FI beyond the node? • Does the design support testing of "bit slices"?	• Does the design include data wrap-around circuitry at major interfaces? • Is a known output defined for every word in a read-only memory (ROM)? • Are sockets provided for microprocessors and other complex components? • If the design incorporates a structured testability design technique (scan path, signature analysis), are all the design rules satisfied?
BIT checklist		
• Can BIT in each item be exercised under control of the test equipment? • Does the BIT use a building-block approach (all inputs to a function are verified before that function is tested)? • Does on-board ROM contain self-test routines?	• Is the test program set designed to take advantage of BIT capabilities? • Does building-block BIT make maximum use of mission circuitry? • Is the self-test circuitry designed to be testable? • Is the predicted FR contribution of the BIT circuitry within stated constraints?	• Are on-board BIT indicators used for important functions? Are BIT indicators designed such that a BIT failure will give a "fail" indication? • Is BIT optimally allocated in hardware, software, and firmware? • Have means been established to identify whether hardware or software has caused a failure indication?

(*Continued*)

Table 7.6   Inherent Testability Checklist (*Continued*)

BIT checklist		
• Does BIT include a method of saving on-line test data for the analysis of intermittent failures and operational failures that are nonrepeatable in the maintenance environment? • Is the additional volume due to BIT within stated constraints? • Does the allocation of BIT capability to each item reflect the relative FR of the items and the criticality of the items' functions? • Are the data provided by BIT tailored to the differing needs of the system operator and the system sustainer? • Does mission software include sufficient hardware error detection capability?	• Is the additional power consumption due to BIT within stated constraints? • Are BIT threshold values, which may require changing as a result of operational experience, incorporated in software, or easily modified firmware? • Is processing or filtering of BIT sensor data performed to minimize BIT false alarms? • Is the failure latency associated with a particular implementation of BIT consistent with the criticality of the function being monitored?	• Is the additional weight due to BIT within stated constraints? • Is the additional part count due to BIT within stated constraints? • Is sufficient memory allocated for confidence tests and diagnostic software? • Are BIT threshold limits for each parameter determined as a result of considering each parameter's distribution statistics, the BIT measurement error, and the optimum FD/false alarm characteristics?
Performance monitoring checklist		
• Have critical functions been identified (by failure mode effects and criticality analysis [FMECA]) that require monitoring for the system operation and users?	• Has the displayed output of the monitoring system received a HE analysis to ensure that the user is supplied with the required information in the best useable form?	• Have interface standards been established that ensure the electronic transmission of data from monitored systems is compatible with centralized monitors?
Diagnostic capability integration		
• Have vertical testability concepts been established, employed, and documented?	• Has a means been established to ensure compatibility of testing resources with other diagnostic resources at each level of maintenance (technical information, personnel, and training)?	• Has the diagnostic strategy (dependency charts, logic diagrams) been documented?

Table 7.6 Inherent Testability Checklist (*Continued*)

Mechanical systems condition monitoring (MSCM) checklist		
• Have MSCM and battle damage monitoring functions been integrated with other performance monitoring functions?	• Are M_P monitoring functions (oil analysis, gear box cracks) in place?	• Have scheduled maintenance procedures been established?
Sensors checklist		
• Are pressure sensors placed very close to pressure-sensing points to obtain wideband dynamic data?	• Has the selection of sensors taken into account the environmental conditions under which they will operate?	• Has the thermal lag between the test media and sensing elements been considered? • Have procedures for calibration of sensing devices been established?
Test requirements checklist		
• For each item, does the planned degree of testability support the level of repair, test mix, and degree of automation decisions?	• For each maintenance level, has a decision been made for each item on how BIT, ATE, and general purpose electronic test equipment will support FD&FI?	• Is the planned degree of test automation consistent with the capabilities of the maintenance technician? • Has a "level of repair analysis" been accomplished?
Test data checklist		
• Do state diagrams for sequential circuits identify invalid sequences and indeterminate outputs? • For computer-assisted test generation, is the available software sufficient in terms of program capacity, fault modeling, component libraries, and postprocessing of test response data? • Is the tolerance band known for each signal?	• If a computer-aided design system is used for design, does the computer-aided design database effectively support the test generation and test evaluation processes? • Are testability features included by the system designer documented in the test requirement document in terms of purpose and rationale for the benefit of the test designer?	• For large-scale integrated circuits (ICs) used in the design, are data available to accurately model the circuits and generate high-confidence tests? • Are test diagrams included for each major test? Is the diagram limited to a small number of sheets? Are inter-sheet connections clearly marked?

Table 7.7   Measures of Testability

Measure	Description
Fraction of faults detectable (FFD)	FFD = FD/FA, where FA = total number of actual faults occurring over time and FD = no. of actual failures correctly identified using defined means
Fraction of faults isolatable (FFI) (also called fault resolution)	$$FFL_L = \left(\frac{100}{\lambda_d}\right)\sum_{i=1}^{N} X_i \sum_{j=1}^{M_i} \lambda_{ij}$$ where: $X_i = 1$ if $M_i \leq 1$; 0 otherwise N = number of unique test responses L = number of mudules isolated to (i.e., ambiguity group size) i = signature index M_i = number of modules listed in signature i j = module index within signature λ_{ij} = FR for jth module for failures having signature i λ_d = overall FR of detected failures = $$\sum_{i=1}^{N} \sum_{j=1}^{M_i} \lambda_{ij}$$
Fault isolation time	Derived from the MTTR. Mean fault isolation time = mean (repair time – [preparation time + disassembly time + interchange time + reassembly time + alignment time + vertification time])
False alarm rate (FAR)	FAR = number of false alarms/total number of faults detected
Maximum ambiguity group size	The largest number of items (modules, subassemblies, etc.) among which the diagnostics cannot distinguish a fault (i.e., the diagnostics can isolate the fault to this size ambiguity group but no lower)

Note that the first two measures in Table 7.7 are interrelated in that before you can isolate a fault, you must first detect it. Therefore, a testability analysis program is designed to analyze the effectiveness of the detection scheme and then to analyze the effectiveness of the isolation scheme. For complex designs, the analysis of testability often requires the use of testability design and analysis tools that provide information on FD&FI, for a given diagnostic approach, or diagnostic capability.

False alarms (in which a failure is "detected" even though none occurred) is a problem related to both testability and a system's diagnostic design. Manifesting themselves in varying degrees in avionics and other types of equipment, false alarms drain maintenance resources and reduce a system's mission readiness. CNDs and RTOKs are the two most common symptoms of false alarms.

False alarms occur for many and varied reasons, including external environmental factors (temperature, humidity, vibration, etc.), design of diagnostics, equipment degradation due to age, design tolerance factors, maintenance-induced factors (e.g., connectors, wire handling, etc.), or combinations of these factors. External environmental factors may cause failure of avionics or other equipment that do not fail under ambient conditions. These factors are believed to be a leading cause of false alarms. When the environmental conditions are removed, the "failure" cannot be found. One solution to the false alarm problem is to use a stress measurement device to record the environmental stresses before, during, and after a system anomaly.

Subsequent diagnosis can use the data to determine what occurred and whether any action (maintenance, modifications, etc.) is needed.

7.5.2 Diagnostics

Defining and developing a product's diagnostic capability depends on a number of factors such as

- The product's performance and usage requirements
- Maintenance support requirements (e.g., levels of maintenance)
- Technology available to improve diagnostics in terms of test effectiveness; reduce the need for test equipment, test manuals, personnel, training, and skill levels; and reduce cost
- The amount of testability designed into the product
- Previously known diagnostic problems on similar systems

Each factor influences the choice of an approach to detecting and isolating faults. As mentioned earlier, BIT is one approach to developing a diagnostics capability. Other approaches may include the use of automatic or semiautomatic test equipment, manual testing using bench-top test equipment, or visual inspection procedures. In nearly all cases, some combination of these approaches is needed. In all cases, trade-offs are required among system performance, cost, and test effectiveness.

Designers and managers must remember that the effectiveness of the diagnostic capability, and the cost of development, is greatly influenced by the amount of testability that has been designed into the system. A lack of test points available to external test equipment, for example, may adversely affect the ability to isolate failures to smaller ambiguity group sizes. The result is higher costs to locate the failure to a single replaceable item. The cost of test development may also increase. BIT design should be supported by the results of an FMEA. An FMEA should be used to define those failures that are critical to system performance and to identify when the effects of a failure can be detected using BIT. Without such information, BIT tests can be developed based only on the test engineer's knowledge of how the system works and not on whether a test needs to be developed for a particular fault. Finally, BIT must be a part of the product design or the risks and consequences shown in Table 7.8 can ensue.

Table 7.8 Risks and Consequences of Not Making BIT as a Part of Product Design

Risks	Consequences
BIT is designed independently of the product	BIT fails to support operational and maintenance needs
BIT is designed after the fact	MTBF of the BIT is less than that of the product
Production personnel are not consulted on BIT	BIT is ineffective in the factory

7.6 Interfaces and Connections

A problematic area in the design for sustainability is interfaces and connections (I&Cs). Without I&Cs, it would be impossible to remove or perform maintenance on individual items. But in disconnecting and reconnecting items, failures can be induced by mismating parts, cross-threading connectors, damaging interface devices (IDs), and so on. Disconnecting and reconnecting items also accounts for much of the time needed to remove and replace items. Finally, in the case of high-voltage electrical or high-pressure hydraulic and pneumatic connections, damage to the item or injury to personnel can result if precautions are not taken.

The risk and the time associated with connecting and disconnecting items can be minimized through proper design for sustainability. Volume II of MIL-HDBK-470A (DoD 1997) provides hundreds of maintenance design guidelines, many of which are related to I&Cs. Table 7.9 summarizes some of the relevant I&C guidelines. Section 7.10 provides a more complete list.

BIT false alarms, CNDs, and RTOKs are typical interface problems that plague many complex products. The cause can often be traced back to poor or incorrect connections. Problems with electronic units can often be fixed by reseating connectors. Connectors can vibrate loose, if not secured by a positive locking device, pins can oxidize, and dirt or other foreign matter can interfere with proper operation. So, ensuring that the proper types of connectors are used, ensuring that connectors can be easily accessed, and developing the proper procedures and training are important actions for the sustainability engineer during design.

7.7 Safety and Induced Failures

Maintaining a product can pose safety risks or result in induced failures. An induced failure is one caused by human error or misuse. Unsafe conditions also can result from human error and misuse. Mislabeling or lack of labeling, poorly written instructions, omission of warnings, inappropriate choices of displays and controls, and so on can also lead to damaged or failed equipment and to injury or death of operators or sustainers. Safety is always important in designing a product. The potential for hazardous conditions, of course, depends on the nature of the product and its intended use. Safety requirements come from a variety of sources:

- The customer
- Building codes
- Government agencies
- Industry standards

Table 7.9   Design Guidelines for I&Cs

• Use integral locking mechanisms and visual indications that show that connectors are properly seated and locked
• Use keying or asymmetrically shaped connectors to ensure proper alignment
• Use corrosion-resistant materials for connectors to reduce or eliminate the need for scheduled inspections or corrosion prevention measures
• Locate and position electrical connectors such that all pin identification for either half can be easily seen
• Provide separation between grouped connectors to allow make or break of any connection
• Use quick disconnects to simplify replacement
• Design electrical connectors so that plugs are cold and receptacles are hot
• Use positive locking, quick disconnect connectors to save labor hours, prevent foreign object damage (FOD), and decrease the chance of personal injury
• Use fiber-optic technologies rather than conventional interconnect concepts to reduce the number of interconnects/interfaces, reduce manufacturing and ownership costs, and significantly improve reliability and maintenance
• Standardize connector and wire types to improve testability and logistic support. Keep the number of "different" standard connectors to a minimum. Use the same connector type keyed differently where possible
• Use torque-set or torque-limiting mechanical connections to prevent failures due to over-torque
• Avoid using cotter pins, safety wire, safety clips, and similar devices to prevent maintenance-induced events leading to ground vehicle accidents or loss of air vehicles
• Locate, position, and orient connectors to prevent the need for sequential installation or removal
• In instances where connector interfaces cannot or are not keyed for a specific orientation, all identification, markings, cautions, and directions should be placed 360° around the interface
• Design interface connectors so that a distinct action is required by an individual to make a disconnection
• Design mating items so they cannot be installed improperly or backward

In the United States, an independent regulatory agency, the Consumer Product Safety Commission (CPSC), is charged with protecting consumers from hazardous products. The CPSC imposes federal regulations only when it believes industry's voluntary efforts are insufficient. In addition, a wide variety of consumer products are tested by Underwriters Laboratory (UL), an independent not-for-profit product safety testing and certification organization.

But product designers should do more than simply ensure that an item meets applicable government standards or can earn the UL Mark. UL approval, for example, is not always an indication that a product is safe. First of all, UL does not always consider factors that can affect the long-term integrity of a product and rarely tests products once they leave the factory. Second, in a number of cases, the CPSC has disagreed with UL's test methods and their findings regarding the safety of a product. In addition, regulations, codes, and UL testing cannot address every aspect of every product, especially products incorporating leading edge technology.

For these and other reasons, an overall system safety program is recommended for all but the most innocuous of products. The principal objective of a system safety program is to ensure that safety, consistent with operational and functional requirements, is designed into systems, subsystems, equipment and facilities, and their interfaces. Such a program helps ensure that the designer will thoroughly consider safety in all design trades and design for safe operation, safe maintenance, and some tolerance to human error.

MIL-STD-882C, System Safety Program Requirements (DoD 1993), has guidelines for developing and implementing a system safety program. This program is sufficiently comprehensive to identify the hazards of a system and to impose design requirements and management controls to prevent mishaps by eliminating hazards or reducing the associated risk to a level acceptable to the managing activity. Four different categories of program elements are addressed in MIL-STD-882C, as shown in Table 7.10. The terms and acronyms are unique to MIL-STD-882C and are therefore included in Table 7.11 to aid the reader.

A system safety program needs to be tailored to the product and program requirements. The requirements of MIL-STD-882C can be tailored by selecting only those elements applicable to a given situation. Table 7.12 provides some guidance in selecting safety tasks.

Table 7.10   MIL-STD-882C Safety Program Elements

Category	Category Description	Elements
Program management and control	Those activities primarily related to management responsibilities dealing with the safety of the program and less to the technical details involved	System safety program A basic system safety program consists of the following safety related elements
		• *System safety program plan.* This plan describes in detail those elements and activities of safety system management and system safety engineering required to identify, evaluate, and eliminate hazards, or reduce the associated risk to a level acceptable to the managing activity throughout the system life cycle. It normally includes a description of the planned methods to be used to implement a system safety program plan, including organizational responsibilities, resources, methods of accomplishment, milestones, depth of effort, and integration with other program engineering and management activities and related systems.

Table 7.10   MIL-STD-882C Safety Program Elements (*Continued*)

Category	Category Description	Elements
		• *Integration/management of associate contractors, subcontractors, and architect and engineering firms.* This element consists of appropriate management surveillance procedures to ensure uniform system safety requirements are developed.
		• *System safety program reviews/audits.* This element is a forum for reviewing the system safety program, to periodically report the status of the system safety program, and, when needed, to support special requirements, such as certifications and first-flight readiness reviews.
		• *System safety group (SSG)/system safety working group (SSWG) support.* This element is a forum for suppliers and vendors to support SSGs and SSWGs established in accordance with government regulations or as otherwise defined by the integrating supplier.
		• *Hazard tracking and risk resolution.* This element is a single closed-loop hazard tracking system to document and track hazards from identification until the hazard is eliminated or the associated risk is reduced to an acceptable level.
		• *System safety progress summary.* This element consists of periodic progress reports summarizing the pertinent system safety management and engineering activity that occurred during the reporting period.
Design and integration	Activities that focus on the identification, evaluation, prevention, detection, and correction or reduction of the associated risk of safety hazards by the use of specific technical procedures	• *Preliminary hazard list (PHL).* Identifies any especially hazardous areas for added management emphasis. The PHL should be developed very early in the development phase of an item.

(*Continued*)

Table 7.10 MIL-STD-882C Safety Program Elements (*Continued*)

Category	Category Description	Elements
		• *Preliminary hazard analysis (PHA).* Identifies safety-critical areas, evaluate hazards, and identifies the safety design criteria to be used.
		• *Safety requirements/criteria analysis (SRCA).* Relates the hazards identified to the system design and identifies or develops design requirements to eliminate or reduce the risk of the hazards to an acceptable level. The SRCA is based on the PHL or PHA, if available. The SRCA is also used to incorporate design requirements that are safety related but not tied to a specific hazard.
		• *Subsystem hazard analysis (SSHA).* Identifies hazards associated with design of subsystems including component failure modes, critical human error inputs, and hazards resulting from functional relationships between components and equipments comprising each subsystem.
		• *System hazard analysis (SHA).* Documents the primary safety problem areas of the total system design including potential safety critical human errors.
		• *Operating and support hazard analysis (O&SHA).* Identifies associated hazards and recommends alternatives that may be used during all phases of intended system use.
		• *Occupational health hazard assessment (OHHA).* Identifies human health hazards and proposes protective measures to reduce the associated risks to levels acceptable to the managing activity.
Design evaluation	Activities that focus on risk assessment and the safety aspects of tests and evaluations of the system and the possible introduction of new safety hazards resulting from changes	• *Safety assessment.* A comprehensive evaluation of the mishap risk that is being assumed prior to the test or operation of a system or at the contract completion.

Table 7.10   MIL-STD-882C Safety Program Elements (*Continued*)

Category	Category Description	Elements
		• *Test and evaluation safety.* Ensures that safety is considered (and safety responsibility assigned) in test and evaluation to provide existing analysis reports and other safety data and to respond to all safety requirements necessary for testing in-house, at other supplier facilities, and at government ranges, centers, or laboratories.
		• *Safety review of engineering change proposals (ECPs) and requests for deviation/waiver.* Performing and documenting the analyses of ECPs and requests for deviation/waiver to determine the safety impact, if any, upon the system.
Compliance and verification	Activities directly related to the actual verification or demonstration that all legal and contractual safety requirements have been compiled with	• *Safety verification.* Verification is conducted to verify compliance with safety requirements by defining and performing tests and demonstrations or other verification methods on safety-critical hardware, software, and procedures.
		• *Safety compliance assessment.* Performing and documenting a safety compliance assessment to verify compliance with all military, federal, national, and industry codes imposed contractually or by law. This element is intended to ensure the safe design of a system and to comprehensively evaluate the safety risk that is being assumed prior to any test or operation of a system or at the completion of the contract.
		• *Explosive hazard classification (EHC) and characteristics data.* Ensures the availability of tests and procedures needed to assign an EHC to new or modified ammunition, explosives (including solid propellants), and devices containing explosives and to develop hazard characteristics data for these items.
		• *Explosive ordnance disposal source data.* Ensures that the following resources are available as needed: source data, explosive ordnance disposal procedures, recommended "render safe" procedures, and test items for new or modified weapons systems, explosive ordnance items, and aircraft systems.

Table 7.11 Definition of Safety Terms and Acronyms

Term	Definition
Fail safe	A design feature that either ensures that the system remains safe or in the event of a failure, forces the system to revert to a state which will not cause a mishap.
Hazard	A condition that is prerequisite to a mishap.
Hazard probability	The aggregate probability of occurrence of the individual events that create a specific hazard.
Hazardous material	Anything that due to its chemical, physical, or biological nature causes safety, public health, or environmental concerns that result in an elevated level of effort to manage.
Mishap	An unplanned event or series of events that result in death, injury, occupational illness, or damage to or loss of equipment or property or damage to the environment. An accident.
Risk	An expression of the possibility of a mishap in terms of hazard severity and hazard probability.
Risk assessment	A comprehensive evaluation of the risk and its associated impact.
Safety	Freedom from those conditions that can cause death, injury, occupational illness, or damage to or loss of equipment or property or damage to the environment.
Safety-critical	A term applied to a condition, event, operation, process, or item of whose proper recognition, control, performance, or tolerance is essential to safe operation or use; e.g., safety-critical function, safety-critical path, or safety-critical component.
Safety-critical computer software components	Those computer software components and units whose errors can result in a potential hazard, or loss of predictability or control of a system.
System safety	The application of engineering and management principles, criteria, and techniques to optimize safety within the constraints of operational effectiveness, time, and cost throughout all phases of the system life cycle.

Table 7.12   Guide to Selecting Safety Tasks

Task	Type of Task	Program Phase
System safety program plan	Management	Generally applicable in all phases of a product's life.
Integration/management of associate contractors, subcontractors, and architect and engineering firms	Management	Most applicable during design and production. May be applicable during operation if other firms involved in engineering changes, maintenance, and so on.
System safety program reviews/audits	Management	Most applicable during design and production. Periodic reviews during operational life may be appropriate.
SSG/SSWG support	Management	Most applicable during design and production. May be applicable during operational use, depending on the nature of the product.
Hazard tracking and risk resolution	Management	Most applicable during design and production. May be applicable during operational use, depending on the nature of the product.
PHL	Engineering	Most applicable during design.
PHA	Engineering	Most applicable during design.
SRCA	Engineering	Most applicable during design (develop prior to start of design).
SSHA	Engineering	Most applicable during design.
SHA	Engineering	Most applicable during design. May be applicable during operational use, depending on the nature of the product.
O&SHA	Engineering	Most applicable during design. May be applicable during operational use, depending on the nature of the product.
OHHA	Engineering	Generally applicable in all phases of a product's life.

(Continued)

Table 7.12 Guide to Selecting Safety Tasks (*Continued*)

Task	Type of Task	Program Phase
Safety assessment	Engineering	May be applicable in all phases of a product's life.
Test and evaluation safety	Management	Most applicable during design and production.
Safety review of ECPs and requests for deviation/waiver	Management	Most applicable during design and production.
Safety verification	Engineering	Most applicable during design. May be applicable during production.
Safety compliance assessment	Management	May be applicable in all phases of a product's life.

7.8 Standardization and Interchangeability

Standardization and interchangeability are important, interrelated, sustainability design factors. Interchangeability is one of the principal means by which standardization is achieved. Good examples of the close relationship between standardization and interchangeability are the standard size base for incandescent lamps and the standard size male plug for electrical appliances. Standardization is a design feature for restricting the feasible variety of items that will meet the hardware requirements. Standardization includes not only parts but also engineering terms, principles, practices, materials, processes, and software. Standardization encourages the use of common items. It is important that sustainability engineers strive for the design of assemblies and components that are physically and functionally interchangeable with other similar assemblies and components of the system. Standardization design will reduce the need for expensive support facilities at all levels of maintenance. Standardization, a major objective of sustainability, helps achieve the following goals:

- Minimizing both the acquisition and support costs of a system
- Increasing the availability of mission-essential items
- Reducing training requirements both in number of personnel and the level of skill required
- Reducing inventories of repair parts and their associated documentation

Despite the advantages offered by standardization, a system should not necessarily be built around a standard item—particularly if the standard item does not meet the required performance and has a record of poor reliability or costly maintenance—or the standard item may satisfy a safety requirement in most environments but not in the unusual environment for which it is being considered. Technological advances may also dictate the development of new material or provide a superior product to replace an existing one.

Interchangeability is the ability to exchange parts or assemblies between similar equipment, without having to alter or physically change the item. This is an extremely important life cycle cost design requirement. Total interchangeability exists when two or more items are physically and functionally interchangeable in all possible applications; that is, when the items are capable of full, mutual substitution in all directions. *Functional* interchangeability is attained when an item, regardless of its physical specifications, can perform the specific functions of another item. *Physical* interchangeability exists when two or more parts or units made to the same specification can be mounted, connected, and used effectively in the same position in an assembly or system. The two broad classes of interchangeability are

- *Universal interchangeable*—Items that are required to be interchangeable in the field even though manufactured by different facilities.
- *Local interchangeable*—Items that are interchangeable with other components made by the same facility but not necessarily interchangeable with those made by other facilities. This may result from different sets of measurement units employed in their design and manufacture.

7.9 Design Tools

To assist in the design of sustainable products, various types of design tools have been developed. These tools can be categorized as analytical, mock-ups, simulation and VR, handbooks and other reference documents, expert systems, and neural networks. These categories are discussed in the following subsections.

7.9.1 Analytical Tools

Most analytical tools available today to assist the designer in designing a sustainable product are related to modeling the human being. Since the late 1970s, more than 50 different human models have been developed. Electronic representations of human forms are used to simulate equipment assembly, operation, and maintenance during the design process. They allow engineers to identify and resolve human interface problems before hardware is built. Early human models used only hands or arms to check clearances for tool manipulation. Today's models create whole-body representations using a basic "link" system resembling a human skeleton to enable posturing of the model within the work environment.

Although a large variety of human models have emerged to support the design effort, few experts agree on how the human form should be configured, what constitutes valid data, what are acceptable levels of accuracy, and what software and communications standards should be adopted. Earlier human models focused on the physical or ergonomic aspects of human/machine interaction. The focus today is on integrating this information with visual and cognitive information processing requirements and with human modeling simulation to create an integrated modeling technology. This provides additional realism not only through accurate replication of human anthropometry, biomechanics and movement but also in simulating purposeful and logical behaviors in response to external stimuli and workload. The purpose of all these models is to integrate human performance analysis with computer-aided design to provide the design team with a high degree of visualization of human performance capabilities and limitations with respect to the product design. Through integration of graphic human models with computer-aided design product models, "rapid prototyping" of human/product simulations or their results can be passed back to equipment designers for resolution of identified problems.

Designing equipment that is easy to operate, assemble, and sustain is often hindered by poor communications between the design team and personnel familiar with the operation, assembly, or maintenance of similar or existing equipment. Improved communication among integrated product development (IPD) team members can be accomplished by simulating equipment operation, assembly, and maintenance using human modeling technology. Human models combine animated 3D human mannequin geometry with equipment geometry to "walk through" designs so that problems can be solved early in the design process. They help to ensure that human-centered design information is readily and accurately documented and preserved to aid in human resources and related logistics planning requirements for system support. The models are used first to influence a product's design for supportability, and then to document the product requirements for human and logistics resources. Another major objective is the development and implementation of design evaluation technology for performance of "design checking" and prescriptive human performance information for recommending corrective action to equipment designers to conform to human performance requirements. The term "human model" in this context refers to the 3D, computer graphic representation of a human form for analysis purposes. It does not address human performance models that are independent of the geometric aspects of the human body (e.g., human error models).

Human modeling systems can support both the design requirements definition and design evaluation when concepts are only represented in 3D computerized form. The human design requirement definition can be accomplished using reach or vision envelopes that describe the minimum conditions a designer must satisfy for physical or visual access. Design evaluations, on the other hand, usually focus on critical task segments in which the human/equipment interface is tested for compliance with stated design requirements and freedom from "won't-fit" or "won't-work" conditions.

Using human modeling in computer-aided design provides important benefits including:

- Eliminating most physical development fixtures by performing evaluations electronically
- Reducing design costs by enabling the IPD team to prototype more rapidly and test a design among themselves
- Avoiding costly design fixes later in the program by considering human factors requirements early in the design effort
- Improving customer communications during product development by using compelling animated graphics to review and confirm equipment function

Application of human modeling technology is likely to impact how engineers design, build, and test products in the future. Those who are responsible for manufacturing planning, tool design, or sustainability engineering will be able to communicate with structural and systems engineering effectively to illustrate assembly or maintenance problems associated with new designs. It is expected that human model applications will spread beyond what is traditionally called engineering and be used by various IPD team members from factory-built units to product support groups.

A variety of suppliers have human modeling software programs. Unfortunately, they have created models that are very different in functionality and in user interface, and in the underlying data driving the mannequins. This diversity has created not only models that look and behave differently but also models that produce distressingly different results when performing the same engineering analysis. For these reasons, the Society of Automotive Engineers (SAE) has formed an ad hoc committee to formulate standards to promote the orderly growth of this technology. The SAE Human Modeling Technology committee has established three major subcommittee activities: user requirements, human model definition, and software standards. A fourth subcommittee activity is being considered on the topic of human performance models that would address human error prediction, human workload, and task time estimation.

7.9.2 Mock-ups

As products became more complex, conceptualizing shape and fit from a 2D drawing became increasingly difficult. As a preproduction version or prototype of the product was constructed, the consequences of inaccurate conceptualization evidenced itself in structural components that would not properly mate, hydraulic lines that did not connect as planned, and so on.

To solve this problem, engineers began using mock-ups of critical sections of the product, often the entire product. Constructed of inexpensive materials, mock-ups are non-functioning, dimensionally accurate, and often full-scale models of the product. Mock-ups allow the fit and mating of components to be checked before constructing any functional hardware. Although being supplanted by computer-aided design and VR, mock-ups are still useful due to their simplicity and relatively low cost.*

7.9.3 Simulation and VR

Simulation, as used here, is a method for representing or approximating an object, event, or environment. VR is a new technology that has been defined as the total or near-total immersion of an observer in a 3D, synthetic environment in which the observer interacts with the environment. Simulation is frequently used to evaluate the maintenance characteristics of a design. Simulation can include physical mock-ups, computer models, or mathematical models. VR is the newest and most technologically advanced form of simulation. Jaron Lanier, founder of VPL Research, initially coined the term "virtual reality" in 1989 (Lanier 1992a and Lanier 1992b). Other related terms include "artificial reality" (Krueger 1991) and "cyberspace." William Gibson coined the term "cyberspace" in his short story Burning Chrome (Gibson 2003). More recent related terms are "virtual worlds" and "virtual environments."

Today, the term "virtual reality" is used in a variety of ways and often in a confusing and misleading manner. Originally, the term referred to "immersive virtual reality." In immersive VR, the user becomes fully immersed in an artificial, 3D world completely generated by a computer. The unique characteristics of immersive VR can be summarized as follows:

■ Head-referenced viewing provides a natural interface for the navigation in 3D space and allows for look-around, walk-around, and flythrough capabilities in virtual environments (VEs).
■ Stereoscopic viewing enhances the perception of depth and the sense of space.
■ The virtual world is presented in full scale and relates properly to the human size.
■ Realistic interactions with virtual objects via data glove and similar devices allow for manipulation, operation, and control of virtual worlds.
■ The convincing illusion of being fully immersed in an artificial world can be enhanced by auditory, haptic (manipulators used to provide force or tactile feedback to humans interacting with virtual or remote environments), and other nonvisual technologies.
■ Networked applications allow for shared VEs.

Currently, the term "virtual reality" is also used for applications that are not fully immersive. The boundaries are becoming blurred, but all variations of VR will be important in the future.

---

* Very sophisticated mock-ups have been constructed. For example, an expensive, full-scale, left half (bisected down the longitudinal axis) of the B-1A bomber was built by Rockwell. The wing was a swept wing. Normally, mock-ups are relatively simple and inexpensive.

These include mouse-controlled navigation through a 3D environment on a graphics monitor, stereo viewing from the monitor via stereo glasses, stereo projection systems, and others. Internet "surfers" are familiar with Apple's QuickTime VR, in which photographs are used to model 3D worlds and provide pseudo look-arounds and walk-throughs of a landscape or object. In general, VR is a method of simulating an environment that

- Is too dangerous for an observer
- Lacks elements, such as an aircraft or other item of study
- Does not exist
- Is not accessible

Three different types of VR have been developed. Although not all these types exactly fit the definition of VR, they do represent variations of the same basic technology. The three types of VR are

- Telepresence, in which observers perceive and interact with a distant environment
- Augmented reality, a combination of a real and synthetic environments, in which a real environment is annotated or augmented with additional details or elements
- VR, in which a synthetic environment is created for the observer (immersive)

Telepresence is used when the environment is dangerous or inaccessible. An example of the former case is disarming a bomb, a hazardous task for a person, even if wearing a helmet, body armor, and other safety devices. A robot equipped with telepresence can be operated by an operator located at a safe distance from the bomb with almost the same feeling of "being there" as if he or she were actually at the site of the bomb. An example of the latter case is controlling robots in earth orbit from a ground station on earth.

In augmented reality, information and details are "added" to the real world, providing guidance, instructions, and so on, to help an observer's understanding or performance. Three examples follow. First, in an augmented reality approach to video-conferencing, a 3D image of a new product still in design could be generated from computer-aided design files and "placed" on the desk or table in front of each conferee. The nomenclature of parts could be "superimposed" on them and would "follow" them no matter how they were moved within the range of the video camera. Another example of the use of augmented reality is the superimposing of the proper locations for drilling holes in an aircraft skin with other information, such as proper hole size. Finally, surfaces or features of an item that are physically occluded can be displayed as an overlay so that an observer can "see" them without disassembling the item.

In a total VR environment, nothing (or very little) but the user is "real." Objects and their physical characteristics, the physical environment, the time of day, and so on, are all generated by a computer and displayed to the user, usually through goggles or a head-mounted display (HMD). The user "sees" and can interact with objects in the environment. Input devices are needed to allow the user to navigate through and interact with the VE. Such devices include data gloves, joysticks, and hand-held wands. In addition, directional sound, tactile and force feedback devices, voice recognition, and other technologies are now used to enrich the immersive experience and to create more "sensualized" interfaces. The HMD was the first device that provided an immersive VR experience. Evans and Sutherland demonstrated a head-mounted stereo display in 1965. Twenty four years later, the first commercially available HMD, the EyePhone, came from VPL Research. A typical HMD houses two miniature display screens and an optical system. The optical system channels the images from the screens to the eyes, thereby presenting a stereo view of a virtual world. A motion tracker continuously measures the position and orientation of the user's head and allows the image-generating computer to adjust the scene representation to the current view. Consequently, the viewer can walk through and observe the surrounding VE. HMDs are often uncomfortable, intrude on the VR experience, and cannot be worn for extended periods of time. To overcome these problems, alternative concepts for immersive viewing of VEs were developed. Two of these alternatives are BOOM (Binocular Omni-Orientation Monitor) and CAVE (Cave Automatic Virtual Environment).

The BOOM from Fakespace is a head-coupled stereoscopic display device. A box houses screens and optical system and is attached to a multilink arm. By looking into the box through two holes, the observer sees the virtual world, and can guide the box to any position within the operational volume of the device. Sensors in the links of the arm that holds the box track the observer's head.

Researchers at the University of Illinois at Chicago developed the CAVE. CAVE immerses observers in an environment created by projecting stereo images on the walls and floor of a room-sized cube. The cube can accommodate several persons who can walk freely inside the CAVE. Observers wear lightweight stereo glasses. A system that tracks the heads of the observers continuously adjusts the stereo projection to the current position of the leading viewer.

VR has definite applications for designing sustainable equipment. For example, based on computer-aided design data files, a virtual copy of the product can be "produced." The maintenance engineer can then enter a VE in which maintenance can be "performed" on the product. The accessibility of components, whether an item fits in an allocated space, and the approximate time required to perform specific maintenance actions all can be evaluated using VR. Virtual copies of support equipment, such as dollies and lifting devices, can be evaluated by "performing" maintenance activities with them. VR maintenance aids could allow technicians to view virtual information panels "superimposed" (using augmented reality techniques) on the actual equipment. In general, the maintenance engineer can use VR to analyze:

- Reachability and access
- Field of view
- Integrated displays
- Attention skills
- Powered hand tools evaluation
- Posture
- Energy expenditure
- Human–machine interface
- Stressor effects on human performance
- Lifting guidelines
- Activity timing
- Cognitive skills, decision making
- Ergonomic analysis of maintenance workstations

In addition to designing for sustainability, VR has many potential training applications. Maintenance and manufacturing procedures, especially procedures that are seldom performed or are difficult to teach using conventional approaches, can be taught using VR. VR could also be used to train individuals in performing hazardous procedures, disposing hazardous materials, or performing life-threatening procedures. For example, surgeons can now "perform" operations without actually using any physical tools or a live patient. As has been the case with previous new technologies, the possible uses of VR cannot be fully appreciated or anticipated. As VR matures, the applications related to design for maintenance will certainly increase in number and in fidelity.

7.9.4 Handbooks and Other Reference Documents

Hard copy handbooks and similar reference documents are considered by some to be passé in today's world of computer-based design and VR. Nonetheless, much of the knowledge gained over the years as well as new information are documented in handbooks, manuals, data books, and so on. Guidance, rules-of-thumb, lessons learned, and similar information, together with explanations, make handbooks and other reference documents important resources for the engineer. Some older documents are being "digitized" for entry into computer databases making it easier to search and update the information. Nearly all new documents are created in digitized form.

7.9.5 Artificial Intelligence

Various forms of artificial intelligence (AI) are beginning to be used in the field of sustainability, particularly in the design of diagnostic tools. Individual AI techniques include expert systems, fuzzy logic, and neural networks. The structural basis and respective advantages and disadvantages for each of these techniques are summarized in Table 7.13.

7.9.5.1 Expert Systems

Expert systems are becoming an important sustainability tool, especially as industry downsizes with a concomitant loss of individual company "maintenance experts." Expert systems are used to "capture" and codify the knowledge of one or more experts in a given field or area of study and to make this knowledge available to nonexperts.

For sustainability, a major use of expert systems is in diagnostic tools. The diagnostic capability of expert systems has been successfully demonstrated in both the medical and maintenance fields. Whether the problem is to identify a specific illness afflicting a patient or to identify the cause of an observed system or equipment failure, expert systems have proved to be efficient and effective.

Another potential use of expert systems comes as a result of "downsizing" and the use of integrated product design teams (IPDTs). As companies have downsized, the number of individuals employed as maintenance engineers has decreased. Many years of corporate experience are being lost and the few remaining maintenance engineers are spread thin. Where IPDTs are used, an engineer who may know very little about sustainability may very well be given the responsibility for that aspect of design. Expert systems can help "replace" the maintenance engineer and assist those given the responsibility for maintenance design. As part of a computer-aided design system, an expert system could guide the designer in equipment placement, selection of fasteners, design of access panels and hatches, and so on.

Two distinct types of expert systems are used: rule based and model based.

7.9.5.1.1 Rule-Based Expert Systems

Rule-based expert systems operate through a set of "IF...THEN" rules processed by an underlying "inference engine." A typical rule-based expert system is composed of four major elements: the inference engine, a knowledge base, a user interface, and an explanation facility. The *inference engine* is that part of the expert system that performs the reasoning. It is analogous to the raw intelligence of a human expert. Many different forms of inference engines exist, but all are designed to perform the same task; that is, to examine the current facts and use available rules to generate new facts. The *knowledge base* is where the information resides within the expert system. It consists of two distinct parts: the rule base "IF <condition> THEN" and the fact base containing simple

Table 7.13 Comparison of AI Techniques

Technique	Basis	Advantages	Disadvantages	Application to Maintenance
Rule-based expert system	"IF...THEN" logic	Audit trail possible.	Difficult to capture "intuitional" rules.	Expert systems for design and for fault diagnosis. Based on knowledge of human "experts."
Model-based expert system	Functional system model	Specific models are available.	Requires the development of a unique model for each problem.	Expert systems for design and for fault diagnosis. Adds model of problem to expert knowledge.
Fuzzy logic	Converts discrete logic into continuous values	Eliminates stepwise approximations. Easy to "Fine Tune."	Each individual output must be "defuzzified."	Expert systems for design and for fault diagnosis. Allows for nondiscrete I/O.
Neural network	Numerous interconnected simple processing modes	Trained by example. Insensitive to "Noise." Able to capture "intuitional" rules.	No theoretical understanding. No practical guidelines. No audit trail possible.	Expert systems for design and for fault diagnosis. Can be "trained" by nonexperts. Can capture intuitional and procedural rules.

statements about the condition of the world as it is applicable to the problem under study. The *user interface* enables the expert system and the user to communicate. The exact form of this interface depends on the intended audience for the expert system. The *explanation facility* presents the user with the expert system's justification for its conclusions (i.e., an audit trail) as necessary.

A typical expert system initially partitions the problem by applying a broad set of inference rules to an initial set of data describing the problem or the symptoms. Each of these inference rules will take the inference engine to a further data acquisition stage (typically another, more directed, questionnaire) or the establishment of a new fact. This process of a directed search with additional data gathering continues until the expert system has reached a leaf node in the resulting decision tree. Some inference engines may resolve an ambiguity, when several inference rules evaluate as

TRUE to a given data set, by selecting the one with the highest associated weighting or confidence factor; others may use a different approach (e.g., fuzzy logic). The rules in the knowledge base, that portion which drives any expert system, are painstakingly constructed by an expert systems specialist interrogating the knowledge expert and subsequently codifying the often imprecise descriptions of their thinking processes into inference rules, possibly with numerical limits. For example, a rule for a medical diagnostic expert system may be stated as follows:

> IF heart rate > 100 beats per minute AND body temperature >101°F THEN recommend that patient be placed in an ice bath.

The fact portion of the knowledge base would simply record the patient's heart rate and temperature. A general approach for the physical development of a maintenance expert system may be

1. Design expert system structure including user interface
2. Knowledge acquisition
3. Rules codification
4. System validation
5. Growth or system enhancement

It may be difficult to capture all of an expert's knowledge in an expert system knowledge base because the expertise is encoded as a causal relationship. "Rational" knowledge, where the solution can be described analytically, is comparatively straightforward to codify into inference rules. "Semirational" knowledge, where the expert can specify suitable ranges for conditions, but cannot (easily) defend the choice of these ranges, is more difficult. This process may take some detective work by the expert system specialist. Unfortunately, however, much of what "makes an expert" occurs at an intuitional or visceral level, where even the expert is unaware of the underlying mechanism behind their decisions and may even be unable to quantify appropriate ranges. This area presents the major challenge and limitation in the design of a rule-based expert system. The following three sections will address some alternative solutions to this problem.

7.9.5.1.2 Model-Based Expert Systems

The second type of expert system—the model-based system—uses a specific functional model to diagnose the observed symptoms and devise a solution to the problem. The knowledge base is usually organized around a functional or representative model of the system, but it is sometimes preferable to use an actual physical model. This model now provides the procedure with a focus of attention directed toward expected goals and guides the process in determining the effects of system/equipment failure symptoms. In the area of testability, a number of detailed models have been developed.

7.9.5.1.3 Fuzzy Logic

Fuzzy logic is essentially an expert system structure tailored to deal with continuous-valued inputs and outputs (I/O) instead of discrete lexical elements. Thus, fuzzy logic can potentially reduce the number of rules required in a system. This is achieved through clever preprocessing of the inputs, where each continuous input value is "fuzzy" or converted from a precise numeric value to a degree-of-membership in a "fuzzy set" as shown in Figure 7.4. Fuzzy logic is attractive because it allows for conflicting "expert opinion," thereby allowing the use of information normally excluded

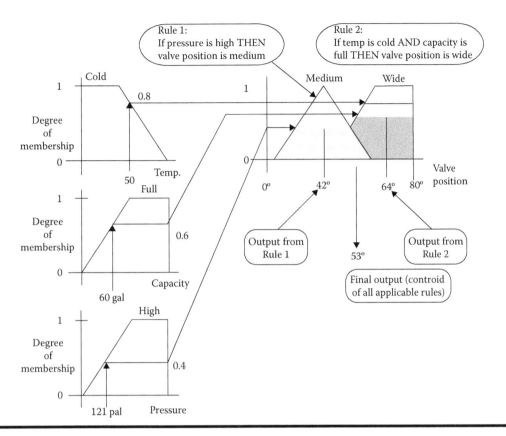

Figure 7.4   Fuzzy logic set membership.

from scientific models. For design, fuzzy logic can be used to define a range of feasible design parameters even when historical data are insufficient to use tractable probability-based approaches.

When an input falls into a region where two or more fuzzy sets overlap, it simply produces a degree-of-membership in each of the overlapping sets. An output term of a fuzzy logic system is itself a fuzzy set, which must converted back to a precise (i.e., "crisp") numeric value. This is done by taking the centroid of the part of the output fuzzy set lying below the degree-of-membership output value. This degree-of-membership can result from a straight mapping of input fuzzy set to output fuzzy set, as shown by Rule 1 in Figure 7.4 or from a logical combination of rules* as used in an expert system (Rule 2 in Figure 7.4). When two or more inference rules trigger on a given output, the "crisp" output is calculated as the centroid of the areas of the contributing rules.

Providing the means for an expert system structure to treat continuous I/O as lexical elements eliminates the stepwise approximation a classical expert system would normally be forced to use in such a situation. This significantly reduces the number of inference rules required and clarifies the program structure. Also, because the mapping between inputs, outputs, and lexical elements is done via simple curve functions, a fuzzy system is easier to "fine tune." Thus, a given fuzzy solution can be taken to other similar domains by rescaling or reshaping the I/O curves while leaving the logical inferences unchanged.

---

* The AND operator selects the smallest degree-of-membership of its operands, whereas the OR operator selects the largest degree-of-membership.

7.9.5.2 Neural Networks

Artificial neural networks consist of a large number of densely interconnected simple processing nodes, each of which produces a nonlinear result of a weighted sum of its inputs (e.g., the output is a binary "1" if the sum exceeds a set threshold). The input stimuli and/or the outputs of other neurons are typically shown in Figure 7.5. Although there are numerous architectures of neural networks, they all work by partitioning the N-dimensional stimulus space into a series of continuous regions and as such, serve as "feature detectors" where the output (1,0) of an output-stage neuron represents the presence or absence of a desired feature. This behavior is especially useful in pattern recognition.

Neural networks, unlike expert systems or fuzzy logic, do not partition the stimulus space based on explicit rules. Rather, they are "trained" with sets of example stimuli and desired outputs. The training procedure gradually adjusts the weighting coefficients on each neuron's input until the global error is minimized. Successive training sets for other stimulus response sets alter the coefficients, but a "memory" of previous training sets remains. Given a sufficient number of training sets, the neural network gradually converges to a stable set.

Neural nets have four significant advantages over expert systems:

1. Although slow to train, neural nets can be trained by someone who is not an expert in the field (the training data sets, however, must be prepared by such an expert), which can translate into time and cost savings.
2. Because the network is trained by example, it can capture the intuitional expertise as well as the procedural aspects.
3. The neural net partitions the stimulus space into contiguous regions, eliminating the gaps, overlaps, and understatement problems inherent in expert systems.
4. Neural networks have been shown to be robust in the face of the noisy data found in nature. They require little or no sensor calibration or special nonlinear quantization schemes.

Several factors, however, mitigate against the use of neural networks. These include

1. The lack of a sound theoretical understanding of neural networks.
2. The absence of practical guidelines for selecting from the multitude of competing architectures often makes the choice one of personal taste.

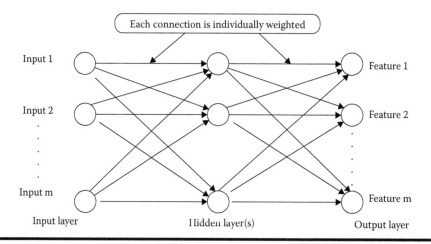

Figure 7.5 Typical neural network configuration.

3. The lack of an established means for determining the correct number of neurons to use in a given architecture for a given problem. Practitioners typically add neurons until they achieve a desired level of network stability.
4. The sensitivity of neural networks to the training data. With too little training, neural nets tend to misidentify stimuli (i.e., mispartition stimuli space). They can also exhibit pattern sensitivity to some data sets. That is to say, the network will not converge to a stable configuration but oscillates between two or more metastable regions.
5. The rapid growth in the number of neurons and divergence that often result from attempts to deal with metastability.
6. The inability of neural networks to provide an audit trail showing how or why a decision was made. This makes them much harder to debug than expert systems and also poses some interesting liability issues.

In summary, neural networks provide several distinct advantages over classical expert systems, most notably, (1) training by example, (2) robust pattern matching in the face of noisy or incomplete data, and (3) the ability to capture an expert's intuitive knowledge. However, they operate "mysteriously," in a field with few landmarks. This makes neural network solutions difficult to develop.

7.10 Design Guidelines

Table 7.14 provides some guidelines for a design for sustainability. These guidelines were developed by Raytheon in 1972 and are listed in Volume II of MIL-HDBK-470A (DoD 1997). Note that even though the Raytheon guidelines were developed years ago, the ones included here are still relevant.

Table 7.14 Design Guidelines

General
• Identify all RUs and assemblies
• Incorporate self-test at the lowest practical economical level of assembly
• Simplify decision making by specifying go/no-go test techniques
• Ensure maintenance procedures call for a logical sequence of tasks

Modularity
• Design modules in uniform sizes and shapes as is practical
• Design modules for complete functions to minimize interconnections and problems of signal tracing and to facilitate troubleshooting
• Design modules to permit operational testing of the modules when removed from the product
• Design low-cost and noncritical modules to be "throw-away"
• Design modules so that they cannot be placed inadvertently in the wrong location
• Use guide pins to facilitate the installation of modules
• Design and locate repairable modules for easy accessibility
• Use quick disconnect devices on modules to facilitate quick and easy removal

(Continued)

Table 7.14   Design Guidelines (*Continued*)

Adjustments and Alignment
• Design equipment to eliminate the need for alignment and adjustments • Design equipment with self-adjusting devices • Design so that disassembly is not required to facilitate adjustments • Alignment and adjustment devices susceptible to vibration or shock should be designed with a positive locking feature • Design so that adjustments can be made by one person • Avoid adjustments where small movements of the adjusting device result in large changes in the affected parameter • Adjustment devices should not be located close to high voltage or hot sections of the equipment

Accessibility
• Design so items can be connected and reconnected without special tools • Design simple connections with few moving parts • Use quick disconnects (self-sealing) in hydraulic/pneumatic systems • Allow adequate space to make connections and disconnections • Design items with safety interlocks to prevent inadvertent disconnect • Design indicators of servicing status of equipment so they are accessible and readable from the ground level for air vehicles and from the operator's position for ground and water vehicles • Equipment bay and compartment structural flanges and stiffeners should be external to the compartment • Mount line replaceable units (LRUs)/subsystems on drawer slides or extender racks to provide easier accessibility during integration, testing, debug, and repair of UUTs • Use a high gloss white to maximize lighting reflectivity, visibility, and rapid leak detection capability in equipment bay and compartment interiors • Provide good access to corrosion-prone structural areas for inspection and treatment • Use transparent windows, quick-opening covers, or openings without any cover to permit quick visual inspections where needed • Use stiffening beads in lieu of flanges to the maximum extent • Locate nutplates or gang channels in accessible areas that do not require extensive disassembly or equipment removal to gain access • Locate identification and modification plates for all major structural assemblies and subassemblies so as to be fully visible and legible when viewed through normal access provisions. Avoid the need to remove equipment or components to view the plates • Provide access to all engine mounts for ground vehicles so that hoisting or crawling under the vehicle is not required • Modularize structural instrument panels, dashboards, and control panels for easy and adequate access to all interfaces, to simplify manufacturing, and to reduce maintenance-induced problems • Provide a sufficient number of compartment fluid drains to ensure fluid drainage

Table 7.14 Design Guidelines (*Continued*)

Human factors (including anthropometric considerations)

- Size structural openings into man-rated fuel cells to enable entry by a 75th percentile male
- In designing the vehicle, system, subsystem, and equipment, attempt to satisfy the personnel spectrum from the 5th percentile female to the 95th percentile male
- Design hinged access doors and panels that can be placed in an opened position so they do not have sharp edges or corners
- Develop decals, placards, and instruction media around an 8th grade reading level and a 10th grade level of comprehension
- Group man–machine interfaces manifold style to enable connect/disconnect in a single action
- Include an integral, highly visible indicator in a man–machine interface connector to denote connector is seated and locked
- Clearly mark all subsystems/LRUs to ease system integration, test, debug, and repair
- Use the English language to report failures rather than alphanumeric codes, lights, indicators, and so on.

Mating, interfaces, and connections

- Use integral locking mechanisms and visual indications that show that connectors are properly seated and locked
- Use keying or asymmetrically shaped connectors to ensure proper alignment
- Use corrosion-resistant materials for connectors to reduce or eliminate the need for scheduled inspections or corrosion prevention measures
- Use positive locking, quick disconnect connectors to save labor hours, prevent FOD, and decrease the chance of personal injury
- Use fiber-optic technologies rather than conventional interconnect concepts to reduce the number of interconnects/interfaces, reduce manufacturing and ownership costs, and significantly improve reliability and maintainability
- Use clamps with torque-set or torque-limiting connections
- Use quick disconnects to simplify replacement
- Use quick-release cables and locate cables to make removal and replacement easy and to avoid having to remove one cable to gain access to another. Provide adequate space for cables, including sleeving and tie-downs, and adequate service loops for ease of assembly/disassembly
- Use torque-set or torque-limiting mechanical connections to prevent failures due to over-torque
- Design mating items so they cannot be installed improperly or backward
- Design interface connectors so that a distinct action is required by an individual to make a disconnection
- Design electrical connectors so that plugs are cold and receptacles are hot.
- Design in-line plumbing connections within a fuel tank or cell so that making/breaking the interface can be done by hand, require no torque, contain integral safety locking mechanisms, and do not require safety wire

(Continued)

Table 7.14   Design Guidelines (*Continued*)

Mating, interfaces, and connections

- Design carry-through bulkheads, major frames, structural ribs, spars, webs, keels, and manufacturing close-outs with constant web thickness to provide flexibility in locating penetration fittings and simplify structural repair. Avoid stepped chemical milling, stepped machining, stepped composite lay-up, and similar manufacturing techniques
- In instances where connector interfaces cannot or are not keyed for a specific orientation, all identification, markings, cautions, and directions should be placed 360° around the interface
- Locate, position, and orient connectors to prevent the need for sequential installation or removal
- Locate and position electrical connectors such that all pin identification for either half can be easily seen
- Locate LRU/subsystem critical nodes (and/or test points) so they are accessible from a connector to prevent the need for internal LRU probing or access
- Avoid using cotter pins, safety wire, safety clips, and similar devices to prevent maintenance-induced events leading to ground vehicle accidents or loss of air vehicles

Standardization and interchangeability

- Hangeability exists—to avoid any potentially dangerous situation
- Ensure items are not physical interchangeability if functional interchangeability is not intended
- Whenever total—functional and physical—interchangeability is impractical, design the items for functional interchangeability, and adapters should be provided to make physical interchangeability possible
- Use identical components, such as pumps, reservoirs, and accumulators, in each individual power subsystem
- Differences should be avoided, where possible, in the shape, size, mounting, and other physical characteristics of functionally interchangeable items
- Modification of parts and units should not change their manner of mounting, connecting, or otherwise alter how they are incorporated in an assembly or system
- To remove latent doubt, provide sufficient information in documented instructions and identification plates to enable the technician to decide positively whether or not two similar items are actually interchangeable
- Design mounting holes and brackets to accommodate parts and units made by different facilities, that is, make them universally interchangeable
- Design exterior structure containing complex integrated antennas or sensors to be interchangeable to enhance repair of battle damage and induced damage
- Design parts and assemblies of a given model product or of models of a product in the same series to be interchangeable or replaceable
- Standardize parts, fasteners and connectors, lines and cables

Table 7.14 Design Guidelines (*Continued*)

Standardization and interchangeability

• Design parts and assemblies of a given model product or of models of a product in the same series to be interchangeable or replaceable
 	
 • Standardize parts, fasteners and connectors, lines and cables, and so on throughout a system, particularly from unit to unit within a given system
 	
 • Design mounting holes and brackets to accommodate parts and units made by different facilities, that is, make them universally interchangeable
 	
 • Provide total interchangeability for all parts and units that
    – are intended to be identical
    – are identified as being identical
    – have the same manufacturer's part number or other identification
    – have the same function in different applications (especially important for parts and units that have a high FR), and so on throughout a system, particularly from unit to unit within a given system
 	
 • Do not develop or identify special tools solely to simplify basic design or development of vehicles, systems, subsystems, or equipment
 	
 • Fully support the development of, or recommendations for, special tools with appropriate analyses, including life cycle costing, to justify the need
 	
 • Ensure that BIT system thresholds are consistent with those across all levels of indenture to prevent excessive numbers of CND and RTOK events from occurring

System/subsystem BIT/built-in test equipment (BITE)

• During design of the BIT, use worst-case stress analysis to ensure that any circuit failures induced by temperature extremes, tolerance build-up, power supply variations, and combinations thereof are identified
 	
 • Limit the amount of data that is recorded to a manageable size by
    – Limiting the number of signals that are monitored
    – Limiting the maximum sampling rate
    – Reducing the time span over which data are accumulated
    – Restricting the type of data accumulated
 • Base the degree of BIT required or proposed on the respective FRs and the appropriate FMECA at all equipment indenture levels
 • Incorporate testability design features as an integral part of equipment preliminary design process
 • Monitor mission-critical functions with BIT
 • Design BIT so it is initiated automatically upon equipment power-up
 • Set BIT tolerances to maximize FD and minimize FAR in the expected operating environment
 • Use concurrent BIT to monitor system-critical functions
 • Design the BIT and BITE so that no fault or failure within the BIT or BITE will degrade, disrupt, or fail the system being monitored

(Continued)

Table 7.14   Design Guidelines (*Continued*)

System/subsystem BIT/built-in test equipment (BITE)
• Design BIT to have a very low FAR (goal of 1% or less)   • Provide for manual control of test sequences so that the test can be selected individually, and appropriate test combinations can be executed at the operator's discretion   • Design the failure detection function to provide the equipment operator with a go/no-go indication of equipment readiness   • Design BIT to have the same level of electromagnetic interference protection as the item being monitored
FI
• Design each FI test to be independent of all other tests   • Design each test so it can be terminated prior to completion and reinitiated at its start point   • Ensure that system user manuals include instructions for faults not covered by BIT such as, system will not power-up or system is being used in an incorrect environment such as, at the wrong altitude, and so on.   • Clearly mark test points and make them easy to access   • Interlock the high power sections of systems and subsystems with visual/audible BIT to ensure safe system activation   • Design feedback loops so that the loop can be broken during test to ensure that faults do not propagate to the point where they cannot be isolated
Safety
• Do not locate equipment servicing points in crew, passenger, or operator areas   • Do not locate heat exchangers using hot liquids as the heat source, inside the compartments used for operator, crew, or passengers   • Route plumbing, lines, or hoses containing hot liquids, toxic gases, or liquids external to operator, crew, or passenger stations   • Use identical types of fluid in all hydraulic subsystems. Brakes may be the exception only if the system is totally separated from and independent of other hydraulic systems   • For vehicles containing two or more systems with different fluids, use different service fittings and different ground power interfaces for each fluid type   • Use cosmetic touch-up and repair materials that are environmentally safe   • Design tires with a color band to provide easy visual indication that maximum wear has been achieved   • Design stored energy devices (e.g., accumulators, nitrogen bottles, gas generators, etc.) that could cause injury, harm, or damage if inadvertently actuated, with integral safing provisions   • Write clear operating or maintenance instructions or procedures that are not easily misinterpreted   • Design items that are not functionally interchangeable so they are not physically interchangeable

Table 7.14   Design Guidelines (*Continued*)

Safety

- Properly locate warning labels and place warnings in procedures in the correct sequence
- Design blind matings with self-guiding features
- Locate high-failure items such that low-failure items do not have to be removed to facilitate maintenance (unnecessarily increasing the removal rate for the latter)
- Design the operation of controls to be consistent with intuition and common practice (i.e., a knob is turned clockwise to increase power)
- Design informational displays to be easy to read and interpret
- Design tasks so they are not physically awkward to perform
- Provide electrical grounds for external metal parts, antenna, and transmission line terminals and control shaft bearings
- Provide safety covers, warning labels, and interlocks for equipment using voltages greater than 70 V
- Use circuit breakers rather than fuses
- Provide guards for high-temperature parts
- Provide protection from moving parts for maintenance personnel
- Round edges and corners and avoid sharp projections, thin edges, and burrs to avoid injuries from cuts or abrasion
- Provide guards around lubrication points that are serviced while the equipment is operating

Nondestructive inspection and nondestructive test

- Avoid reliance on extensive interpretation by nondestructive inspection equipment operators to detect structural flaws
- Do not use nondestructive inspection technologies to maintain or protect the reliability of an item
- Derive nondestructive inspection and nondestructive test requirements from the FMECA and the associated RCM analysis and documentation

Handling

- Provide handles on items that
  - Are difficult to grasp, carry, or remove
  - Are frequently carried or handled
  - Weigh more than 10 pounds
  - Have fragile components that might be used as handles
- Locate single handles over the center of gravity of the item
- Place handles so they do not interfere with operation or maintenance of the item
- Provide provisions for mechanical handling of items weighing more than 90 pounds

(*Continued*)

Table 7.14   Design Guidelines (*Continued*)

Handling
• Provide hoist lugs with "LIFT HERE" markings located adjacent to the lugs for items weighing more than 150 pounds
• Handles, lugs, and other handling gear should be permanent parts of the item's case

References

Gibson, William. 2003. *Burning Chrome*. New York: Harper Collins.

Krueger, Myron. 1991. *Artificial Reality II*. 2nd ed. Reading, MA: Addison-Wesley Professional.

Lanier, Jaron. 1992a. "The Future of Virtual Reality." *Journal of Communication* 42 (4): 150–72.

Lanier, Jaron. 1992b. "Virtual Reality: The Promise of the Future." *Interactive Learning International* 8 (4): 275–9.

U.S. Department of Defense (DoD). 1997. "Designing and Developing Maintainable Products and Systems." Volume I. MIL-HDBK-470A, August 4, 1997, 4–13.

U.S. Department of Defense (DoD). 1993. "Military Standard System Safety Program Requirements." MIL-STD-882C, January 19, 1993.

Chapter 8

Sustainability Analysis

8.1 Equipment Downtime Analysis

Equipment DT analysis is a commonly used analysis technique to evaluate the expected time for which piece of equipment is not available (i.e., it is down) due to maintenance or a supply backlog. This value is the sum of elapsed maintenance time (EMT), awaiting parts (AWP) time, and awaiting maintenance (AWM) time. It is a primary measure of merit that considers reliability, maintainability, support system attributes, and operational environment. The results of this analysis may be used to calculate other equipment measures of merit, such as mission capable rate and equipment availability. The results of the analysis indicate those areas driving nonavailability of the equipment and are used to evaluate alternative design and support concepts based on total system DT.

Equipment DT is derived by using reliability and maintainability parameters and support parameters. The DT is the sum of EMT, AWP time, and AWM time, and it can be expressed as follows:

$$DT = EMT + AWP + AWM$$

This parameter indicates repair time for corrective (unscheduled) maintenance. The EMT is a function of failure rate (FR), maintenance action rate, maintenance action to failure ratio, and MTTR. The AWP time combines mean operating hours between demands, not-repairable rates for the equipment, and expected available inventory to determine the expected length of time for which a part is not available. The AWM time is the expected length of time for which equipment cannot be worked on due to any other considerations such as unavailable personnel, administrative delays, logistics delays other than spare parts (e.g., SE) delays, and weather delay. This is usually derived from field data.

Equipment DT analysis is typically performed at the total system level to provide the operator with information that can be used for alternative design or support system concept comparisons, operations or mission planning, and readiness capability assessment. Individual subsystems and lower indenture equipment items can also be evaluated using this analysis approach to identify the effects of individual equipment modifications or high-driver contributors to overall system DT.

Equipment DT analysis may be used any time during the program or product life cycle with the depth of the analysis increasing as the system becomes more completely defined and parametrically described. Early use of DT analysis will provide criteria to influence design for supportability, whereas later use will point out corrective actions that can be taken through changes in the design or support system.

Equipment DT analysis results in a figure of merit called equipment DT, measured in hours, days, or some other time cycle appropriate for the equipment evaluated. It can be used to identify areas driving system nonavailability, to compare alternate design or support system concepts, and as input to other equipment capability measures.

8.2 Sustainability Design Analysis

Sustainability design evaluation is the process of analyzing maintenance implications of a proposed or evolving design and providing feedback to the designer in a timely manner. A major goal of this evaluation is to ensure that sustainability is designed into the product from the start.

The process starts with a set of standards available to the designer and sustainability engineer. These standards normally consist of a preliminary "use study," a maintenance concept, qualitative and quantitative sustainability requirements, and lessons learned. In-process evaluations refine the maintenance concepts that will later form the basis for maintenance elements of logistics support analysis. The depth of this analysis depends on the phase that the design program is in at the time and the complexity of the equipment being designed. More complex equipment will need extensive evaluation to ensure that all maintainability requirements are being met. The design guidelines discussed in Chapter 7 provide a basis for evaluating a design for sustainability.

8.3 FMEA

The FMEA (also referred to as failure mode, effects, and criticality analysis, or FMECA, when criticality of failures is also determined) is used as a reliability analysis and design tool. However, results of an FMEA are also a key input to the design for sustainability. The FMEA helps establish necessary sustainability design characteristics based on potential failure modes and their effects on subsystems, equipment, and product operation. Results of an FMEA are used to determine placement and nature of test points, develop troubleshooting schemes, establish design characteristics relative to the ease of maintenance, and develop FD&FI strategies. Although FMEA is most often used on products, it can be and is more frequently being applied to processes.

Some of the prime outputs of an FMEA, from a sustainability viewpoint, include the following:

- Identification of single point failures
- Fail-safe design deficiencies
- False alarm occurrences
- Operator/maintenance person safety considerations
- Potential failure detection methodology, including the following:
    - Protective and warning devices
    - Failure override features
    - BIT provisions

The FMEA should describe the means by which the occurrence of a specific functional failure (failure mode) is detected and localized by the operator or the maintenance person. The FMEA outputs are very important to the design of a diagnostic system, which may include BIT. By identifying both local and next higher level effects of each potential system failure mode, methods for identifying, annunciating, and isolating failure modes that affect system operation can be devised. Any applicable warning devices, BIT indications, or other indications that make evident that an item has failed or malfunctioned should be clearly identified. If no such indication exists, the situation should be flagged in the FMEA as a potential sustainability problem. Proper recognition of an item failure or malfunction requires that normal, abnormal, and incorrect indications are known. A normal indication is one that is obvious to an operator or maintenance person when the item is operating normally. Abnormal and incorrect indications are those that are evident when an item malfunctions or fails.

An FMEA can be used to identify potential false alarms. False alarms often occur when a system's BIT detects and annunciates a failure during operation that cannot be repeated or duplicated later at the initial maintenance level. A false alarm can occur when the failure is an out-of-tolerance condition that exceeds the preset BIT limits that define "good" indications of system operation. The FMEA can be used to identify failure modes that result in an out-of-tolerance condition rather than in a hard failure. This knowledge can then be used to design the BIT so that it recognizes such a condition and only declares a failure if the condition persists over a specified period (time, cycles, etc.) of operation. This is one example of how an FMEA can be used to avoid false alarms. Another example is when the out-of-tolerance condition has no effect on system operation. Without this knowledge, the BIT could be designed to declare a failure, resulting in an unnecessary mission abort. With such knowledge, the BIT can be designed to ignore (intentionally override) this condition, allowing the mission to be completed.

Finally, the FMEA can be used to identify failures that are undetectable but have no effect on the mission. In such cases, the consequences of a second failure can be analyzed. For cases in which the mission would be jeopardized by the second failure, the FMEA can be used to determine whether or not a failure indication would now be evident to the operator, maintenance person, or BIT.

Figure 8.1 illustrates the steps in an FMEA. As mentioned earlier, when the criticality of each failure mode is also determined the analysis is known as an FMECA. Figure 8.2 illustrates a typical FMEA work sheet. Finally, Figure 8.3 shows the abbreviated results of an FMEA performed on a subsystem. The example in this case is a solid rocket motor.

8.4 Testability Analysis

Testability analysis is important at all levels of design and can be accomplished in a variety of ways. For instance, when designing complex ICs, such as application-specific ICs, or ASICs, it is important to develop test vectors that detect a high percentage of "stuck at" faults (i.e., signal stuck at logic "1" or "0"). This is almost always determined via logic simulation wherein a model of the design is developed in an appropriate fault simulation language. Once the model is compiled and ready to be simulated, a set of test vectors are applied to the model. The fault simulation program then produces a list of faults detected by the test vectors, as well as reporting the percentage (or fraction) of faults detected. Many such programs also identify specific signals that were not detected such that adjustments can be made either in the design or in test vectors themselves in order to increase FD percentage.

For nondigital electronics, FD efficiency is typically determined with the aid of an FMEA. An FMEA identifies faults that result in an observable failure, and can therefore be detected. The test engineer must then develop a test that will verify operation and detect any malfunctions as identified in the FMEA. FD percentages are then determined by summing the number of faults identified in the FMEA that are detected versus the total number identified as being detectable.

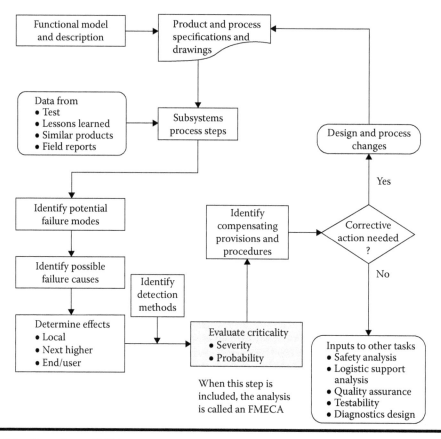

Figure 8.1 Steps in an FMEA.

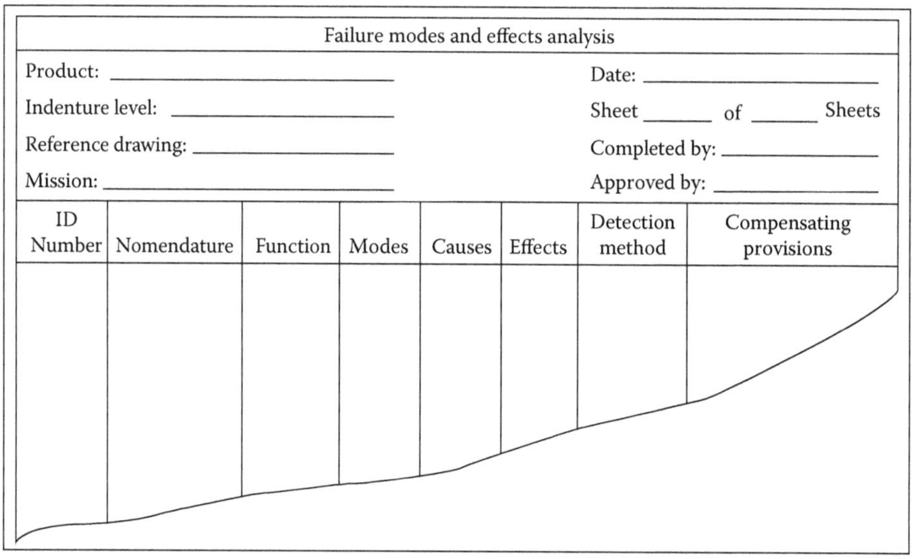

Figure 8.2 Typical FMEA work sheet.

FMEA			
Subsystem: Rocket motor	Prepared by: A.N. Engineer		Date: January 3, 1993.
Component or item	Failure mode	Cause of failure	Effects
1. Motor case	1. Rupture	1. Poor workmanship 2. Defective materials 3. Damage during shipping 4. Damage during handling 5. Damage during storage 6. Overpressurization	1. Destruction of missile
2. Propellant grain	1. Cracking 2. Voids 3. Bond separation	1. Abnormal stresses from cure 2. Excessively low temperature 3. Aging	1. Excessive burn rate 2. Overpressurization 3. Motor case rupture during operation
3. Liner	1. Separation from case 2. Separation from motor grain or insulation	1. Inadequate cleaning of case after fabrication 2. Unsuitable bonding material 3. Bonding process inadequate or not in control	1. Excessive burn rate 2. Overpressurization 3. Motor case rupture during operation

Figure 8.3   Abbreviated results from an FMEA of a solid propellant rocket motor.

This process can occur at all levels of design. The fault grading methods described in the first paragraph of this section are primarily applied at the IC and printed circuit card levels. In addition to determining FD percentage, a testability analysis should be performed to determine FI effectiveness of designed tests. For digital electronics, many of the tools used to grade test vectors also provide statistics on FI percentages, typically in a fault dictionary. During fault simulation, response of the circuit is determined in the presence of faults. These responses collectively form the fault dictionary. Isolation is then performed by matching the actual response obtained from the circuit or test item with one of previously computed responses stored in the fault dictionary. Fault simulation tools can determine from the fault dictionary the percentage of faults that are uniquely isolatable to an ambiguity group of size n ($n = 1, 2, 3, \ldots$). These tools can be used to verify FI goals or requirements via analysis, prior to actual testing. For nondigital circuits, hybrid circuits, or even digital systems above the printed circuit card level, analysis of FI capability can be performed with the aid of a diagnostic model and a software tool that analyzes that model. Examples are dependency modeling tools such as the system testability analysis tool (STAT) or the system testability and maintenance program (STAMP).* These tools, and others like them, can be used to determine FI capability of a design based on design topology, order of test performance, and other factors such as device reliability. Statistics such as percentage of faults isolatable to an ambiguity of group size n are provided as is the identification of which components or modules are in an ambiguity group for a given set of tests. However, test effectiveness and model accuracy are the responsibility of the test designer.

In the development of any product and prior to its release to the customer, the product is tested to

- Verify that the hardware and software meet product performance specifications
- Validate that the design is reliable and maintainable
- Improve product quality by uncovering design and manufacturing process problems, determining the root causes of problems, and subsequently introducing fixes

* The STAT is a registered trademark of DETEX Systems, Inc., and STAMP is a registered trademark of the ARINC Research Corporation.

Although a product should not be developed without performing some kind of testing, it is often difficult to determine how much testing should be done given the constraints of limited test samples, time, and budget. Given these factors, it becomes important to develop a well-coordinated and conceived product development and evaluation test plan. This plan should include input from all disciplines, including sustainability.

8.4.1 Objectives of Testing

Testing is performed to meet two objectives:

1. Validate and refine the design (and the design approaches and tools)
2. Verify the level of performance achieved (i.e., determine if a specification has been met)

In both cases, several tests may be required to meet these objectives. A sustainability test objective may be to validate that a product's subassemblies can be removed and replaced by a person using a defined set of tools. Another test may be performed to determine if the specification for MTTR is being achieved. In either case, planning must be accomplished early during development to determine if a formal sustainability test must be performed. If not, then a well-coordinated data collection program must be initiated that solicits information important to sustainability. For example, if a reliability growth test is to be conducted, then data should be collected on FD&I times and diagnostic efficiency when failures occur that must be fixed. In addition, data should be collected on ease of maintenance during removal and replacement of failed items. Sustainability testing should be planned using standardized methods, such as those discussed in Section 8.4.3 (verification, demonstration, and evaluation).

8.4.2 Testing in General

In general, testing can be grouped into five basic areas: (1) Functional, (2) performance, (3) verification, (4) demonstration, and (5) evaluation. Table 8.1 summarizes these five types of testing. Formal/specific sustainability testing involves the latter three categories of testing (verification, demonstration, and evaluation). They are discussed in Section 8.4.3.

Functional testing and performance testing are performed throughout various phases of product development and may include the use of models, simulations, test beds, and prototypes or development models of the product. Whereas such testing is almost always performed as part of the product design and development process, testing of sustainability features of product design, such as diagnostics, must also be planned in a similar fashion. In fact, diagnostic and other sustainability performance testing must be an integral part of all types of testing. This is important to evaluate performance, uncover deficiencies, and implement corrective action while the product is still being developed.

Sustainability testing should be a part of the overall test plan of the product. This plan should reflect an integrated approach to testing. That is, whenever possible each test should be developed to serve as many objectives as possible, thereby reducing total test cost. For example, data from a functional test can be used in the evaluation of reliability. Any maintenance actions that result from failures during that test can be used in evaluating at least certain aspects of maintainability. Without an integrated approach to testing, certain risks and consequences can ensue, as shown in Table 8.2. Because testing budgets may be limited, it may be necessary to validate, refine, and demonstrate a product's sustainability using the results of tests performed for other reasons. If this is the case, the sustainability engineer must be involved in the test planning process such that

Table 8.1 Types of Testing

Type	Description
Functional	Verify that a product, or product function, is behaving as intended. Typically involves applying a known stimulus or set of stimuli to the test item and comparing item response to a known response or set of responses.
Performance	Verify that the level of performance of product functions meet the requirements. Goes beyond functional testing. It is insufficient, for example, to verify that for a given input signal the product provides the right kind of output signal; the characteristics of the signal (amplitude, noise level, and so on) and reliability, maintainability, safety, and so on of the product must meet the requirements.
Verification	Determine the accuracy of and update the analytical data obtained from engineering analysis. Performed continuously throughout product development. Typically performed prior to any planned demonstration or evaluation test to provide assurances that the maintainability of the product can be achieved and demonstrated. All kinds of test data collected, such as from a functional test of diagnostics, should be used for verification of maintainability analyses and requirements.
Demonstration	Determine whether specified maintainability requirements have been achieved. Usually a formal process conducted by product developer and end customer. Such testing will involve development of a formal test plan, using defined methods of analysis to determine compliance.
Evaluation	Determine, at all levels of maintenance and product design, the impact of operational, maintenance, and support environments on the maintainability parameters of the product. Should involve performance of defined maintenance tasks in the product's actual use environments. Should be integrated with testing designed to evaluate other product parameters, as should other forms of testing. Can benefit from VR technology by allowing the testing of some maintenance tasks in a simulated usage environment, rather than the actual one. Some obvious cost savings are possible with this approach.

provisions can be made to collect sustainability-related information that will assist in evaluation of product sustainability.

8.4.3 Verification, Demonstration, and Evaluation*

Although all categories of testing can provide insight into sustainability, the three basic kinds of formal/specific testing are verification, demonstration, and evaluation. The first two types are performed on early models or prototypes and are used during design and development. Evaluation normally uses production versions of the product. Figure 8.4 provides a time-phased chart for these three kinds of testing. Note that the figure reflects the phases of a product's life cycle.

---

* The information in this section is adapted from MIL-HDBK-470A (DoD 1997).

Table 8.2 Risks and Consequences of a Nonintegrated Testing Approach

Risks	Consequences
Critical tests are omitted	Design shortcomings may appear after the customer assumes ownership of the product.
Tests are duplicated	Development costs increase and schedules are affected.
Test resources are inadequate	Tests are delayed, results are incomplete, results are inaccurate or invalid, faults are missed, and product performance suffers.
Test schedules are not coordinated	Inadequate time for testing, tests occur in wrong sequence, tests compete for critical test equipment, test requirements are not met, and so on.
Schedules are milestone oriented	Test results seem to confirm progress but do not result in needed product design improvements.

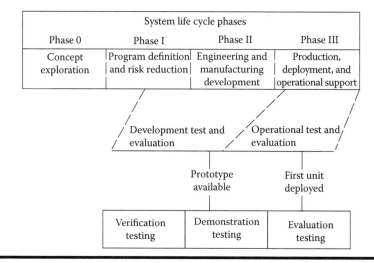

Figure 8.4 Time phasing of sustainability testing.

The following terms are used and discussed in this section:

- Sustainment task: The effort necessary for retaining an item in, changing to, or restoring it to a specified condition
- Sustainability model: A quantifiable representation of a test or process the purpose of which is to analyze results to determine specific relationships of a set of quantifiable sustainability parameters
- Verification: The effort performed from system concept through the hardware development phase to determine the accuracy of and update the analytical (predicted) data obtained from sustainability engineering analysis, identify sustainability design deficiencies, and gain progressive assurance that sustainability of the system or item can be achieved and demonstrated in subsequent phases

- Demonstration: The effort (often performed jointly by the system developer and the system procuring activity) to determine whether specified sustainability requirements have been achieved
- Evaluation: The procuring activity effort to determine, at all levels of maintenance, impact of the operational, maintenance, and support environment on the sustainability parameters of the system or item and to demonstrate depot-level maintenance tasks
- Sustainability concept: A description of the planned general scheme for maintenance and support of an item in the operational environment
- Sustainability environment: The climatic, geographical, physical, and operational conditions (e.g., combat, mobile, continental) under which an item is sustained.

8.4.3.1 Sustainability Testing: Verification

Three basic methods are used for verification and validation: (1) Inspection, (2) analysis, and (3) test. Inspection is best suited to the physical aspects of a product, such as finish and form. Although inspection might be used in determining accessibility, it is not particularly useful in verifying and validating maintainability of a product. Analytical methods can be quite useful and, in the case of enormously expensive products, may be the only affordable method of performing verification and validation. Many of the analytical methods used for verification and validation are the same ones used in the design of the product. Finally, testing is currently the most definitive method of performing verification and validation.

Verification testing is conducted in the earliest possible phase of overall system life cycle. Other sustainability testing (i.e., demonstration and evaluation) follow verification testing during subsequent phases of system life cycle. The means by which verification is accomplished depends on the sustainability characteristic or data element to be verified. For qualitative elements, such as accessibility, clearances for use of tools, available work space, or safety concerns, maintenance studies can be planned and executed. These may include the use of mock-ups or, in the future, use of virtual mock-ups using VR techniques, as was discussed in Section 7.9.3. Other verification methods include design reviews, other special studies such as maintenance task analysis or simulation studies, or review of historical information on like systems. The verification process should be continuously performed throughout the system life cycle, and therefore data obtained from sustainability demonstration testing and OT&E should also be used to verify both quantitative and qualitative features.

The process of executing a verification program should follow the following general procedural elements:

- Identify possible data sources and develop data collection and analysis plan
- Evaluate both qualitative and quantitative data
- Compare results with requirements

Quantitative data analysis typically relates to collecting information on maintenance task times, such as removal and replacement times, and developing an approach to statistically compare actual data with predicted values.*

---

* See Downs, W. R., "System Maintainability Verification—The Paired Time Comparison Method," Proceedings, 1979 Annual Reliability and Maintainability Symposium, for an example of a method used for verification.

The effort to verify sustainability parameters that are developed during phase 0 of the life cycle, shown in Figure 8.4 (e.g., predicted values of MTTR), is incremental in nature commencing with initial design and continuing through hardware development from components to the CI. The basic objectives of verification are as follows:

- ■ To verify and update the contractor's sustainability model.
- ■ To ensure economical correction of design deficiencies and to provide assurance that sustainability requirements will be achieved and demonstrated. Verifications such as limited low confidence maintainability tests and time-motion measurements should be performed early in the design process.
- ■ To provide progressive assurance that sustainability requirements can be achieved and demonstrated and that elements of the integrated support plan directly related to sustainability are valid.

Maximum use should be made of data resulting from maintenance performed in conjunction with tests such as development, prototype, mock-up, qualification, and reliability tests. In this respect, collection of maintenance task data must be planned for and coordinated with other disciplines. Further, specific maintenance tasks used in developing the sustainability model and prediction must be clearly defined such that when failures occur during development testing that result in a specific maintenance task, maintenance time can be compared with that used in the prediction model. This comparison must be done for both M_P (as applicable) and M_C tasking.

Development of a predicted value for sustainability is based on time estimates of individual tasks associated with maintenance of a system and its individual parts. Reliability data, such as failure frequency and failure mode and effects, play a role in sustainability prediction as well since it is the failure modes, their effects, and frequency of occurrence that ultimately define individual maintenance tasks to be performed. In developing a sustainability prediction, one must assume that a person with a particular skill level is assigned to maintain a particular subsystem within the system of interest. Also, estimates of FI time and FD capability must also be developed based on the testability design and assumed effectiveness of any BIT features incorporated at various system levels. Time estimates derived in this manner are often based on experience and knowledge of the individual performing the estimate.

A prediction can be biased, especially if factors such as the maintenance environment are not well understood early in the design phase. It is important then to verify the maintenance task database that is used to obtain predicted values as a means of improving the sustainability design aspects of a system and to improve the chances of performing a meaningful and successful sustainability demonstration.

8.4.3.2 Sustainability Testing: Demonstration

Sustainability demonstration is an analysis used to determine whether specific maintainability and testability requirements are achieved. It helps to determine ease of access, spares variability, test equipment readiness, testability provisioning, technical orders accuracy, and training requirements. Table 8.3 summarizes common maintainability parameters that can be demonstrated. Table 8.4 provides guidance in selecting a demonstration plan, and Table 8.5 provides a summary of testability demonstration plans.

Demonstrations are conducted to provide assurance that a specified maintainability index (e.g., MTTR, DT, and mean corrective time M_{ct}) will be attained during operation of the product. In a demonstration test, maintenance tasks are performed at a specified level of maintenance (e.g., organizational, intermediate, depot) by personnel having the skill levels available or required

Table 8.3 Common Maintainability Indexes That Are Demonstrated

Specified Maintainability Index	Test Calculations for Sample Maintenance Tasks	Accept/Reject Criteria
Mean corrective maintenance time (μ_c)	$$\bar{X}_c = \frac{\sum_{i=1}^{n_c} X_{c_i}}{n_c}$$	Accept if μ_c (specified) $\geq \bar{X}_c + \dfrac{\phi \hat{d}_c}{\sqrt{n_c}}$ Reject if μ_c (specified) $< \bar{X}_c + \dfrac{\phi \hat{d}_c}{\sqrt{n_c}}$
Mean M_P time (μ_{pm})	$$\bar{X}_{pm} = \frac{\sum_{i=1}^{n_{pm}} X_{pm_i}}{n_{pm}}$$	Accept if μ_{pm} (specified) $\geq \bar{X}_{pm} + \dfrac{\phi \hat{d}_{pm}}{\sqrt{n_{pm}}}$ Reject if μ_{pm} (specified) $< \bar{X}_{pm} + \dfrac{\phi \hat{d}_{pm}}{\sqrt{n_{pm}}}$
Mean maintenance time (including both M_P and M_C actions; $\mu_{p/c}$)	$$\bar{X}_{p/c} = \frac{f_c \bar{X}_c + f_{pm} \bar{X}_{pm}}{f_c + f_{pm}}$$	Accept if $\mu_{p/c}$ (specified) $\geq$ $\bar{X}_{p/c} + \phi \sqrt{\dfrac{n_{pm}(f_c \hat{d}_c)^2 + n_c(f_{pm} \hat{d}_{pm})^2}{n_c n_{pm}(f_c + f_{pm})^2}}$ Reject if $\mu_{p/c}$ (specified) $<$ $\bar{X}_{p/c} + \phi \sqrt{\dfrac{n_{pm}(f_c \hat{d}_c)^2 + n_c(f_{pm} \hat{d}_{pm})^2}{n_c n_{pm}(f_c + f_{pm})^2}}$
Maximum value of a percentile of M_C times (M_{max_c})	$$M'_{max_c} = \exp\left[\frac{\sum_{i=1}^{n_c} \ln X_{c_i}}{n_c} + \phi \sqrt{\frac{\sum_{i=1}^{n_c} \ln X_{c_i} - \dfrac{\left(\sum_{i=1}^{n_c} \ln X_{c_i}\right)^2}{n_c}}{n_c - 1}} \right]$$	Accept if M_{max_c} (specified) $\geq M'_{max_c}$ Reject if M_{max_c} (specified) $< M'_{max_c}$

Source: U.S. Department of Defense (DoD), "Designing and Developing Maintainable Products and Systems," Volume I, MIL-HDBK-470A, August 4, 1997. With permission.

Notes: c = M_C; d = sample variance; n = number of maintenance tasks; pm = M_P; ϕ = the value, corresponding to risk, that is obtained from a normal distribution table for a one-tail test.

Table 8.4 Selecting Maintainability Demonstrations

Test Characteristic	Program Constraints		
	Calendar Time Required	Test Facility Limitations	Desired Confidence in Results
Fixed sample size-type or sequential-type tests.	Much less than that required for reliability demo. Time required is proportional to sample size number. Sample size may vary depending on program.	None.	Fixed sample size tests and sequential-type tests gives demonstrated maintainability to desired confidence.
Test plan risks for consumer and producer ([1 – consumer risk] = confidence). Risks can be tailored to program.	Lower producer and consumer risks require larger sample sizes than higher risks.	Must be able to simulate operational maintenance environment, scenario, skills, and levels available.	Higher confidence levels require more samples than lower confidence levels.
The following demonstrations apply to testability:			
Fixed sample size–type tests.	Calendar time much less than that required for reliability demonstration. Time required is proportional to sample size. May vary depending on program.	Same as that required for maintainability demonstration.	Provides producer's risks of 10%. Provides consumer assurance that designs with significant deviations from specified values will be rejected.
Preset risks for consumer and producer ([1 – consumer risk] = confidence).	Risks inversely proportional to sample size used.	None.	Higher confidence levels require more samples than lower confidence levels.

in the fielded maintenance environment. Time required to perform each maintenance task is recorded. Depending on the sustainability index being demonstrated and the test plan chosen, once a statistically significant number of tasks are performed the collected data are used to determine if sustainability is acceptable or not. In addition to the quantitative data collected during the demonstration test, qualitative information, such as the adequacy of test support documentation or ease of maintenance (accessibility, safety, etc.), is also collected and reviewed.

The testability aspects of a design, such as BIT effectiveness, are not easily demonstrated in a formal demonstration test. In fact, demonstrations are inadequate for assessing testability because failure mechanisms that cause transient or intermittent behavior are not easily simulated in a laboratory environment (where many demonstrations take place). Also, the number of failures induced in a demonstration is small compared to the overall number of failures that may occur during fielded operation

Table 8.5   Testability Demonstration Plan Summary

Test Variable	Distribution Assumptions	Sample Size	Procedure	Consumer/ Producer Risk
FFD	None	Same as maintainability demonstration	FMEA on maintainability demonstration samples selected	• 10% Producer • 30%–40% Consumer
FFI; to given level of ambiguity	None	Same as maintainability demonstration	FMEA on maintainability demonstration samples selected	• 10% Producer • 30%–40% Consumer
FAR	None	Actually occurring false alarms over a given period of operating time	Collect data on false alarms during maintenance demonstration	• 25% Consumer • Producer risk is sample size dependent

Note: Since each plan demonstrates a different testability parameter, usually all three plans (FFD, FFI, and FAR) are used.

and are therefore insufficient to really demonstrate the diagnostic capabilities of a system design. Consequently, a well-planned verification program that optimizes naturally occurring failures during development and subsequent testing is needed to assess the diagnostic characteristics of the design.

8.4.3.2.1 Demonstration Test Specification

Demonstration test specification is defined as a set of numerical requirements and associated risk levels that govern the design and decision criteria of the demonstration test. For test plans to be described, this specification involves decisions regarding the following:

■ Type of sustainability index to be specified
■ Acceptable and unacceptable values of the index
■ Associated risk levels

For example, the test specification hypothesis might be as follows:

H_0: Mean corrective maintenance time = 40 minutes
H_1: Mean corrective maintenance time = 80 minutes
$\alpha = 0.20, \beta = 0.10$

Terms H_0 and H_1 are for null and alternative hypotheses, respectively. Values α and β are producer and consumer risks, respectively. For the defined specifications, a demonstration test must be designed such that the probability of rejecting a system whose mean M_C time is 40 minutes is .20, whereas the probability of accepting a system whose mean M_C time is 80 minutes is .10. This is presented as follows in the form of probability equations:

$$P\left(\text{reject} \mid M_{ct} = 40 \text{ minutes}\right) = .20$$

$$P\left(\text{accept} \mid M_{ct} = 80 \text{ minutes}\right) = .10$$

Typically maintainability index is specified in the procurement specification. It should be a measure directly influenced by equipment design, allowing the producer to plan for high assurance of a pass decision while bearing the responsibility for a reject decision. The index should also be appropriate for and measurable in the demonstration-test environment. If a demonstration of the chosen index is required, adequate sampling and statistical evaluation procedures, such as those described in this chapter, should be available for demonstrating conformance to the requirement. Finally, the specified index and risk values should not lead to sample sizes that exceed available test resources.

8.4.3.2.2 Choosing a Demonstration Test Method

Table 8.6 provides a matrix of available test methods. For each test method, the table lists the index for which the plan is designed to demonstrate a summary of assumptions, the required number of samples,

Table 8.6   Test Method Matrix

Test	Test Index	Assumptions	Sample Size	Sample Selection	Specification Requirement
1-A	Mean	Lognormal distribution and prior knowledge of variance	See test method	Naturally occurring failures or stratification	H_0, H_1, α, β
1-B	Mean	No distribution assumption, prior knowledge of variance	See test method		H_0, H_1, α, β
2	Critical percentile	Lognormal distribution, prior knowledge of variance	See test method		H_0, H_1, α, β
3	Critical maintenance time or worker hours	None	See test method		H_0, H_1, α, β
4	Median	A specific variance; lognormal	20		ERT
5[a]	CMDT/flight	None	See test method	Naturally occurring failures	Operational ready rate (ORR) or availability (A), NCMDT/NOF, DDT/NOF, α, β

Table 8.6 Test Method Matrix (*Continued*)

Test	Test Index	Assumptions	Sample Size	Sample Selection	Specification Requirement
6[b]	MH rate	None	See test method	Naturally occurring failures	MH rate Δ MR
7[c]	MH rate	None	See test method	Naturally occurring failures or stratification	μ_R, α
8	Mean and percentile dual percentile	Lognormal; none	See test method	Naturally occurring failures or simple random sampling	Mean, M_{max}; dual percentile
9	Mean (corrective task time, previous maintenance time, DT) M_{max} (90 or 95 percentile)	None	30 minimum	Naturally occurring failures or stratification	μ_c, μ_{pm}, $\mu_{p/c}$, M_{max_c}
10	Median (corrective task time, previous maintenance time), $M_{max_{ct}}$ (95 percentile), $M_{max_{pm}}$ (95 percentile)	None	50 minimum		$M_{max_{ct}}$, $M_{max_{PM}}$
11	Mean (M_P task time), M_{max} (M_P task time at any percentile)	None	All possible tasks	All	μ_{pm}, $M_{max_{pm}}$

[a] Test method 5 is an indirect method for demonstrating ORR or availability.

[b] Test method 6 is intended for use with aeronautical systems and subsystems.

[c] Test method 7 is intended for use with ground electronic systems where it may be necessary to simulate faults.

the method by which samples are selected, and the demonstration specification parameters. Definitions of individual terms found in the "Specification Requirement" column of Table 8.6 can be found in the corresponding discussion given below in Section 8.4.3.3 (Evaluation) where the method is presented.

A number of factors influence the choice of a sustainability demonstration test method. These factors includes the index to be demonstrated, any assumptions about the statistical nature of the index as related to test method requirements, the means by which sample maintenance tasks are selected, the number of maintenance tasks that must be demonstrated to obtain a statistically significant number of data samples, and the individual producer's and consumer's risk for some of the tests. Guidance and discussions of some of these factors are provided in Sections 8.4.3.2.3 through 8.4.3.2.7 that follow to aid the user in making informed decisions when specifying and executing a sustainability demonstration.

8.4.3.2.3 Choosing a Sustainability Index

Historically, the specified index is also the one that is demonstrated. However, it is important to provide some guidance on choosing such an index, as this can affect which of the test methods (outlined in Table 8.1) is chosen for sustainability demonstration testing. The principal criterion in selecting the index (and therefore for a product specification) is consistency with mission objectives and operational constraints. This criterion generally means that equipment DT is the time measurement of the index, since operational effectiveness cannot be achieved unless DT is controlled. If the need for a piece of equipment or system is not critical and manpower control is important, a labor-hour index may be most appropriate. M_P labor hours per operating hour are preferable to DT due to M_P without fear of operational demand during the maintenance action.

By the same reasoning, M_C is more crucial than M_P, especially if the latter can be scheduled to take place during known periods of nonuse. For equipment operated or needed continuously, such as a radar, total maintenance time is of prime importance. For equipment demanded at random times, such as a missile defense system, the approach might be to use separate controls for M_C and M_P. The choice of the statistical measure to be used often depends on the mission objective. If there is an A_o requirement for the system, then the equation for A_o is used to determine maintainability requirements:

$$A_o = \frac{\text{MTBM}}{\text{MTBM} + \text{MDT}}$$

Inherent availability may also be a requirement, in which case MTBF and MTTR are substituted for MTBM and MDT, respectively, in the aforementioned equation. When either availability expression is appropriate, a mean value becomes the index to be demonstrated. There may be, however, an availability requirement where a maximum DT or MTTR is required or more appropriate. Such a requirement would apply to critical equipment aboard an aircraft, for example, where the aircraft may have to be available for a new mission within 2 hours after completing a mission. In this case, a requirement of .95 probability, for instance, of completing the necessary maintenance within 100 minutes would be more consistent with the operational objective than a mean-value index. Of course, the maintenance level for which the requirement is developed also plays a role. For instance, maximum time to repair may not be appropriate or needed at the intermediate or depot level of maintenance assuming that an adequate amount of spares are available at the next lower level of maintenance (i.e., organizational or intermediate, respectively).

No matter what indices are specified in the requirements or as the index to be demonstrated, values for such indices must be realistic and based on current knowledge of the state of the art, past history of similar products or services, and engineering judgment. Whether historical data or prediction, or both, is used for assessing realism, careful judgment is required. If an allocation leads to an M_{ct} value of 20 minutes but a 30-minute value was observed for the most similar existing item, can it be concluded that 20 minutes is achievable? In all such cases, the following questions should be considered:

- How similar are the items?
- How similar will the maintenance environment be?
- Since the observed 30-minute value is necessarily based on a sample, what is the lower confidence limit associated with such a mean-value estimate?
- How much sustainability improvement can reasonably be expected?
- Is there any margin for increasing the 20-minute specified value?

8.4.3.2.4 Demonstration Environment versus Requirements

Past history shows that demonstrated and predicted values of both reliability and maintainability often do not correlate well with actual field experience. In the case of demonstrated maintainability values, this most likely stems from the fact that the demonstration environment is often not the fielded environment. Studies have shown that the closer the demonstration-test environment is to the expected field environment, the more meaningful is the demonstration test; therefore, every effort should be made to achieve such similarity.

In most cases, it is likely that demonstration environments continue to differ from the field environment. Because of this, when sustainability demonstration environments are based on operational requirements, applicability of these requirements to the demonstration environment needs to be considered. As a general principle, the specified value based on operational goals and conditions must be suitably adjusted to reflect the maintenance environment–governing demonstration. Often, it is difficult to adhere to this principle. With an avionics system, for example, a certain amount of time will be spent in the field to access the equipment in the aircraft, and the time to locate the malfunction and complete repairs and checkout is a function of this accessibility. If the demonstration test is not to take place in the aircraft (and this is often the case), there is the question of whether the specified value should be adjusted and by how much. It might be possible to construct a simulation of the actual condition, thus eliminating the need for adjustment. However, this type of simulation is often expensive and therefore not practicable. Tables 8.7 and 8.8 list various factors to be considered in evaluating the applicability of a specified sustainability index.

8.4.3.2.5 Maintenance Task Sampling

It is necessary to choose a specified number of maintenance tasks for the demonstration test. In general, there are two basic approaches for sample selection:

1. Observe maintenance tasks as they occur naturally in an operational or simulated situation.
2. Induce faults in the system and observe the maintenance actions to correct these faults.

For the fault-inducement approach, a decision must be made on the type of sampling procedure to be used. This is generally made between stratified sampling and simple random sampling. Guidelines are presented in Section 8.4.3.2.7 for evaluation of applicability of the two basic approaches, obtaining maintenance task samples, and choosing the appropriate sampling design and procedure.

Table 8.7   Factors Affecting the Suitability of a Specified Index for Demonstration

Physical equipment	Support items
Stage of completion	Tools
Similarity to production items	General and special test equipment
Physical location	Spares availability
Interfacing equipment	Technical manuals
Test location and facility	Operational factors
Lighting factors	Mode of equipment operation
Weather factors	Procedures for instituting maintenance
Space factors	
Test team	
Organization	
Training and experience	
Indoctrination	

Table 8.8   Causes of Discrepancies between Test and Field Results

Causes of Optimistic Test Results

1. Demonstration maintenance technicians are not representative of typical field maintenance personnel because they have more education and training or greater knowledge of the equipment's design.
2. The monitoring situation imparts to the technician an urgency not normally encountered in the field.
3. Known probable tasks are rehearsed beforehand.
4. Necessary SE is readily available.
5. Observed times are not contaminated with such factors as administrative or logistic delays, as field results sometimes are.
6. Difficult to isolate faults such as intermittent failures, and degradation failures are not simulated during demonstration.

Causes of Pessimistic Test Results

1. Technicians are not familiar with the equipment and have not acquired the necessary experience for rapid FI.
2. Field and procedural modifications to reduce maintenance time have not yet been made.
3. Initial manuals may be incomplete or require revision.
4. The monitoring situation can adversely affect a technician's performance.

8.4.3.2.5.1 Natural versus Induced Failures — It is important that sample selection is done early in the development program, especially if the choice is a naturally occurring failure or a combination of natural and induced failures. The natural failure approach is dependent on whether the program schedule allows enough time to obtain the required number of maintenance tasks, where allowable time is related to reliability. Given the MTBF, or θ, of a system, the average number of operating hours needed to yield n failure occurrences is $n\theta$. Therefore, for items with large MTBFs (i.e., hundreds of hours) and a required sample size of, say, 30–50 tasks, the number of required equipment operating hours can easily exceed 10,000 (e.g., 50 samples from an item with an MTBF of 200 hours). Because of time requirements of this magnitude, most maintainability demonstrations are based on the fault-inducement approach, which allows demonstration to be completed in a few days.

When the reality of cost and schedules dictate the use of induced failures for maintainability demonstration, the natural failure approach is preferred. A disadvantage that always exists with inducing failures is that there is no guarantee such faults are representative of the faults that will be seen in operation. This disadvantage is amplified when considering demonstration of diagnostic features of a design. Because of these problems, the following general recommendations are made concerning sample selection:

- If the schedule allows natural failures, then natural sampling is preferred.
- If the complete demonstration cannot be completed with only naturally occurring failures, a combination of natural and induced approaches should be used. One possibility is to take advantage of other development tests, such as the reliability demonstration test, and correct faults that occurred in these tests. Close coordination between the sustainability demonstration test and any other test is required.
- If natural failure testing cannot be conducted, any natural failures that do occur during the induced failure test should be included in the sample.

8.4.3.2.5.2 Failure Inducement Approach — An initial step in developing a sample set of tasks for demonstration is to develop a hypothetical maintenance task population. The two basic approaches to identifying maintenance task groups are simple random sampling and stratification. For discussion purposes, comments are restricted to stratification, as they also generally apply to simple random sampling when task selection by failure inducement is considered.

The first task in stratification is choosing criteria by which to stratify. This involves the characteristic by which to stratify, the number of strata, and boundaries defining an individual strata. The major objective here is to divide the equipment for which sustainability is to be demonstrated into a subset of homogeneous groups. To accomplish this, the maintenance tasks within each group, or stratum, should require approximately the same amount of maintenance time or the same number of worker hours, whichever is most appropriate. Blind application of this requirement, however, is not recommended. Repairing an electronic assembly within a system may take approximately the same amount of time as repairing a motor within the same system; however, differences between the two types of maintenance actions would make it unnatural to place them in the same stratum. Therefore, it is reasonable to make sure that there are similarities among the tasks assigned to a stratum. As is evident, engineering judgment must always play a role when grouping elements of this nature. The following approach is presented in four steps as additional guidance to stratum development:

1. First divide the equipment or item by physical entities, such as equipment within a system or units within the equipment. These first-level breakdowns are called blocks.
2. For each block, subdivide to the highest system level at which maintenance will be performed. If the block is the highest level, no further subdivision is necessary. If the equipment

is under test and the organizational maintenance philosophy is unit replacement, subdivide into units. These elements of the subdivision are called subblocks.

3. For each subblock, list the associated maintenance tasks and estimated maintenance task times or worker hours. For a subblock that is a LRU, or equivalent, removal or replacement may be the only task listed. However, if LRU adjustment or some further tasks such as crystal replacement are possible, they would also be listed as subblock tasks.

4. Group together the tasks in each subblock that require essentially similar actions and are expected to have similar maintenance times or worker hours, whichever index applies. The use of historical and predicted data and previous development tests should be used as inputs for time estimates. These groups then form part of the initial set of strata. The initial set of strata may have to be revised when actual tasks to be induced and sample-size requirements are considered.

To minimize any biasing problems due to task rehearsals and the problem of not being able to physically induce a selected fault, it is necessary to select a much larger number of possible tasks than required by the demonstration method. Most of the methods are based on having a demonstration population of four times the specified sample size. This number should then be allocated to the individual groups using the relative frequency of occurrence method. Further allocation within modules of a group is also recommended.

The entire process of selecting maintenance tasks is summarized in a 17-step approach shown in Table 8.9.

There are alternative approaches for choosing the maintenance tasks to be demonstrated, such as the symptom matrix approach referenced in Table 8.6. However, this approach requires a much more detailed analysis of system design.

Table 8.9　Steps in Selecting Tasks for Demonstration

Step	Description
1	Identify the major units comprising the equipment.
2	Subdivide each unit to the functional level at which maintenance for the demonstration is to be performed in accordance with the approved maintenance plan. This level may be an assembly, a module, a printed circuit card, or a piece part.
3	For each functional level of maintenance, identify the type of maintenance task or tasks to be performed and the estimated mean maintenance time for the task or tasks. The maintenance tasks and estimated maintenance time would be derived from a maintenance engineering analysis, a maintainability prediction effort, or historical data. The same maintenance task, such as "remove and replace" of a module, may result from different faults within the module.
4	Determine the FR (failures per 10^6 hours) for each module, printed circuit card, and so on, for which the maintenance task is identified. The FRs used should be the latest available from an associated reliability program.
5	Determine the quantity of items in each major unit associated with each task.
6	Determine the duty cycle for each item associated with each task (e.g., operating time of a receiver to the operating time of the radar, engine operating hours to aircraft FHs).

Table 8.9   Steps in Selecting Tasks for Demonstration (*continued*)

Step	Description
7	Group together the identified maintenance tasks that have both the following: 1. Similar maintenance actions. Note: A maintenance action is an element of a maintenance task. Although the estimated maintenance time for different maintenance tasks may be similar, the actions may be different (i.e., one task may involve significant diagnostics and another minimum diagnostics but significant access time). 2. Similar estimated maintenance times. The maintenance times in each group should be within a range that shall not exceed the smallest value in the group by more than 50%. Task grouping should be limited to within major units.
8	Determine the total FR for each task grouping.
9	Determine the relative frequency of occurrence for each task grouping by dividing the sum of total FR into individual total FR for each group.
10	Select a fixed sample. A sample of maintenance tasks equal to four times the sample size specified for the selected test method or as specified or agreed upon with the procuring activity should be allocated among the task groups in accordance with the relative frequency of occurrence of the task group.
11	The maintenance tasks to be demonstrated are allocated among the task groups in accordance with the relative frequency of occurrence of maintenance for the group. The maintenance task to be demonstrated is then randomly selected from the maintenance tasks allocated to the group or modules, assemblies, and so on, within the group. The maintenance task to be demonstrated is not returned to the sample pool and is therefore demonstrated only once.
12	Variable sample/sequential test: When variable sample size/sequential test methods are employed, a simple random sampling of the total population of maintenance tasks using a random number table based on a uniform distribution from 0 to 1 is used. Determine from relative frequency of occurrence the cumulative range of frequency of occurrence for each task group. A maintenance task is selected from that group whose cumulative range of frequency of occurrence includes the number selected from the random number table. The number selected from the random number table is then returned to the table before selecting a second number. The specimen task demonstrated is also returned to the sample pool.
13	Identify the maintenance task of interest.
14	Determine the failure modes that will result in the maintenance task of interest.
15	Determine the effect of each identified failure mode.
16	Determine the relative frequency of occurrence of each failure mode.
17	Simple random sampling: Determine the cumulative range of frequency of occurrence for each failure mode. Using a random number table, a number is selected and the failure mode to be induced is that whose cumulative range of frequency of occurrence includes the number selected. The number selected from the random number is returned to the table before selecting a second number. The specimen demonstrated is also returned to the sample pool.

8.4.3.2.6 Statistical Demonstration Plans

The matrix presented in Table 8.6 summarizes the major characteristics of each test method as well as the quantitative requirements that must be specified for each. The data analysis method included with each test method provides the decision criteria for acceptance or rejection of the item being demonstrated.

Each of the test plans includes an equation or other directions for determining a minimum sample size for maintenance tasks. Any departure from the minimum sample size requirements can affect the statistical validity of test procedures. Some test plans require a prior estimate of the variance of the distribution of interest for calculation of sample size. Such prior estimates are typically obtained from data on similar systems provided similarities in maintainability design, skill levels of maintenance personnel, test equipment, manuals, and the maintenance environment are considered in the estimation process. To preserve desired risk values in cases where the variance is predicted, the 85th to 95th upper confidence bound on predicted or estimated variance should be used. Average values of the variance range from 0.5 to 1.3.

Because of the difficulty of obtaining prior information and estimates of variance and due to the fact that the mean M_C time (M_{ct}) and maximum M_C time ($M_{max_{ct}}$) have historically been the maintainability requirement most often cited in a procurement specification, test plan 9 (mean corrective task time, mean M_p time, MDT) in Table 8.6 is the most chosen method for sustainability demonstration. Note that this method does not rely on any assumptions regarding the distribution of maintenance times. Despite this fact, there are examples when one or more of the other test methods have been employed and, therefore, all methods are given equal consideration.

8.4.3.2.7 Task Selection

We have seen methods of determining which tasks to be sampled under the fault inducement approach, which are applicable to each of the test methods presented in this section. When the demonstration is a requirement of the development program, the procuring activity historically has had the option of surveillance over and/or participation in the random selection of tasks comprising the demonstration population down to and including the specific faults to be simulated or induced. It is recommended to continue this practice. Further details on this and other management aspects of sustainability demonstration are presented in Section 8.6.8. In all cases, whenever a chosen task results in events detrimental to the safety of personnel or property, appropriate redesign action must take place. In the event that secondary failures result from an induced fault, they should be documented and their impact on item maintainability assessed. A report of such findings is typically made to the procuring activity or demonstration authority.

Two basic types of tests may be used for statistical maintainability demonstration: (1) Sequential and (2) nonsequential tests. In sequential testing, testing continues until a decision to accept or reject the hypothesis under consideration can be made. One drawback of sequential testing is that the length of the test cannot be determined in advance. However, sequential testing accepts very low MTTRs or reject very high MTTRs very quickly. A nonsequential, or fixed, sample size is best when maintainability must be demonstrated with a given confidence level.

Whenever sequential test plans are used (e.g., see test plan 1), care must be exercised in selecting and sampling tasks to ensure a true simple random sample is obtained. Departures from simple random sampling, such as proportionate stratified sampling, can affect the validity of the test procedures presented here. However, this effect is considered minimal for sample sizes required by test procedures that are not sequential tests. In short, simple random sampling must be used for sequential test methods.

8.4.3.3 Sustainability Testing: Evaluation

Evaluation is conducted to analyze the impact of the actual operational, maintenance, and support environment on the sustainability parameters of a system, to evaluate the correction of any deficiencies exhibited during demonstration, and to demonstrate maintenance tasks when needed. For the military enterprise, an evaluation is managed and conducted during OT&E as part of total system evaluation. Evaluation testing is similar to demonstration testing with the following notable exceptions:

- All evaluation items are products or product equivalent models.
- The evaluation is conducted in the actual operational and maintenance environment unless otherwise specified.
- All maintenance tasks are performed by personnel, either a procuring agency (e.g., government or civil service) or contractor, who would normally perform maintenance on the system in the fielded environment at the specified maintenance level.
- Maintenance tasks to be evaluated are those resulting directly from and incidental to actual operation and maintenance. These tasks should be supplemented by fault simulation only to evaluate specific tasks or special tasks that do not occur by chance during the evaluation phase.

In general, maintainability index to be evaluated is the primary consideration in selecting a sustainability test method. Considerable savings in sample size can be obtained by use of sequential test procedures in preference to fixed sample size tests. As a general rule, however, the sequential test should be used only when prior knowledge (e.g., from prediction methods) indicates that the system may be much better (or worse) than the specified value.

There are 11 evaluation test methods (summarized in Table 8.6):

1. Test on the mean maintainability index value
2. Test on a required critical percentile value and a design goal value
3. Test on critical maintenance time or worker hours
4. Test on median equipment repair time (ERT)
5. Test on chargeable maintenance DT (CMDT)
6. Test on man-hour (MH) rate (worker hours per flight hour [FH])
7. Test on MH rate (worker hours per operating hour)
8. Test on a combined mean and percentile requirement: A dual requirement for the mean and either 90th or 95th percentile of maintenance times
9. Test on mean maintenance time (corrective, preventive, or a combination of corrective and preventive)
10. Test on percentiles and corrective M_p time
11. Test for M_p time

Each of these evaluation test methods is discussed in more detail in Tables 8.10 through 8.22.

The following symbols and notations are common to test methods 1–3 discussed in Tables 8.10 through 8.12:

X = The random variable that denotes the maintenance characteristics of interest (e.g., X can denote M_C time, M_p time, fault location time, worker hours per maintenance task, etc.).

X_n = The nth observation or value of the random variable X.

n = Sample size. $\overline{Y}$

$\overline{X}$ = The sample mean (i.e., $\overline{X} = \dfrac{1}{n}\sum_{i=1}^{n}(X_i)$).

E (random variable) = The expected value of the variable.

$\sigma^2 = E[(\ln X - \theta)^2]$ = The true variance of $\ln X$.

$\mu = E(X)$ = The true mean of X.

$d^2 = \text{Var}(X) = E[(X - \mu)^2]$ = The true variance of X.

$\hat{d}^2$ = The sample variance of X (i.e., $\hat{d}^2 = \dfrac{1}{n-1}\sum_{i=1}^{n}(X_i - \overline{X})^2 = \dfrac{1}{n-1}\left(\sum_{i=1}^{n}X_i^2 - n\overline{X}^2\right)$.

$\tilde{d}^2$ = The prior estimate of variance of maintenance time.

X_p = The $(1-p)$th percentile of X (i.e., $X_{.05}$ = the 95th percentile of X).

$\tilde{M} = X_{.50}$ = Median of X.

$Y = \ln X$ = Natural logarithm of X.

$\overline{Y}$ = Sample mean of Y.

$\theta = E(\ln X)$ = The true mean of $\ln X$.

$\tilde{\sigma}^2$ = The prior estimate of the variance of the logarithm of maintenance times.

s^2 = Sample variance of $\ln X$.

Z_p = The standardized normal deviate exceeded with probability p (i.e., $\displaystyle\int_{Z_p}^{\infty} \dfrac{1}{\sqrt{2\pi}}\, e^{\left(\frac{-z^2}{2}\right)}dz = p$).

$Z_\alpha\, Z_{(1-\beta)}$ = Standardized normal deviate exceeded with probabilities α and $(1-\beta)$, respectively.

α = The producer's risk; probability that the equipment will be rejected when it has a true value equal to the desired value (H_0).

β = The consumer's risk; probability that the equipment will be accepted when it has a true value equal to the maximum tolerable value (H_1).

H_0 = The desired value specified in the contract or specification and expressed as a mean, critical percentile, or critical maintenance time.

H_1 = The maximum tolerable value. Note: $H_0 < H_1$.

When X is a lognormally distributed random variable,

$$f(x) = \dfrac{1}{\sigma x\sqrt{2\pi}}\, e^{-1/2\sigma^2(\ln x - \theta)^2},\ 0 < x < \infty.$$

If $Y = \ln X$, the probability density of Y is normal with mean θ and σ^2 variance $Y \sim N(\theta, \sigma^2)$.

Properties of the lognormal distribution:

Mean $= \mu = e^{\left(\theta + \frac{\sigma^2}{2}\right)}$

Variance $= d^2 = e^{(2\theta + \sigma^2)}\left(e^{\sigma^2} - 1\right)$

Median $= \tilde{M} = e^{\theta}$

Mode $= M = e^{(\theta - \sigma^2)}$

$(1 - p)$th percentile $= X_p = e^{\left(\theta + Z_p^{\sigma}\right)}$

The following symbols are common to test method 4 shown in Table 8.13:

$X_{ci} =$ Maintenance DT per M_C task (of the ith task).

$X_{pm_i} =$ Maintenance DT per M_P task (of the ith task).

$n_c =$ Number of M_C tasks sampled.

$n_{pm} =$ Number of M_P tasks sampled.

$\beta =$ Consumer's risk.

$\phi =$ The value, corresponding to risk, that is obtained from a table of normal distribution for a one-tail test.

$f_c =$ Number of expected M_C tasks occurring during a representative operating time (T).

$f_{pm} =$ Number of expected M_P tasks occurring during a representative operating time (T).

$T =$ Item representative operating time period.

$DT =$ Total maintenance DT in the representative operating time (T).

$\bar{X}_c, \bar{X}_{pm}, \bar{X}_{p/c} =$ MDTs of sample (corrective, preventive, and combined preventive/corrective maintenance times, respectively).

$M'_{max_c} =$ Sample calculated maximum M_C DT.

$\mu_c =$ Specified mean M_C time.

$\mu_{pm} =$ Specified mean M_P time.

$\mu_{p/c} =$ Specified mean maintenance time (taking both corrective and M_P times into account).

$M_{max} =$ A requirement levied in terms of a maximum value of a percentile of task time (i.e., 95% of all corrective task times must be less than 60 minutes) usually taken as the 90th or 95th percentile.

$M_{max_c} =$ Specified M_{max} of M_C DTs.

$\theta_c = E(\ln X_c) =$ Expected value of the logarithm of M_C tasks.

$\log X_{c}, \log X_c =$ Log to the base 10 of X_c, X_c.

$\ln X_c, \ln X_c =$ Natural logarithms of $X_c, X_c,$ respectively.

$\tilde{M}_{ct} =$ Median value of M_C tasks.

$\tilde{M}_{pm} =$ Median value of M_P tasks.

The following definitions apply to test method 5:

A = Availability; a measure of the degree (expressed as a probability) to which an aircraft is in the operable and committable state at the start of a mission, when the mission is called for at an unknown (random) point in time. For this test method, availability is considered synonymous with operational readiness. The aircraft is not considered to be in an operable and committable state when it is being serviced and is undergoing maintenance.

TOT = Total active time in hours.

Active time = The time during which an aircraft is assigned to an organization for the purpose of performing organizational mission. It is the time during which

1. The aircraft is flying or ready to fly.

2. Aircraft maintenance is being performed.

3. Aircraft maintenance is delayed for supply or administrative reasons.

DUR = Daily utilization rate; the number of flying hours per day.

AFL = Average flight length; flying hours per flight.

NOF = Number of flights per day.

DT = Downtime; time (in hours) during which the aircraft is not ready to commence an assigned mission (i.e., have the flight crew aboard the aircraft).

CMDT = Chargeable maintenance DT; time (in hours) during which maintenance personnel are working on the aircraft, except when the only work being done would fall under the nonchargeable maintenance DT (NCMDT) category.

NCMDT = Nonchargeable maintenance DT; time (in hours) during which the aircraft is not available for immediate flight but the only maintenance being performed is not chargeable. It would include the following:

1. Correct maintenance or operational errors not attributable to technical orders, contractor-furnished training, or faulty design

2. Miscellaneous tasks such as keeping of records or taxiing or towing the aircraft to or from somewhere other than the work center area

3. Repair of accident or battle damage

4. Modification tasks

5. Maintenance caused by test instrumentation

DDT = Delay DT; DT (in hours) during which maintenance is required but no maintenance is being performed on the aircraft for supply or administrative reasons. It would include the following:

1. Supply DDT

    a. Not operationally ready supply time

    b. Item obtainment time from locations other than the work center area

2. Administrative DDT

   a. Personal breaks such as coffee or lunch

   b. No maintenance people available for administrative reasons α = Producer's risk; the risk that a producer (or supplier) must take that the hypothesis that a true mean = μ_0 will be rejected even though it is true. The desirable value of α must be determined by judgment and agreed upon by the procuring activity and the systems developer. All other things being equal, a smaller value of α requires a larger sample size.

M = The maximum mean CMDT per flight.

M_0 = The required mean CMDT per flight.

$M - M_0$ = Difference between the maximum mean (M) of the parameter being tested and the specified mean (M_0). This value must be determined in conjunction with a value for β, that is, the consumer's risk. The value M is a value, greater (or worse) than the specified mean, that the consumer is willing to accept, but only with a small risk or probability (β). If the true mean is in fact equal to the value of M selected, the hypothesis that the true mean = M_0 will be accepted, although erroneously, β percent of the time.

β = The consumer's risk. The risk that the consumer is willing to take of accepting the hypothesis that the true mean = M_0 when in fact the true mean = M. All other things being equal, a smaller value of β requires a larger sample size.

σ = The true standard deviation of the parameter (CMDT per flight) being tested; this value, unless it is a specification requirement, will not be known, but an estimate must be made. (It is assumed that both M and M_0 will have the same value of σ.) The developer's maintainability math model, previous models, or previous data may be used. All other things being equal, a larger value of σ requires a larger sample size.

The following symbols are common to test methods 8, 9, 10, and 11, which are shown in Tables 8.17, 8.18, 8.19, and 8.20, respectively:

X_{ci} = Maintenance DT per M_C task (of the ith task).

X_{pm_i} = Maintenance DT per PM task (of the ith task).

n_c = Number of M_C tasks sampled.

n_{pm} = Number of PM tasks sampled.

β = Consumer's risk.

ϕ = That value, corresponding to risk, that is obtained from a table of normal distribution for a one-tail test.

f_c = Number of expected M_C tasks occurring during a representative operating time (T).

f_{pm} = Number of expected PM tasks occurring during a representative operating time (T).

T = Item representative operating time period.

DT = Total maintenance DT in the representative operating time (T).

$\bar{X}_c, \bar{X}_{pm}, \bar{X}_{p/c}$ = MDTs of sample (corrective, preventive, and combined preventive/corrective maintenance times, respectively).

M'_{max_c} = Sample calculated maximum CM DT.

μ_c = Specified mean M_C time.

μ_{pm} = Specified mean M_P time.

$\mu_{p/c}$ = Specified mean maintenance time. (Taking both corrective and P_M times into account.)

M_{max} = A requirement levied in terms of a maximum value of a percentile of task time (i.e., 95% of all corrective task times must be less than 60 minutes) usually taken as the 90th or 95th percentile.

M_{max_c} = Specified M_{max} of CM DTs.

$\theta_c = E(\ln X_c)$ = Expected value of the logarithm of M_C tasks.

$\log X_{c_i}$, $\log X_c$ = Log to the base 10 of X_{c_i}, X_c, respectively.

$\ln X_{c_i}$, $\ln X_c$ = Natural logs of X_{c_i}, X_c, respectively.

$\tilde{M}_{ct}$ = Median value of M_C tasks.

$\tilde{M}_{pm}$ = Median value of PM tasks.

8.4.4 Dependency Analysis

Dependency analysis has gained in popularity recently, and it is therefore prudent to provide some additional information on this testability technique. Dependency analysis starts with the creation of a dependency model of the item to be analyzed. The model is designed to capture the relationship between tests or test sites within a system, and the components and failure modes of components that can affect the test. As an example, consider the simple functional block diagram shown in Figure 8.5. The dependency model for this system, in the form of a tabular list of tests and their dependencies, is provided in Table 8.21.

Figure 8.5 has been labeled to identify each potential test site within the system; in this example exactly one test is being considered at each node. The dependency model shown in Table 8.21 is a list of "first-order dependencies" of each test. For example, the first-order dependency of test T3 is C2 and T2. This would indicate that T3 depends on the health of the component C2 and any inputs to C2, which is T2 in this case. For this simple system it is also obvious that T3 also depends on C1 and T1, but these are considered higher-order dependencies. Each of the tools mentioned previously (i.e., STAT and STAMP mentioned at the beginning of Section 8.4 on testability) determine all higher-order dependencies based on a first-order dependency input model.

Dependency modeling is attractive due to its applicability to any kind or level of system. Note in the aforementioned example that neither the nature nor the level of the system is required to process the model. Consequently, this methodology is applicable to almost any type of system technology and any level (i.e., component to system).

Based on the input model, the analysis tools can determine the percentage of time isolation to an ambiguity group of "n" or fewer components will occur. In addition, each of the tools discussed will also identify which components or failures will be in the same ambiguity group with other components or failures. Furthermore, any test feedback loops that exist, including those components contained within the feedback loop, will also be identified. Note that the ambiguity group sizes and statistics are based on a binary test outcome (i.e., test is either good or bad), and in most cases the tools assume that the test is 100% effective. This means that if the model indicates that a particular test depends on a specified set of components, the tools assume that should the test pass all components within the dependency set are

Table 8.10   Test Method 1

Test Method	Purpose	Assumptions	Hypotheses	Sample Size	Decision Procedure
1	Test on the mean: Demonstrate maintainability when requirement is stated in terms of required mean value (μ_1) and a design value goal (μ_0); or when the requirement is stated in terms of μ_1 and μ_0 is chosen by the contractor). The test plan is subdivided into two basic procedures: Test plan A and test plan B. Test A makes use of the lognormal assumption for determining the sample size, whereas test B does not. Both are fixed sample tests that employ the central limit theorem and the asymptotic normality of the sample mean for their development.	Test A: Maintenance times can be adequately described by a lognormal distribution. The variance σ^2 of the logarithms of maintenance times is known from prior information, or reasonably precise estimates can be obtained. Test B: No specific assumption concerning the distribution of maintenance times is necessary. The variance d^2 of maintenance times is known from prior information, or reasonably precise estimates can be obtained.	H_0: mean $= \mu_0$ H_1: mean $= \mu_1$, ($\mu_1 > \mu_0$). Note that μ_0 is normally the specified maintainability index value and that μ_1 is typically the maximum acceptable value of the specified index.	For a test with producer's risk α and consumer's risk β, the sample size for test A is given by $$n = \frac{(Z_\alpha \mu_0 + Z_\beta \mu_1)^2}{(\mu_1 - \mu_0)^2} (e^{\tilde{\sigma}^2} - 1)$$ where $\tilde{\sigma}^2$ is a prior estimate of the variance of the maintenance times and Z_α and Z_β are standardized normal deviates. The sample size for test B is given by $$n = \left(\frac{Z_\alpha + Z_\beta}{\frac{\mu_1 - \mu_0}{\tilde{d}}} \right)^2$$ where $\tilde{d}^2$ is a prior estimate of the variance of the maintenance times. Z_α and Z_β are standardized normal deviates. Minimum sample size of 30 for both test A and test B.	Obtain a random sample of n maintenance times, X_1, $X_2 \ldots, X_n$, and compute the sample mean: $$\bar{X} = \frac{1}{n} \sum_{i=1}^{n} X_i$$ Sample variance: $$\hat{d}^2 = \frac{1}{n-1} \left(\sum_{i=1}^{n} X_i^2 - n\bar{X}^2 \right)$$ Test A: Accept if $\bar{X} \le \mu_0 + Z_\alpha \dfrac{\hat{d}}{\sqrt{n}}$. Test B: Accept if $\bar{X} \le \mu_0 + Z_\alpha \dfrac{\hat{d}}{\sqrt{n}}$, reject otherwise.

Table 8.11 Test Method 2

Test Method	Purpose	Assumptions	Hypotheses	Sample Size	Decision Procedure
2	Test on critical percentile: Demonstrate maintainability when requirement is stated in terms of a required critical percentile value (T_1) and a design goal value (T_0; or when the requirement is stated in terms of T_1 and T_0 is chosen by the system developer). If the critical percentile is set at 50%, then this test method is a test of the median. The test is a fixed sample size test. The decision criterion is based on the asymptotic normality of the maximum likelihood estimate of the percentile value.	Maintenance times can be adequately described by a lognormal distribution. The variance σ^2 of logarithms of maintenance times is known from prior information, or reasonably precise estimates can be obtained.	H_0: $(1-p)$th percentile, $X_p = T_0$ or $P[X > T_0] = p$ H_1: $(1-p)$th percentile, $X_p = T_1$ or $P[X > T_1] = p$, $(T_1 > T_0)$.	To meet specified α and β risks, the sample size to be used is given by the following formula: $$n = \left(\frac{2+Z_p^2}{2}\right)\tilde{\sigma}^2\left(\frac{Z_\alpha + Z_\beta}{\ln T_1 - \ln T_0}\right)^2$$ (round up to the next integer) where $\tilde{\sigma}^2$ is a prior estimate of σ^2, the true variance of logarithms of maintenance times. Z_p is the standardized normal deviate corresponding to the $(1-p)$th percentile.	Compute the following: $$\bar{Y} = \frac{1}{n}\sum_{i=1}^{n}\ln X_i$$ $$s^2 = \frac{1}{n-1}\left[\sum_{i=1}^{n}(\ln X_i)^2 - n\bar{Y}^2\right]$$ $$X^* = \ln T_0 + Z\alpha s\left[\frac{1}{n} + \frac{Z_p^2}{2(n-1)}\right]^{1/2}$$ Accept if $\bar{Y} + Z_p s \le X^*$, reject otherwise.

Table 8.12   Test Method 3

Test Method	Purpose	Assumptions	Hypotheses	Sample Size	Decision Procedure
3	Test on critical maintenance time or worker hours: Demonstrate maintainability when the requirement is specified in terms of a required critical maintenance time (or critical worker hours; X_{P1}) and a design goal value (X_{P0}; or when the requirement is stated in terms of X_{P1} and X_{P0} is chosen by the system developer). The test is distribution free and is applicable when it is desired to establish controls on a critical upper value on the time or worker hours to perform specific maintenance tasks. In this test both the null and alternate hypotheses refer to a fixed time and the percentile varies. It is different from test method 2 where the percentile value remains fixed and the time varies.	No specific assumption is necessary concerning the distribution of maintenance time or worker hours.	$H_0: T = X_{P_0}$ $(p_1 > p_2)$ $H_1:$ $T = X_{P1}$ for a specified α and β.	The normal approximation to the binomial distribution is used to find n and c when p_0 is not a small value. Otherwise, the Poisson approximation is used. The equations for n and c are as follows: For $.20 < p_0 < .80$ ($p_1 = 1 - Q_1$); $$n = \left[\frac{Z_\beta \sqrt{p_1 Q_1} + Z_\alpha \sqrt{p_0 Q_0}}{p_1 - p_0} \right]^2$$ (Use next higher integer value.) $$c = n \left[\frac{Z_\beta p_0 \sqrt{p_1 Q_1} + Z_\alpha p_1 \sqrt{p_0 Q_0}}{Z_\alpha \sqrt{p_0 Q_0} + Z_\beta \sqrt{p_1 Q_1}} \right]$$ (Use next lower integer value.) For $p_0 < .20$, n and c can be found from the following two equations: $$\sum_{r=0}^{c} \frac{e^{-np_0}(np_0)^r}{r!} \geq 1 - \alpha$$ $$\sum_{r=0}^{c} \frac{e^{-np_1}(np_1)^r}{r!} \leq \beta.$$	Random samples of maintenance times are taken, yielding n observations $X_1, X_2, ..., X_n$. The number of such observations exceeding the specified time T is counted. This number is called r. Accept H_0 if $r \leq c$. Reject H_0 if $r > c$.

Table 8.13 Test Method 4

Test Method	Purpose	Assumptions	Sample Size	Method	Decision Procedure
4	Test on the median (ERT): This method provides demonstration of maintainability when the requirement is stated in terms of an ERT median, which is specified in the detailed equipment specification.	This method assumes the underlying distribution of M_C task times is lognormal.	The required sample size is 20. This sample size must be used to make use of the equation described in this test method.	Sample tasks are selected in accordance with the stratification procedure outlined in the chapter. The duration of each task is recorded and used to compute the following statistics: $$\log \text{MTTR}_G = \frac{\sum_{i=1}^{n_c}(\log X_{c_i})}{n_c}$$ $$S = \sqrt{\frac{\sum_{i=1}^{n_c}(\log X_{c_i})^2}{n_c} - (\log \text{MTTR}_G)^2}$$ (Note: All logarithms in the aforementioned two equations are to be taken to the base 10.) where MTTR_G is the measured geometric MTTR. It is equivalent to $\tilde{M}_{ct}$ used in other plans included in this chapter.	The equipment under test will be considered to have met the maintainability requirement (ERT) when the measured MTTR_G and standard deviation (S) as determined using a previous equation satisfies the following expression: Accept if $\log \text{MTTR}_G \leq \log \text{ERT} + 0.397(S)$ where $\log \text{ERT} =$ logarithm of the ERT

Table 8.14 Test Method 5

Test Method	Purpose	Assumptions	Sample Size	Method	Decision Procedure
5	Test on CMDT per flight: Because of the relatively small size of the demonstration fleet of aircraft and administrative and operational differences between it and fully operational units, ORR or availability cannot be demonstrated directly. However, a contractual requirement for chargeable DT per flight can be derived analytically from an operational requirement of ORR or availability. This chargeable DT per flight can be thought of as the allowable time (hours) for performing maintenance given that the aircraft has levied on it a certain availability or operational ready requirement. The requirement for chargeable DT per flight will be established using the procedure presented here.	This method requires no assumption as to the probability distribution of chargeable DT per flight. The method is valid only if the central limit theorem applies, which means that the sample size (NOF) must be large enough for this theorem to apply. The sample size must be at least 50, but the actual size is to be determined using appropriate equations.	Since the central limit theorem is applied, the expected distribution of the means will take on a normal distribution. If the true mean is equal to M_0 and a particular α is desired, the upper distribution (the mean of the distribution will equal M_0) will apply. It is on this basis that an acceptance rule is generated to the effect that if $\bar{X}$ is found to be equal to or less than the value $M_0 + \frac{Z_\alpha \sigma}{\sqrt{n}}$ the item is to be accepted.	The requirement for CMDT per flight, which will be demonstrated, is $$\frac{CMDT}{NOF} =$$ $$\frac{24(AFL)}{DUR} - \frac{A(24)(AFL)}{DUR} - \frac{NCMDT}{NOF} - \frac{DDT}{NOF}.$$ Values for DUR and AFL should be those planned for the aircraft during operational use. Values for $\frac{NCMDT}{NOF}$ and $\frac{DDT}{NOF}$ are functions of the operational environment. They should be provided to the system developer in the RFP or, if not, they must be provided by the developer in his or her proposal. The value for availability or ORR should be provided in the RFP.	The CMDT (X_i) after each flight will be measured and, at the end of the test, the total chargeable DT will be divided by the total NOF to obtain ($\bar{X}$) the sample mean CMDT and the sample standard deviations of CMDT. $$\bar{X} = \frac{\sum_{i=1}^{NOF} X_i}{NOF}$$ $$s = \sqrt{\frac{\sum_{i=1}^{NOF}(X_i - \bar{X})^2}{NOF - 1}} =$$ $$\sqrt{\frac{1}{(NOF-1)}\sum_{i=1}^{NOF} X_i^2 - (NOF)\bar{X}^2}.$$ Accept if $\bar{X} \le M_0 + \frac{Z_\alpha S}{\sqrt{NOF}}$. Reject if $\bar{X} > M_0 + \frac{Z_\alpha S}{\sqrt{NOF}}$.

Table 8.15   Test Method 6

Test Method	Purpose	Assumptions	Sample Size	Method	Decision Procedure
6	Test on MH rate: Demonstrate MH rate (worker hours per FH). Method intended for use with aeronautical systems and subsystems. It is based on a determination during phase II test operation of the total accumulated chargeable maintenance worker hours (MMH) and the total accumulated demonstration FHs.	Normally, all maintenance performed by approved test maintenance personnel during phase II and documented in appropriate maintenance reports will be the source of data for identifying chargeable MMH. The procuring activity may elect to terminate the demonstration prior to phase II completion if sufficient data are collected to project that the requirement will be met. The MH rate requirement must pertain to the aircraft configuration provided in the contract. For phase II flights conducted with a configuration other than this, an appropriate amount of chargeable MHs will be included in calculating the total chargeable MHs. This amount will be based on the predicted MH rate associated with the equipment not installed.	Not applicable.	The demonstrated MH rate is calculated as follows: $$\frac{\text{Total chargeable MMH}}{\text{Total demonstration FH}}$$ Care must be exercised in assuring that the predicted MH rate pertains to flight time and not equipment operating time. Appropriate ratios of equipment operating time to flight time must therefore be developed.	If the demonstrated MH rate is less than or equal to the MH rate requirement plus a maximum value (ΔMR), by which the demonstrated MH rate will be permitted to differ from the required MH rate, then the requirement has been met. ΔMR will be provided, by the procuring activity, as a percentage of the system MH rate requirement and will be determined based on such considerations as expected phase II duration and prior experience with similar systems. It is recognized that this demonstration method is nonstatistical in nature and does not allow the determination of quantitative producer's and consumer's risk levels. It is for this reason that ΔMR is provided (in a subjective manner) to minimize producer's risk.

Table 8.16 Test Method 7

Test Method	Purpose	Assumptions	Sample Size	Method	Decision Procedure
7	Test on MH rate (using simulated faults): Demonstrate MH rate (worker hours per operating hour). It is based on the predicted total FR of the equipment and the total accumulative chargeable maintenance MHs and the total accumulative simulated demonstration operating hours. The method is intended for use with ground electronic systems where it may be necessary to simulate faults.	None.	30 or more.	The demonstrated MH rate is calculated as follows: $$\frac{\text{Total chargeable MH}}{\text{Total operating time}} = \frac{\sum_{i=1}^{n} X_{C_i} + (\text{PS})}{T}$$ Where: X_{C_i} = Worker hours for M_C task i. n = Number of M_C tasks sampled; n must not be less than 30. MTBF = MTBF of the unit. PS = Estimated average total MHs that would be required for PM during a period of operating time equal to $n.(\text{MTBF})$ hours $$\frac{\sum_{i=1}^{n} X_{C_i}}{n} = \bar{X}_c = \text{Average number of } M_C$$ MHs per M_C task. T = Operating time.	If the MH rate requirement = μ_R: Accept if $$\bar{X}_c \leq \mu_R(\text{MTBF}) - \left(\frac{\text{PS}}{n}\right) + Z_\alpha \frac{\hat{d}}{n}$$ where α denotes producer's risk.

Table 8.17 Test Method 8

Test Method	Purpose	Assumptions	Sample Size	Method	Decision Procedure
8	Test on a combined mean/percentile requirement: Demonstrates maintainability when the specification is couched in terms of a dual requirement for the mean and either 90th or 95th percentile of maintenance times when the distribution of maintenance time is lognormal.	For use as a dual mean and 90th or 95th percentile requirement, the mean must be greater than 10 and less than 100 units of time; the ratio of the 90th percentile maximum value to the value of the mean must be less than 2; the ratio of the 95th percentile maximum value to the value of the mean must be less than 3. Maximum ratio of percentile to mean: 90th percentile value—2; 95th percentile value—3	Not applicable.	Sample tasks are to be selected with respect to the procedure defined for variable sample/sequential tests. The same sample tasks may be used simultaneously in the demonstration of both the mean and M_{max} requirements. Table 8.18, Table 8.19, and Table 8.20 (which are based on the sequential probability ratio test of proportion) define the accept/reject criteria for the values of required mean, M_{max} when defined as the maximum 90th percentile value, and M_{max} when defined as the maximum 95th percentile value, respectively. The number of observations greater than and less than the required values of the mean and M_{max} must be cumulated separately and compared to the decision values shown in the tables applicable to the two requirements. When one plan provides an accept decision, attention to that plan is discontinued. The second plan continues until a decision is reached. The equipment is rejected when a decision to reject on either plan has occurred regardless of the status of the other plan.	The equipment is accepted only when an accept decision has been reached on both plans. If no accept or reject decision has been made after 100 observations, the following rules apply: Plan A_1—Accept only if 29 or less observations are more than the value of the required mean. Plan B_1—Accept only if 5 or less observations are more than M_{max_C}. Plan B_2—Accept only if 2 or less observations are more than M_{max_C}.

Table 8.18 Test Method 9

Test Method	Purpose	Conditions for Use	Sample Size	Quantitative Requirements	Accept/Reject Criteria
9	Test for mean maintenance time (corrective, preventive, combination of corrective and preventive) and M_{max}: Demonstrates the following indices of maintainability: Mean M_C time (μ_c), mean pPM time (μ_{pm}), mean maintenance time (includes PM and M_C actions; $\mu_{p/c}$), and M_{max} (percentile of repair time).	Procedures are based on the central limit theorem. No information relative to the variance (d^2) of maintenance times is required. So it may be applied whatever the form of the underlying distribution, provided the sample size is adequate. Note: The procedure of this method for demonstrating M_{max_c} is valid for those cases where the underlying distribution of M_C task times is lognormal.	Minimum sample size of 30 M_C tasks is required for demonstration of M_C indices and demonstration of PM indices, where sampling is permitted. Actual sample size (if sample size greater than 30 is required) must be determined for each equipment to be demonstrated, and this is usually approved by the procuring activity.	Application of this plan requires identification of the index or indices of interest and specification of quantitative requirements for each. When demonstration involves μ_c or μ_{pm}, or a combination of both, consumer's risks need to be specified. When demonstration involves M_{max_c}, the percentile point that defines the specified value of M_{max_c} is specified.	Consult a table of the normal distribution function for values of φ (for a one-tailed test) corresponding to the specified level of consumer risk β: 1. μ_c: Accept/reject value (A/RV) is $\bar{X}_c + \dfrac{\varphi \hat{d}_c}{\sqrt{n_c}}$, where $\hat{d}_c$ = standard deviation of sample of M_C tasks. Accept if μ_c (specified) ≥ A/RV; reject if μ_c (specified) < A/RV 2. μ_{pm}: A/RV is $\bar{X}_{pm} + \dfrac{\varphi \hat{d}_{pm}}{\sqrt{n_{pm}}}$ where $\hat{d}_{pm}$ = standard deviation of sample of PM tasks. Accept if μ_{pm} (specified) ≥ A/RV; reject if μ_{pm} (specified) < A/RV 3. $\mu_{p/c}$: A/RV is $\bar{X}_{p/c} + \varphi\sqrt{\dfrac{n_{pm}(f_c \hat{d}_c)^2 + n_c(f_{pm} \hat{d}_{pm})^2}{n_c n_{pm}(f_c + f_{pm})^2}}$. Accept if $\mu_{p/c}$ (specified) > A/RV; reject if $\mu_{p/c}$ (specified) < A/RV. 4. M_{max_c}: A/RV is M'_{max_c}. Accept if M'_{max_c} (specified) ≥ M'_{max_c}; reject if M_{max_c} (specified) < M'_{max_c}.

Table 8.19 Test Method 10

Test Method	Purpose	Conditions for Use	Sample Size	Quantitative requirements	Accept/Reject Criteria
10	Tests for percentiles and maintenance time (M_C/M_P): This method uses a test of proportion to demonstrate achievement of $\tilde{M}_{ct}$, $\tilde{M}_{pm}$, M_{max_c}, $M_{max_c'}$ and $M_{maxp_{pm}}$ when the distribution of M_C and M_P repair times is unknown.	This method is intended for use in cases where no information is available on the underlying distribution of maintenance task times. The plan holds the confidence level at 75% or 90% as may be desired.	Requires a minimum sample[a] size (N) of 50 tasks.	Application of this method requires specification of $\tilde{M}_{ct}$, $\tilde{M}_{pm}$, $M_{max_{ct}}$ (95th percentile), or $M_{max_{pt}}$ (95th percentile) and selection of 75% or 90% confidence level.	The item under test is accepted when the number of observed task times, which exceed the required value of each specified index, is less than or equal to the following values corresponding to each index for the specified confidence level. Test for the median: The acceptance level for the median M_C and M_P tasks is 22 at a confidence level of 75% and 20 at 95% with a sample size (N) of 50 tasks. Test for M_{max_c} and $M_{max_{pm}}$: The acceptance level for M_{max_c} and $M_{max_{pm}}$ is 1 at a confidence level of 75% and 0 at 95% with a sample size (N) of 50 tasks.

a Sample tasks are selected in accordance with the stratification procedures outlined in this chapter. The duration of each task will be compared to the required value(s) of the specified index or indices ($\tilde{M}_{ct}$, $\tilde{M}_{pm}$, $M_{max_{ct}}$ and $M_{max_{pm}}$) and recorded as greater than or less than each index.

Table 8.20   Test Method 11

Test Method 11: Test for PM Times

This method provides maintainability demonstration when the specified index involves μ_{pm} and/or $M_{max_{pm}}$ and when all possible PM tasks must be performed.

- Conditions of use: All possible tasks are to be performed and no allowance need be made for underlying distribution.

- Quantitative requirements: Application of this plan requires quantitative specification of the index or indices of interest. In addition, the percentile point defining $M_{max_{pm}}$ must be stipulated when $M_{max_{pm}}$ is of interest.

- Task selection and performance: All PM tasks will be performed. The total population of PM tasks will be defined by properly weighing each task in accordance with relative frequency of occurrence as follows: Select the particular task for which the equipment operating time to task performance is greatest and establish that time as the reference period. Determine the frequency of occurrence (f_{pm}) of all other tasks during the reference period; where the frequency of occurrence of a given task is a fractional number, the frequency is set at the nearest integer.

- Accept/reject criteria:

  Test for μ_{pm}: The mean is computed as follows:

  $$\mu_{pm}\,(\text{Actual}) = \frac{\sum_{i=1}^{k} f_{pm_i}(X_{pm_i})}{\sum_{i=1}^{k} f_{pm_i}}$$

- Where f_{pm_i} is the frequency of occurrence of the ith task in the reference period, k is the number of different PM tasks, and Σf_{pm_i} is the total number of PM tasks in the population. Accept if μ_{pm} (required) $> \mu_{pm}$. (actual).

  Reject if μ_{pm} (required) $< \mu_{pm}$ (actual).

- Test for $M_{max_{pm}}$: The PM tasks are ranked by magnitude (lowest to highest value). The equipment is accepted if the magnitude of the task time at the percentile of interest is equal to or less than the required value of $M_{max_{pm}}$.

good. Conversely, a failed test makes all the components within the considered dependency set suspect. Therefore, the accuracy of the model, in terms of what components and component failure modes are actually covered by a particular test, is the responsibility of the model developer. The coverage is very much dependent on test design and knowledge of the system's functional behavior.

Even before intimate knowledge of what tests are to be performed is obtained, such as in the early stages of system development, a model can be created that assumes a test at every node, for instance. The system design can be evaluated as to where feedback loops reside; which components are likely to be in ambiguity; and where more visibility, in terms of additional test points, need to be added to improve the overall testability of the design. Once the design becomes more developed and knowledge of each test becomes available, the dependency model can be refined. Given that the analyst is satisfied with the model results, each of the tools discussed can be used to develop optimal test strategies based on system topology and one or more weighting factors.

**Table 8.21   First-Order Dependency Model
for a Simple System**

Test	First-Order Dependencies
T1	None
T2	C1, T1
T3	C2, T2
T4	C3, T2

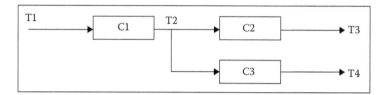

Figure 8.5   Test dependencies on a simple system.

Typical weighting factors are test cost, test time, component FRs, time to remove an enclosure to access a test point, and so on.

One of the drawbacks faced in the past of dependency modeling was the time taken to create a model. However, translation tools exist today and are continuously being developed that can translate a design captured in a CAD format, such as the electronic data interchange format, to a dependency model compatible with the specific dependency analysis tool being used. The analyst is still responsible for verifying the accuracy of the model, however, as in some cases not all dependencies are 100% correctly translated. Despite this fact, the amount of time that can be saved by translation outweighs any additional time it may take to verify the model.

The two tools mentioned at the beginning of Section 8.4, STAT and STAMP, provide the same basic kinds of outputs as just discussed. Each tool has other features that may be attractive depending on the system being analyzed, CAD tools being used in the design process, and so on. Therefore, more information should be gathered on these and other similar tools prior to making a final decision on which one to acquire.

The key point to remember about any of these tools is that model accuracy is most important. Therefore, it is important to understand how the system behaves in the presence of a failure and which tests can be developed to detect such behavior. Thus, to gain the most benefit from the model development process, experts in design and test should be involved in this process.

8.4.5 Other Types of Testability Analyses

Other types of testability analyses that do not require the use of a software tool are ad hoc procedures, such as reviewing a design against a known set of testability design practices. Grumman, and later Raytheon, developed such a procedure for the U.S. Air Force Rome Laboratory (Rome, New York) that rates a design based on the presence or absence of design features that increase or decrease ease of testing. The result is a score that is subjectively evaluated as indicating the design is anywhere between untestable without redesign to very testable. Used in conjunction with a design guide, also developed as part of the process by the aforementioned companies, this method can be very effective in making the test engineer's job easier and less costly. Testability analysis is a combination of applying any of the techniques mentioned in Section 8.4 to a system design and should be tailored according to the design level and design technology.

8.5 Human Factors Analysis

One of the most basic sustainability requirements is that the system is easy to maintain by human personnel. Sustainability analysis of a system typically involves maintenance tasks that deal with the repair or removal and replacement of a part or subassembly. Maintenance tasks usually involve disassembly, which is needed to access the target component; component repair or replacement; and subsequent reassembly. Thus, human factors analysis is performed to identify problems related to the interaction between maintenance personnel and the design model in performing each maintenance task. This analysis is used to verify that humans can perform each required maintenance task and its associated motions and manipulations. It deals more with the qualitative requirements than the quantitative requirements. Also, it is extremely important that this analysis is done while the product is still in the early design stages, that is, before any "metal is bent."

Human factors problems may involve the limited strength of maintenance personnel; limited or no work clearance required to carry out the task, that is, accessibility problems; and problems related to visual requirements of the maintenance person performing the task. Human factors analysis involves three types of analysis: (1) Strength, (2) accessibility, and (3) visibility analyses, as described in Table 8.22.

In the past, human factors analyses were very time consuming. They required the construction of expensive physical mock-ups to perform the analysis and also, unfortunately, the analysis was not done until the final stages of design, when modifications were very costly. However, there are a variety of modern, animated CAD tools and new VR techniques available today to assist the maintainability engineer in effectively and efficiently performing these analyses. Furthermore, when problems are discovered in the course of the human factors analysis, the proposed design modifications are quickly verified for their effectiveness using these same tools and techniques.

8.6 Managing the Analysis

The sustainability analyses presented in this chapter should include the development of some type of program plan. This plan should be an integral part of the overall test plan for the development of the project. The plan should include sections tailored according to the specific requirements of a program. Each of these sections should be in some way identified as being applicable to the verification, demonstration, and evaluation phases of the program:

Background information: Includes a description of quantitative and qualitative maintainability requirements; the maintenance concept; maintenance environment; applicable levels of maintenance; where the testing is to be conducted; test facilities requirements; participating agencies; modes of operation of the items of interest, including configuration and mission requirements; the specific items that are subject to verification, demonstration,

Table 8.22 Three Major Types of Human Factors Analysis

Analysis Type	Description	Comment
Strength	Used to determine the feasibility of disassembly and assembly sequences. Determine whether or not the maintenance person is able to carry out a maintenance activity that requires a certain level of human strength. That is, to evaluate the ability of the maintenance person to carry, lift, hold, twist, push, and pull objects in a standard body position (i.e., standing, bending, sitting, squatting, lying, etc.).	Strength analysis can be one of the most important criteria for the evaluation of a maintenance task.
Accessibility	Performed to identify design problems related to the inability of maintenance personnel to access the work area, that is, to detect possible collisions during the maintenance activity.	Based on size of men and women at a given percentile of the population.
Visibility	Determine visibility of work or operation area to operators and maintainers.	For some maintenance activities, it is important (e.g., for safety considerations) that the maintenance person is able to fully observe the work area.

and evaluation; and data required for the completion of verification/demonstration/evaluation.

Item interface: A description of the adequacy or inadequacy of item support elements and an estimate of their effect on item maintainability. These elements would include maintenance planning; support and test equipment; supply support; transportation, handling, and storage; technical data; facilities; and personnel and training.

Test team: The test team should be described in the plan. The description should include organization and degree of participation of procuring activity personnel and system developer personnel, including managerial, technical, maintenance, and operation personnel. The plan should also include test team member qualifications, quantity, sources, training requirements, and indoctrination requirements.

Support material: This section should cover SE, tools and test equipment, technical manuals to be used (or required), spares and consumables requirements/needs, safety equipment needs, and calibration equipment requirements.

Preparation plan: This section includes a description of and schedule for organization and assembly of the test team; training of personnel; preparation of facilities; and availability, assembly, checkout, and preliminary validation of support material.

Implementation: This section describes the test objectives of each test phase; schedule of tests; procedures for selection of maintenance tasks when faults are to be simulated; any special maintenance tasks, such as those requiring unique skills, equipment, and test methods to be performed, including method of demonstration; test method; data acquisition methods; data analysis methods and procedures; specific data elements; type and schedule of reports to be generated, if any; and the maintenance tasks to be verified, demonstrated, and evaluated.

Retesting requirements: This section provides a provisional schedule for special or repeat testing required to investigate any deficiencies or trouble areas. Deficiencies should be corrected in any item that has failed to meet the acceptance criteria. Retesting should include the corrected portions of the item and any other portions of the item affected by the correction. The maintenance tasks to be demonstrated should be designated by the procuring activity.

8.6.1 Sustainability Test Procedures

In designing sustainability test procedures, both qualitative and quantitative requirements should be verified, demonstrated, and evaluated. Typically, qualitative sustainability requirements to be verified, demonstrated, and evaluated are described in a checklist prepared by the system developer and coordinated with the procuring activity, when applicable. These checklists permit observation, analysis, and identification of sustainability characteristics incorporated or omitted. Quantitative requirements are verified, demonstrated, and evaluated by actual demonstration of maintenance tasks.

8.6.2 Maintenance Tasks

As implied, verification, demonstration, and evaluation is accomplished by performing maintenance tasks at a specified maintenance level. Generation of specific maintenance tasks during each maintainability test phase can take several forms. The means by which the tasks are generated should be considered, planned, and documented in the maintainability test plan. Actual operation of the item in the specified test, operational, and maintenance environment is always the preferred method of maintenance tasks generation (i.e., maintenance is performed as a result of naturally occurring failures). However, it is not always certain that a sufficient number of failures or maintenance tasks occur during the test period to satisfy any minimum sample requirements for the test method employed. So this method of maintenance task generation must be considered early in the development stages to ensure that a sufficient number of test or operational hours are planned, both through tests dedicated for maintainability and other forms of testing, to make this approach feasible. Close coordination with the entire development team is required for this approach to maximize all test time planned.

In lieu of the naturally occurring failure approach to maintenance task generation, there is the fault or failure simulation approach. In this approach, failures are introduced by way of faulty parts, deliberate misalignment, open leads, shorted parts, and so on. As part of this approach, a maintenance task sampling plan must be prepared. When done as part of demonstration testing, actual task selection should not be made by the test team until immediately prior to the demonstration.

8.6.3 Administrative Tasks

A test team consisting of members of the procuring activity, if any, and the system developer should be formed to manage the test program. The team members should be empowered to make decisions for their respective organizations. Each member of the team may have advisors from their organizations who are knowledgeable in various aspects of the demonstration and requirements of the verification, demonstration, and evaluation plan. Responsibilities of the team are described in the maintainability test plan and should typically include the following:

- Maintain surveillance over maintenance and inspection operations. Any apparent discrepancies in maintenance task accomplishment and documentation observed by any member of the team should be promptly brought to the attention of the remaining test team members.
- Evaluate and validate maintenance and operational data to determine applicable labor hours, flying hours, operating time, maintenance time, DT, item status, and so on.
- Assure that the demonstration item selected is adequately prepared in accordance with applicable technical manuals and that no maintenance has been deferred that will compromise the successful completion of the next scheduled operation or mission prior to being placed in an operation ready status.
- Decide if resulting failures, maintenance time, elapsed DT, maintenance labor hours, and so on, should be chargeable in cases where operator or maintenance crew errors are committed.
- Rule on questions of whether or not the verification, demonstration, and evaluation plan has been adhered to.
- Rule on controversial issues that are not specifically covered by applicable specifications or other pertinent documentation. Further, determine those matters requiring contractual interpretation or resolution by the appropriate procuring authority and system developer organizations. For these matters, the test team majority and minority statements should be submitted to the procuring activity or other applicable authorities' contracting officer for resolution.
- Prepare and submit demonstration status reports to the procuring activity and the system developer.
- Analyze data and determine the extent of achievement of specified maintainability requirements.
- Prepare and submit final results of each of the maintainability test phases to the procuring activity and the system developer within the time period indicated in the test plan.
- Ensure that the following conditions have been fulfilled prior to the start of the demonstration and evaluation test phase:
  - Each test item complies with the established configuration or all deviations reported have been accepted by the procuring activity.
  - All required technical manuals have been updated as necessary.
  - All support resources are available in the type and quantity specified in the verification, demonstration, and evaluation test plan.
  - All operator or maintenance crew personnel are properly trained and meet established skill-level requirements.
  - All records of approved changes in personnel requirements, operating and maintenance manuals, data-handling procedures, and analysis techniques have been incorporated into the final revision of the verification, demonstration, and evaluation test plan.

8.6.4 Other Duties

In addition to the duties listed in Section 8.6.3 for the test team, administrative duties usually accompany the implementation of a sustainability test plan. For instance, the designated test team should have a test director who has the authority to decide in all cases of deadlock between the members of the team. This person is usually designated by the procuring activity. Other such requirements or "rules of conduct" are as follows:

Instrumentation failures: Any failures of test instrumentation used to instrument the demonstration item for test purposes or failures induced by such test instrumentation installation or operation and all associated maintenance are not chargeable maintenance tasks.

Maintenance due to secondary failures: If secondary failures result from a chargeable primary failure, the total resultant maintenance time to restore the items are chargeable as a single maintenance task, except when the secondary failure results from the method used to simulate a fault rather than from the fault itself. If the reason for the secondary failure is removed (corrected), time charged for the secondary failure can be deleted.

Inadequate technical manuals or support equipment: If in the accomplishment of a maintenance task a technician finds the applicable technical manuals or SE to be inadequate, this finding should be brought to the attention of the test team, and if the inadequacy is verified this portion of the demonstration can be terminated. In such cases, times measured are not chargeable. Action must then be taken to correct the inadequacies of the technical manuals or SE after which the same maintenance task is repeated.

Cautions: If an item is damaged or maintenance errors induced by item design complexity, poor design practice, or following improper procedures that allow improper maintenance without proper caution in the technical manuals, the failure and resultant maintenance times are chargeable. In these cases, action is taken to correct the improper procedures or deficiencies and the corrective action verified. When this action is completed, the maintenance time saved can be deleted.

Personnel number and skill: Each task should be performed by the prescribed number of personnel with the prescribed skills. If personnel are required on an intermittent or sequenced basis, the labor hours assessed against the maintenance task will include the required standby time only if the standby time is of a type or duration that prevents standby personnel from performing other productive tasks.

Cannibalization: The maintenance associated with the removal or reinstallation of the item or SE assemblies or components for cannibalization purposes are not chargeable unless the deficiency can be directly related to lack of recommendations for proper level of support spares or expendables. If the system developer takes action to correct the deficiency, the time charged can be deleted.

8.6.5 Data Collection

Data collection is important in identifying weaknesses in the sustainability design of a system and in the subsequent correction of such weaknesses. For verification, demonstration, and evaluation, a sound data collection system must exist and be coordinated with other disciplines and tests. The data system should be accessible by all members of the test team, including the procuring activity, and should include information on all mission debriefings, failures, and maintenance data. The descriptions of all maintenance tasks must be adequate to enable determination of which maintenance task was performed. It is important to include in the maintainability database or maintenance-related data records all direct maintenance DT or labor hours that are not specifically determined to be nonchargeable. This information will then feed into the quantitative calculations of all applicable maintainability metrics. Maintenance times that may not be charged could result from the following causes:

- Maintenance and operational errors not chargeable to technical manuals, system developer–furnished training, or faulty design
- Miscellaneous tasks such as keeping of records and taxiing and towing of aircraft to or from an area other than the assigned work center area
- Repair of accident damage
- Documented delay DT (supply or administrative), which is clearly outside the responsibility of the system developer
- Modification tasks
- Maintenance of test instrumentation exclusive of normal configuration
- Maintenance time accountable to test instrumentation installation (other than normal configuration) accrued during maintenance task performance

In any case, it is extremely important to establish up front in the program which maintenance tasks will be chargeable and which ones will not be. Doing so will avoid confusion and arguments later on between test team members.

8.6.6 Parameter Calculations

All data acceptable to the test team during each applicable test phase (i.e., verification, demonstration, etc.) is used in calculating the maintainability parameters of interest.

8.6.7 Documentation Tasks

After each phase of sustainability testing, a final report should be developed that documents, as a minimum, the following information:

- Summary of data collected and location of data files
- Factors that influence the data
- Analysis of the data
- Results of the sustainability testing phase and certification that the specified objectives and requirements have or have not been met
- Assessment of integrated logistic support factors, such as technical manuals, personnel, tools and test equipment, SE, maintenance concept, and provisioning for the effect of these factors on quantitative and qualitative demonstrated maintainability parameters
- All noted deficiencies
- All recommendations to correct deficiencies and to make improvements

8.6.8 Sustainability Demonstrations

Despite the fact that sustainability demonstrations are quite successful, testability-related problems, especially those associated with BIT, have continued to plague the sustainability performance of many complex systems. Metrics such as CND rate, RTOK, and FAR continue at unacceptable values in actual operations of complex products resulting in too many resources being spent on maintenance.

Why does testability performance in the field continue to fall short of both expectations and demonstrated values when maintainability demonstrations are usually successful? Specifically, current demonstration techniques are inadequate to demonstrate testability metrics such as fraction of faults detected and FI resolution. Most maintainability demonstrations are performed in laboratory environments using fault insertion methods. Furthermore, the faults selected for insertion represent a small percentage of those likely to occur during fielded operation. The number of faults selected for insertion is limited because some faults would result in equipment damage and others cannot be easily inserted. Only hard faults, such as open leads or shorted components that are relatively easy to detect, isolate, and repair, are selected. Also, many of the faults that result in CNDs or RTOKs are not easily simulated in a demonstration test. Finally, it is not possible to simulate failures or intermittent conditions that can be considered false alarms, thus eliminating the ability to demonstrate any specified FAR for BIT.

Given these facts, effective demonstration of testability is probably not possible in the near future and should not be considered as part of future development programs. However, maintainability demonstration, as described in this chapter, is still useful. The need to demonstrate ease of maintenance and the adequacy of logistical support services such as technical manuals, SE, sparing levels, and training is still extremely important to maintainability. Furthermore, if the diagnostic system designed into a system cannot detect and isolate even those hard failures induced as part of a maintainability demonstration, then this is an indication that a redesign is warranted.

If it is impossible to adequately demonstrate testability characteristics of the design in terms of the aforementioned metrics, the question still remains: How can customers provide some assurance that the diagnostic system will meet their needs in terms of overall system performance? The key is to do a better job early on in development of determining exact system diagnostic needs and then to develop a process by which higher-level requirements are allocated properly to subsystems. Further, a single individual must be given overall authority for testability. This person must be given equal status in the decision-making process, so that testability needs and requirements do not take a backseat to other performance needs. In this manner, any design decision must consider the impact on testability prior to finalizing any approaches.

8.6.9 Defining Needs

In addition to making sure that testability receives equal consideration, the design team must determine several items that contribute to an effective testability design. For instance, the team must define what constitutes a failure. In particular, failures that can affect BIT performance, such as drift, must be clearly defined. Definitions are a problem in design that has plagued BIT performance in the field. The BIT algorithms that are too sensitive may detect and report failures that only occur intermittently due to environmental or other factors; but it may be impossible to duplicate in the maintenance environment the conditions that caused the failure. A formal process must also be in place to ensure that test verticality is maintained from one maintenance level to the next.

Another area requiring clear definition is which failures need to be reported by BIT. Should all BIT failures be reported or only those that degrade safety or mission capability? All reportable BIT indications should be carefully reviewed to define failure state and appropriate action.

Testability needs should also be determined from field and manufacturing data on like systems. However, many data collection systems do not adequately report testability problems. Therefore, the data collection system must be devised to collect data such as CNDs. These data can then be analyzed to determine the root causes of such behavior so that corrective actions can be implemented in next-generation designs.

8.6.10 Using Test Programs to Verify Testability Design Attributes

Although a formal test program to demonstrate testability features is impractical, full use must be made of all forms of other testing, including reliability demonstration tests and other development tests, to improve the testability of a system. This requires, however, that the diagnostic system, such as BIT, is available prior to the occurrence of testing. Once again, close coordination between individuals responsible for the diagnostic design and other disciplines within the IPDT is absolutely essential.

During testing that includes diagnostics, all failures and diagnostic system response to those failures, as well as the ability to detect and isolate faults using test SE, must be recorded and analyzed to identify problems and develop corrective actions. This process includes collecting diagnostic performance data on both hardware and software faults. A training program should be instituted that disseminates to all individuals responsible for data collection the importance of testability information and how to properly record such data. This form of verification and evaluation, as opposed to dedicated demonstration testing, is much more effective for testability since testing provides a means for testing diagnostics for long periods of time without the need for unique diagnostic tests and extra assets.

8.7 Sustainability Prediction, Allocation, and Assessment

Sustainability predictions are estimates of design performance. Allocation is the process of apportioning the predictions to lower levels of assembly. Assessment is the procedure that provides assurance that a specified prediction will be attained during the fielded operation of the system.

8.7.1 Sustainability Prediction

Sustainability predictions are estimates of design performance from a maintainability perspective. They are a means for comparing design options, assessing the feasibility of achieving maintainability requirements, and assessing progress in achieving the maintainability requirements. However, predictions are imprecise with the degree of imprecision being determined by the validity of assumptions, amount of available performance data, applicability of the method, and so on. Predictions should, therefore, never be used as the sole basis for programmatic or engineering decisions.

Sustainability prediction is a useful tool for determining where to place the most emphasis in designing for maintainability. Each subsystem, equipment, and component can be evaluated in terms of FR, maintenance time required, and complex maintenance tasks. The designer is thus provided with the necessary visibility into the attributes of subsystems, equipment, and components to

- Identify design weaknesses from a maintainability perspective
- Support trade-off studies
- Determine if the design is ready to proceed to the next phase of development

Sustainability prediction is an iterative estimate of the future observed maintainability characteristics of the product. Prior to starting the prediction process, all assumptions must be specifically defined and evaluated as to their validity and applicability. Rationale must be provided for all assumptions.

Predictions are also useful in logistical planning. Early estimates of maintenance time, labor hours, and other maintainability metrics can be used to make preliminary assessments of SE, spare parts, personnel, training, and other logistics resources needed to maintain the system in operational use. As predictions are refined using test and demonstration data, the estimates of logistics resources can be revised. Although other factors determine the needed types and amounts of logistics resources, maintainability predictions are important for this purpose and, in the early stages of a program, may be the only basis on which to plan logistics. By beginning the process of identifying logistics resources early in the program, a "fully operational" status can be rapidly achieved after the fielding of a new system.

A variety of methods and metrics are used to predict maintainability. Each prediction method is designed for a specific application. All methods depend on at least two basic input parameters. These two common parameters are as follows:

1. The FRs of components at the specific level of assembly of interest
2. Repair times* required at the maintenance level involved

Historically, the most commonly used methods for maintainability predictions were those found in MIL-HDBK-472, "Maintainability Prediction" (DoD 1966). Of the five different maintainability methods discussed in this handbook (see Table 8.23), only procedure V was included in MIL-HDBK-470A (DoD 1997), which replaced MIL-HDBK-472. The reason for including only procedure V was that all commercial computer software development to date has concentrated on procedure V to the exclusion of the other four maintainability prediction procedures.

8.7.2 Sustainability Allocation

Sustainability allocation is the process of apportioning system-level maintainability requirements to lower levels of assembly. In other words, the system requirements are apportioned to each subsystem; each subsystem's requirements are apportioned to components and equipment within the subsystem; and, finally, the component and equipment requirements may be apportioned to modules.

Sustainability allocation requires a detailed analysis of system architecture and knowledge of characteristics of various types of systems, subsystems, and so on. Allocations are made primarily for M_C requirements. Historically, system-level requirements have been difficult to fully assess without a prototype or first-production version of the system. So allocations have been used to assess the progress being made toward achieving the system-level maintainability requirement.

Allocations are a natural management tool. The customer, prime contractor, and subcontractors and suppliers can use them to

■ Derive "not-to-exceed" maintainability values (i.e., maximum MTTR) for the system's lower-level indentures of assembly.

---

* Most maintainability experts agree that repair times exhibit the skewed characteristics of the lognormal distribution. Thus, repair times are usually assumed to be lognormally distributed.

Table 8.23   Maintainability Prediction Methods from MIL-HDBK-472

Procedure	Description
I	Typically intended to be used to predict flight-line maintenance of airborne electronic and electromechanical systems involving modular replacement. It uses a calculation procedure based on a list of elemental activities for which normalized distributions and occurrence probability are given. The parameters used in this method are the distribution of DTs for various elemental activities, maintenance categories, repair times, and system DT.
II	Typically intended to be used to predict the maintainability of shipboard and shore electronic equipment and systems. It could be used to predict the maintainability of mechanical systems provided that task times and functional levels are established. Procedure II contains two different approaches: (1) Part A and (2) part B. Part A: The parameter used is M_C time expressed as an MTTR in hours. Part B: The parameters used are active maintenance in terms of mean M_C time in labor hours, mean M_P time in labor hours, and mean active maintenance time in terms of mean MH s per maintenance action.
III	Typically intended to be used to predict the mean and the maximum active M_C DT for Air Force ground electronic systems and equipment. It can also be used to predict M_P DT. The parameters used in this method are mean and maximum active corrective DT (at the 95th percentile), mean and maximum preventive DT, and MDT.
IV	Typically intended to be used to predict the mean or total corrective and PM DT of systems and equipment. The parameters used in this method are mean system maintenance DT, mean M_C DT per operational period, total CM per operational period, and total M_P DT per operational period.
V	Developed much later than the other four procedures and by far the most versatile. Typically used to predict the maintainability parameters of avionics, ground and shipboard electronics at the organizational, intermediate, and depot levels of maintenance. It presents a tabulation of time standards in relation to illustrations of what each time represents. The parameters used include MTTR, M_{max} (ϕ), MMH/repair, MMH/OH, and MMH/FH.

Source: U.S. Department of Defense (DoD), "Maintainability Prediction," MIL-HDBK-472, May 24, 1966. With permission.

■ Provide designers and maintainability engineers with a standard for monitoring and assessing compliance with stated maintainability objectives.
■ Identify areas needing additional emphasis (regarding maintainability) and areas where improvements in maintainability will have the greatest effect on the system.

Sustainability allocations provide a budget of maintainability values that, if met, ensure with a high degree of confidence that the system-level requirements are achieved. This budget is the standard against which subsequent maintainability predictions and demonstrated (i.e., measured) values are compared. The allocation of maintainability requirements must be completed and the results made available to designers and any subcontractors early in the program. Allocation is an

Table 8.24   Typical Types of "In-Place" Repair and Maintenance

Type of Maintenance Action	Performed on
Repair	• Hydraulic, pneumatic, lubrication, and fuel lines • Electrical cables and wiring • Structural components • Control cables
Calibration and adjustments	Subsystems, components, or items
Fueling and servicing (includes lubrication)	Product, components, or items

iterative process. The feasibility of achieving the initial set of allocated values must be evaluated and, if the allocated values are not reasonable, the allocation must be revised.

One final note regarding allocations is thus: As discussed thus far, and shown in the specific methods discussd in Sections 8.7.2.1 through 8.7.2.4, maintainability values allocated to subsystems, components, and so on, are expressed in the same term as that used for the product (e.g., MTTR). However, an item may simply be removed and replaced to repair the product. Repair of the item itself would then be done off the product. For example, if an aircraft (a product) had an engine failure internally, the engine would be removed and replaced. It then would be sent to the engine shop or the engine manufacturer for repair. For complex products that are mobile (wheeled and tracked vehicles, aircraft, railroad engines and cars and, to a lesser extent, ships), many repairs consist of removing and replacing the failed item or component. Table 8.24 shows the types of repairs and maintenance that are made on the product (i.e., in-place repairs).

Methods for allocating sustainability values are as follows:

■ The FR complexity method (FRCM)
■ Variation of the FRCM
■ Statistically based allocation method
■ Equal distribution method

8.7.2.1 FRCM

In this method, the most stringent maintainability requirements (i.e., the lowest MTTR values) are allocated to the subsystems and components having the lowest reliability and, conversely, the least stringent maintainability requirements are allocated to the subsystems and components having the highest reliability. The assumption is that the most complex items have the highest FRs. For this reason, the method is referred to as the FRCM. The procedure for the method is as follows:

Step 1: Determine N_i, the number of each item in the product for which allocation is being made.

Step 2: Identify λi, the FR for each item (constant FR is assumed).

Step 3: Multiply λi by N_i to find C_{fi}, item i's contribution to total FR.

Step 4: Express each item's MTTR, M_i, as the product of (λ_h/λ_i) and M_H, where H is the item having the highest FR.

Step 5: Multiply each result from step 4 by the corresponding λ_i. The result is C_{Mi}.
Step 6: Using step 5, solve for MTTR of the item having the highest FR:

$$\text{MTTR}_{\text{Product}} = \frac{\sum_i C_{Mi}}{\sum_i C_{fi}} \quad \text{where} \quad C_{Mi} = M_i C_{fi}$$

Step 7: Solve for the MTTR of the other items by multiplying the MTTR found in step 6 by λ_h/λ_i.

Table 8.25 illustrates an example of sustainability allocation using FRCM for the subsystems shown in Figure 8.6. The same method is used to allocate MTTRs found for subsystem B to its components.

8.7.2.2 Variation of the FRCM

A method used by Blanchard and Fabrycky (1998) is a variation of the FRCM. In this approach, an initial MTTR is assumed for each item and the product-level MTTR M_{Product} is calculated. If the result is equal to or less than the required M_{Product}, the allocation is complete. If it is not, then new values of each item's MTTR are selected and the process is repeated until the calculated M_{Product} is equal to or less than the required M_{Product}. The initial values for the items' MTTRs can be selected based on similar items already in use or on engineering estimates.

Table 8.25   Allocation Using FRCM

Item	Step 1: Determine No. of Items per Product (N_i)	Step 2: Identify FR λ_i ×10^{-3} failures per hour	Step 3: Calculate Contribution to Total FR C_{fi} = $N_i \lambda_i \times 10^{-3}$ failures per hour	Step 4: Express each MTTR (M_i) as (λ_H/λ_i) × M_H	Step 5: Calculate Contribution to System MTTR ($C_{Mi} =$ $M_i C_{fi}$)
A	1	5	5	M_a	$5\,M_a$
B	1	1.111	1.111	$4.5\,M_a$	$5\,M_a$
C	1	0.833	0.833	$6\,M_a$	$5\,M_a$
			$\sum C_{fi} = 6.944$		$\sum C_{Mi} = 15\,M_a$

Step 6: Solve for M_a

MTTR$_{\text{Product}} = \sum C_{Mi}/\sum C_{fi} \rightarrow 1.44 = 15\,M_a/6.944 \rightarrow M_a = 0.67$ hours

Step 7: Solve for M_b and M_c

$M_b = 4.5\,M_a = 3$ hours; $M_c = 6\,M_a = 4$ hours

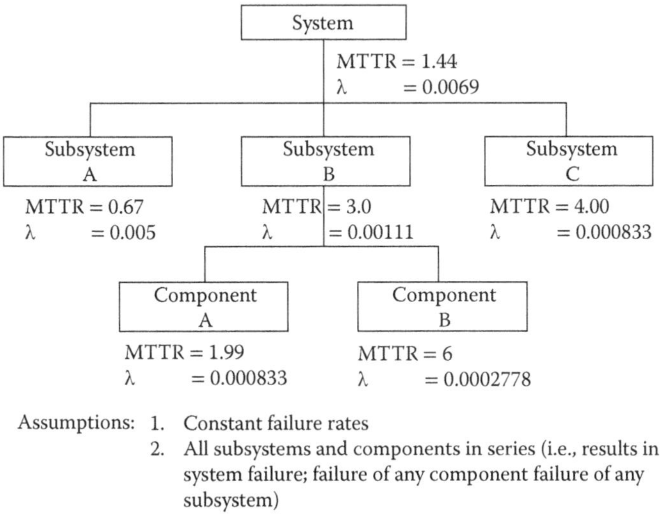
System
MTTR = 1.44
λ = 0.0069
Subsystem A
MTTR = 0.67
λ = 0.005
Subsystem B
MTTR = 3.0
λ = 0.00111
Subsystem C
MTTR = 4.00
λ = 0.000833
Component A
MTTR = 1.99
λ = 0.000833
Component B
MTTR = 6
λ = 0.0002778
Assumptions: 1. Constant failure rates
2. All subsystems and components in series (i.e., results in system failure; failure of any component failure of any subsystem)

Figure 8.6 Example of sustainability allocation.

8.7.2.3 Statistically Based Allocation Method

A well-documented, statistically sound methodology for performing a sustainability allocation may be found in IEC 706-6, "Guide on Maintainability of Equipment" (ANSI 1994). The key underlying assumption is that within a product item maintainability is inversely proportional to item complexity. This method is based on the frequently used assumption that maintenance times, especially the active M_C part of them, which is generally under the control of the supplier, can be adequately described by a lognormal distribution with mean active M_C time (MACMT) and 95th fractile maximum active M_C time (ACMT95; also called $M_{max}(95)$). Active M_C times longer than ACMT95 are also determined so as to provide the complement to the accumulated MACMT specified for the item.

8.7.2.4 Equal Distribution Method

Equal distribution is applicable when items have equal, constant FRs. Equal distribution simply allocates the product-level value of maintainability to each lower indenture item. As shown in Table 8.26 for the product depicted in Figure 8.6, using the product-level MTTR for each item does indeed result in an allocation that supports the product-level requirement. The assumption underlying this method is that repair times are unrelated to FR (i.e., MTTR is not affected by complexity). The method is identical in principle to the equal distribution method used for reliability allocations.

8.7.3 Sustainability Assessment

Periodically, assessments are made of the quantitative and qualitative sustainability characteristics of a design. In making these assessments, terms and parameters must be carefully defined. Maintainability terms and parameters are many and varied to the extent that sustainability engineers sometimes use different terms for the same measure. Therefore, all parties involved in the task must understand and track, for example, how the time intervals are measured and which type of

Table 8.26 Example of the Equal Distribution Method

Item	No. of Items per Product (N_i)	Item FR $\lambda_i \times 10^{-3}$	Contribution to Total FR C_{fi} = $N_i\,\lambda_i \times 10^{-3}$	MTTR (M_i) (Each Set Equal to Product-Level M)	Contribution to System MTTR ($C_{Mi} = MC_{fi}$)
A	1	5	5	1.44	7.2
B	1	1.11	1.11	1.44	1.6
C	1	0.833	0.833	1.44	1.2
			$\sum C_{fi} = 6.943$		$\sum C_{Mi} = 10$
Check					
MTTRProduct = $\sum C_{Mi} / \sum C_{fi}$ = 10/6.943 = 1.44 hours					

activities (preventive, corrective, or both) are included and which are excluded. If units or categories get mixed, the results of the computations will be inconsistent and have no effective meaning or use.

Sustainability assessments are conducted to provide assurance that a specified maintainability index (i.e., MTTR, MDT, mean MHs [MMH] per operating hour [MMH/OH], etc.) will be attained during a fielded operation. In a demonstration test, maintenance tasks are performed at a specified level of maintenance (e.g., organizational, intermediate, or depot levels) by personnel having the skill levels available or required in the fielded maintenance environment. The time required to perform each maintenance task is recorded. Depending on the maintainability index being demonstrated and the test plan chosen, once a statistically significant number of tasks are performed the collected data are then used to determine if the maintainability is acceptable or not. In addition to the quantitative data collected during the demonstration test, qualitative information, such as the adequacy of test support documentation or ease of maintenance (accessibility, safety, etc.), is also collected and reviewed.

The testability aspects of a design, such as BIT effectiveness, are not easily demonstrated in a formal demonstration test. In fact, demonstrations are inadequate for assessing testability because failure mechanisms that cause transient or intermittent behavior are not easily simulated in a laboratory environment (where many demonstrations take place). Also, the number of failures induced in a demonstration is small compared to the overall number of failures that may occur during fielded operation and are therefore insufficient to really demonstrate the diagnostic capabilities of a system design. Consequently, a well-planned verification program that optimizes naturally occurring failures during development and subsequent testing is needed to assess the diagnostic characteristics of the design.

8.8 Quantitative Measures of Sustainability

Quantitative sustainability requirements are associated with those design characteristics that are controllable by the designer. They are determined through an analysis of customer needs and constraints. Customer-imposed constraints include the following:

- Expected operating time (or cycles) per unit of calendar time
- Maximum DT or maintenance time, or required availability

- Operational environment and mission profile
- Skill types and skill levels of maintenance personnel
- Existing types of diagnostics and other maintenance SE available to support the product and the customer
- Turnover rate of personnel

Quantitative maintainability requirements may be expressed using many different metrics and may be established at any or all levels of maintenance. For example, they may be structured as functions of time or labor hours, or in terms of FD&FI. Examples of quantitative maintainability requirements include the following:

- Active maintenance in terms of M_C time in labor hours
- Mean M_p time in labor hours
- Mean active maintenance time in terms of mean labor hours per maintenance action
- Unit removal and installation times
- Inspection times
- Turnaround time
- Reconfiguration time
- The MTTR
- Mean time to restore system
- Maximum time to repair (at a specified confidence level ϕ; $M_{max}(\phi)$)
- The MMH per repair
- The MMH/OH
- Mean time to restore functions
- Direct MHs per maintenance action
- Mean equipment corrective maintenance time to support a unit hour of operating time
- Maintenance ratio (MR; also called direct maintenance MHs per flying hour)
- Mean time to service
- Mean time between preventive maintenance (MTBPM)
- The MMH per flying hour (MMH/FH)
- Probability of FD
- Proportion of faults isolatable
- Proportion of faults and percentage of time detected for failure modes to be detected or isolated by automatic or BITE
- Maximum FAR for automatic or BITE

Sustainability models and maintenance activities block diagrams are essential elements of sustainability analysis. Models and diagrams are developed and used as the basis for allocation and prediction processes. Models may also be used as graphical representations of maintenance tasks and may be used to assess compatibility with maintenance human resource requirements. Finally, models and diagrams are used to augment systems engineering trade-off studies. Models and maintenance activities block diagrams may be based on system engineering models and are developed for alternative system concepts or configurations or for proposed design changes. The models and maintenance activities block diagrams must be well documented and used consistently throughout the design process. As an example, Figure 8.7 shows the maintenance activities block diagram for the example maintenance tasks provided in Table 8.27.

In Figure 8.7, note the activities that can be done simultaneously (in parallel). Although it might be possible for one person to perform the task by doing each activity serially, two people make the job easier and, during application of electrical power and operational checkout, safer. Also note that both people are not needed during the entire maintenance action. The individual not performing activities 5, 6, 7, and 8 can perform other work in the vicinity of the aircraft.

At a minimum, sustainability models should consider the following:

- Operational maintenance concept
- Safety considerations

Table 8.27   Example Maintenance Tasks for the Block Diagram Shown in Figure 8.7

Task description	Replace a receiver/transmitter (R/T) on an aircraft
Number of maintenance personnel	2
Equipment/parts	Allen wrench, new R/T unit, maintenance stand, electrical power cart
Activity	
1	Get Allen wrench from toolbox (or tool bin).
2	Get maintenance stand and position adjacent to R/T equipment bay.
3	Move electrical power cart to aircraft and connect to aircraft power connection.
4	Open access panel Tridair fasteners using Allen wrench.
5	Disconnect electrical cables with BNC (Bayonet Neill–Concelman or British Naval Connector) connectors.
6	Unscrew ATR (aircraft transmitter/receiver) latches securing unit to rack and remove R/T.
7	Install new R/T and secure in position with ATR latches.
8	Reconnect electrical cables with BNC connectors.
9	Start power cart.
10	Apply power to and run operational checks on R/T unit using established procedures; then turn power off.
11	Turn off power cart.
12	Close access panel; resecure Tridair fasteners using Allen wrench.
13	Disconnect power cart and move away from aircraft.
14	Move maintenance stand away from aircraft and return to storage site.
15	Return Allen wrench to toolbox (or tool bin).

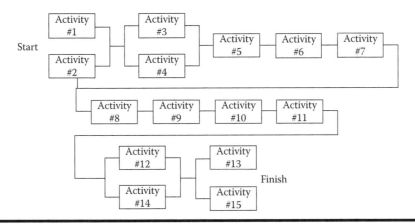

Figure 8.7   Maintenance activities block diagram.

- Applicable levels of system hierarchy
- List of LRUs
- Organizational policies

8.9 Qualitative Measures of Sustainability

Any sustainability requirement that cannot be categorized as a quantitative requirement is, by definition, qualitative. Qualitative sustainability requirements encompass a wide variety of desired outcomes considered to be essential in ensuring the product is maintainable. For example, the following qualitative requirements are often used:

- Minimize the number of new skills.
- Minimize the number of new SE.
- Minimize the number of new tools.
- Minimize the use of safety wire.
- Minimize the number and variety of fastener types.
- Use standard fasteners whenever possible.
- Use self-locking fasteners to the maximum possible extent.

Such qualitative requirements can provide guidance to the designer, but they are not useful for contractual purposes (i.e., at what point is the number of new skills minimized?). So, measurable design rules must be established to ensure that the design reflects the qualitative requirements. Now it might be tempting to say that the minimal number of new tools, skills, and so on, is zero. Or that using self-locking fasteners to the maximum extent means that only self-locking fasteners can be used. Such extreme interpretations would drive up costs and might be impractical.

So how would qualitative requirements be written as quantitative requirements? Using some of the examples of qualitative requirements already given, design rules that are quantifiable might be as follows:

- No less than 80% of all maintenance actions will be performed using only those tools in the customer's standard tool kit and no torque wrenches will be used.
- No safety wire or lockwire shall be used.

- Existing skill levels must be used at all maintenance levels.
- No more than 15% of all access panels will be designed to require more than 4 fasteners per side (or a total of 12 per perimeter).

Such design rules would be based on previous products, an assessment of the product, practical considerations, cost, and engineering judgment and experience. Many of the design guidelines are qualitative in nature. When applying these guidelines to a particular program or product, they should be tailored and rewritten as quantitative requirements.

References

American National Standards Institute (ANSI). 1994. "Guide on Maintainability of Equipment." Part 6: Section 9. "Statistical Methods in Maintainability Evaluation," Annex A, "Maintainability Allocation." (International Electrotechnical Commission) IEC 706-6. 1430 Broadway, New York, NY.

Blanchard, B. S., and W. J. Fabrycky. 1998. *Systems Engineering and Analysis*. Englewood Cliffs, NJ: Prentice-Hall, Inc.

U.S. Department of Defense (DoD). 1966 (24 May). "Maintainability Prediction." MIL-HDBK-472.

U.S. Department of Defense (DoD). 1997 (4 August). "Designing and Developing Maintainable Products and Systems." Volume I. MIL-HDBK-470A.

Chapter 9

Assessing Sustainability Capability Maturity

9.1 Overview of the Sustainability Maturity Model

The model presented in this chapter, the sustainability maturity model, describes components of the capability of an entity to be sustainably mature. It is neither a sustainability engineering process nor a life cycle model. It is a model to be used for assessment. The model defines an extensive set of sound sustainability engineering process activities and constituent practices, and it provides a measurement method for quantifying the capabilities and sustainability engineering maturity of an enterprise based on these practices. The model also offers an orderly process improvement path for enhancing sustainability engineering capabilities and maturity within an organization, such as the military maintenance, repair, and overhaul enterprise.

The foundation for sustainability maturity model is the Software Engineering Institute's (SEI) Capability Maturity Model Integration (CMMI®) (Software Engineering Institute 2006). The model is based on generic sustainability engineering process and management activities that span the full life cycle of an entity. This model was selected because it is one of the first comprehensive models of its type to be released to the public, and it has widespread government and industry interest, participation, sponsorship, and support. A good reference text for the CMMI is Chrissis et al. 2011.

A stated design goal of CMMI is to capture the key concepts of the activities and emerging engineering standards from several organizations, such as the EIA, IEEE, INCOSE, and ISO. The CMMI architecture is an adaptation of the ISO Software Process Improvement Capability determination (SPICE) Baseline Practices Guide along with the applicable references. It includes all the basic activities (BAs) that are required to perform good technical sustainability engineering activities as well as the management functions that are required to empower, enable, and track these sustainability engineering activities, such as planning, monitoring, control, and training. These processes also provide the basis for assessment and a stepwise improvement path. This path can be used to systematically improve sustainability engineering effectiveness and maturity within the military sustainment enterprise.

9.1.1 Purpose and Description of the CMMI Model

The purpose of CMMI is to help organizations improve their development and maintenance processes for both products and services. The CMMI is a collection of best practices that is generated from the CMMI Framework. The CMMI Framework supports the CMMI Product Suite by allowing multiple models, training courses, and appraisal methods to be generated that support specific areas of interest.

The CMMI is a process improvement model that provides a set of industry-recognized best practices to address productivity, performance, costs, and stakeholder satisfaction in the systems engineering and software development process. The CMMI does the following:

- Helps an organization examine the effectiveness of its processes
- Establishes priorities for improvement
- Helps implement these improvements

The CMMI provides an integrated, consistent, and enduring framework for enterprise-wide process improvement and can accommodate new initiatives as future needs are identified. It is unlike single-discipline or stove-pipe models that can result in confusion and higher costs when implemented together.

The CMMI is for organizations that provide systems and software engineering products and services to projects that transform customer needs, expectations, and constraints into products and for organizations that support these products throughout their life cycle. Organizations that manufacture, code, analyze, maintain, or document products need this model.

9.1.2 CMMI Terminology

Process descriptions in a wide range of references including standards published by IEEE, INCOSE, EIA, and ISO generally use conventional terminology such as processes with process activities, subprocesses with subprocess activities, and process outcomes or work products. The CMMI avoids the use of the term process. Instead, it uses the terms goals and practices, which are grouped to satisfy goals. The idea is to focus on areas of assessment and to group activities for that purpose. Although process areas (PAs) may look similar to processes, they are not intended to contain all the process components necessary to design products as are the processes presented in Chapter 5.

Sustainability engineering is a set of processes. These processes comprise four process categories: (1) Engineering, (2) project management, (3) support, and (4) process management. The engineering category includes PAs that are concerned with the technical and engineering aspects of a project. The project management category is similar to the set of sustainability engineering management processes discussed in Section 5.4. It consists of PAs that are primarily concerned with the technical management infrastructure needed to perform development process on a project. The support category focuses on QA and causal analyses. The process management category contains PAs primarily associated with providing an enterprise infrastructure that supports sustainability engineering activities across projects and processes that are usually concentrated at a higher-level organization within an enterprise rather than the project level.

Each of the process categories has a set of functionally similar sustainability engineering activities associated with it, which is called a PA. There are a total of 22 PAs, which make up the 4

Table 9.1   Process Categories and PAs

Category	PA
Engineering	REQM
	RD
	TS
	PI
	VER
	VAL
Project Management	PP
	PMC
	SAM
	IPM
	RSKM
	QPM
Support	CM
	PPQA
	MA
	CAR
	DAR
Process Management	OPF
	OPD
	OT
	OPP
	OID

process categories. Table 9.1 outlines the 4 categories and their associated 22 PAs. Each PA consists of a set of BAs, typically 6–9 BAs per PA. These BAs are the specific constituents of a PA, which together achieve the purpose of the PA. The BAs can also be thought of as building blocks. They can be arranged to fit various sustainability engineering life cycles, by varying the chronological order and frequency of performance of each BA. The BAs are the lowest level of sustainability engineering processes for purposes of assessment.

One other distinction between processes discussed in Chapter 5 and those presented in this chapter is as follows: The organizational improvement processes in the process management category are not used in Chapter 5. Table 9.2 lists differences in the processes between the two chapters of this book.

9.1.3 Capability Maturity Levels

A capability maturity level consists of a generic goal and its practices as they relate to a PA, which can improve an organization's processes associated with that PA. As organizational entities satisfy

Table 9.2 Processes Presented in Chapter 5 versus Processes Presented in Chapter 9

Processes in Chapter 5		Processes in Chapter 9	
ID	*Process Name*	*ID*	*Process Name*
SEP 1	Problem evaluation		
SEP 2	Project definition		
SEP 3	Supply	SAM	
SEP 4	Acquisition	SAM	
SEP 5	Work directives		
SEP 6	Life cycle model management		
SEP 7	Technical measurement	MA	
SEP 8	Project/phase closeout		
SEP 9	Stakeholder requirements	RD	
SEP 10	Requirements analysis	TS	
SEP 11	Logical architecture design	TS	
SEP 12	Physical architecture design	TS	
SEP 13	Document the design	TS	
SEP 14	Technical planning	PP	
SEP 15	Process implementation strategy	PP	
SEP 16	Technical assessment		
SEP 17	Sustainability engineering technical reviews		
SEP 18	Requirements management	REQM	
SEP 19	Interface management		
SEP 20	Decision analysis and resolution	DAR	
SEP 21	Data management		
SEP 22	Configuration management	CM	
SEP 23	QA	PPQA	
SEP 24	Risk management	RSKM	
SEP 25	Supplier performance management		
SEP 26	Design implementation		
SEP 27	Product integration	PI	

Table 9.2   Processes Presented in Chapter 5 versus Processes Presented in Chapter 9 (*Continued*)

Processes in Chapter 5		Processes in Chapter 9	
ID	*Process Name*	*ID*	*Process Name*
SEP 28	Verification	VER	
SEP 29	Product sustainability engineering analysis		
SEP 30	Testing		
SEP 31	Validation	VAL	
SEP 32	Product support readiness		
SEP 33	Product deployment		
SEP 34	Operations		
SEP 35	Maintenance		
SEP 36	Disposal		
		OPF	Organizational process focus
		OPD	Organizational process definition
		OT	Organizational training
		OPP	Organizational process performance
		OID	Organizational innovation and deployment
		CAR	Causal analysis and resolution
		QPM	Quantitative process measurement

the generic goal and its generic practices at each capability level, they reap the benefits of process improvement for that PA. The 6 capability levels, designated 0 through 5, are as follows:

0: "Incomplete"
1: "Performed" (informally)
2: "Managed" (planned and tracked)
3: "Defined" (well defined)
4: "Quantitatively managed and controlled"
5: "Optimizing" (continuously improving)

Table 9.3 lists the common features of each level. Tables 9.4 and 9.5 expand on Table 9.1 by providing the capability maturity level of each of the 22 PAs.

Table 9.3 Capability Levels and Common Features

Level	Capability	Common Features
5	Continuously improving	Improving organizational capability Improving process effectiveness
4	Quantitatively controlled	Establishing measurable quality goals Objectively managing performance
3	Well defined	Defining a standard process Performing the standard process
2	Planned and tracked (managed)	Planning performance Disciplined performance Verifying performance Tracking performance
1	Performed informally	Base practices performed
0	Not performed	Base practices not performed

Table 9.4 Capability Levels and PAs

Capability Level	Capability Name	PAs
5	Continuously improving	OID CAR
4	Quantitatively controlled	OPP QPM
3	Well defined	RD TS PI VER VAL OPF OPD OT IPM RSKM DAR

Table 9.4 Capability Levels and PAs (*Continued*)

Capability Level	Capability Name	PAs
2	Planned and tracked	REQM
		PP
		PMC
		SAM
		MA
		PPQA
		CM
1	Performed informally	—
0	Not performed	—

Table 9.5 Categories, PAs, and Their Associated Maturity Levels

Category	PA	Maturity Level
Engineering	REQM	3
	RD	2
	TS	3
	PI	3
	VER	3
	VAL	3
Project management	PP	2
	PMC	2
	SAM	2
	IPM	3
	RSKM	3
	QPM	4
Support	CM	2
	PPQA	2
	MA	2
	CAR	5
	DAR	3
Process management	OPF	3
	OPD	3
	OT	3
	OPP	4
	OID	5

9.1.3.1 Capability Level 0: Incomplete Process

An incomplete process is a process that is either not performed or partially performed. One or more of the specific goals of the PA are not satisfied, and no generic goals exist for this level since there is no reason to institutionalize a partially performed process.

9.1.3.2 Capability Level 1: Performed Process

A capability level 1 process is characterized as a performed process. A performed process is a process that satisfies the specific goals of the PA. It supports and enables the work needed to produce work products.

Although capability level 1 results in important improvements, such improvements can be lost over time if they are not institutionalized. The application of institutionalization (the CMMI generic practices at capability levels 2–5) helps to ensure that improvements are maintained.

9.1.3.3 Capability Level 2: Managed Process

A capability level 2 process is characterized as a managed process. A managed process is a performed (capability level 1) process that has the basic infrastructure in place to support the process. It is planned and executed in accordance with policy; uses skilled people who have adequate resources to produce controlled outputs; involves relevant stakeholders; is monitored, controlled, and reviewed; and is evaluated for adherence to its process description. The process discipline reflected by capability level 2 helps to ensure that existing practices are retained during times of stress.

9.1.3.4 Capability Level 3: Defined Process

A capability level 3 process is characterized as a defined process. A defined process is a managed (capability level 2) process that is tailored from the organization's set of standard processes according to the organization's tailoring guidelines and contributes work products, measures, and other process improvement information to the organizational process assets.

A critical distinction between capability levels 2 and 3 is the scope of standards, process descriptions, and procedures. At capability level 2, the standards, process descriptions, and procedures may be quite different in each specific instance of the process (e.g., on a particular project). At capability level 3, the standards, process descriptions, and procedures for a project are tailored from the organization's set of standard processes to suit a particular project or organizational unit and, therefore, they are more consistent except for the differences allowed by the tailoring guidelines.

Another critical distinction is that at capability level 3 processes are typically described more rigorously than at capability level 2. A defined process clearly states purpose, inputs, entry criteria, activities, roles, measures, verification steps, outputs, and exit criteria. At capability level 3, processes are managed more proactively using an understanding of the interrelationships of process activities and detailed measures of a process, its work products, and its services.

9.1.3.5 Capability Level 4: Quantitatively Managed Process

A capability level 4 process is characterized as a quantitatively managed process. A quantitatively managed process is a defined (capability level 3) process that is controlled using statistical and other quantitative techniques. Quantitative objectives for quality and process performance are established and used as criteria in managing the process. Quality and process performance is understood in statistical terms and is managed throughout the life of the process.

9.1.3.6 Capability Level 5: Optimizing Process

A capability level 5 process is characterized as an optimizing process. An optimizing process is a quantitatively managed (capability level 4) process that is improved based on an understanding of the common causes of variation inherent in the process. The focus of an optimizing process is on continually improving the range of process performance through both incremental and innovative improvements.

9.1.4 Application to IPTs

The process activities presented in the Section 9.2 are equally applicable to functionally organized programs and to programs organized according to an IPT approach. In the IPT approach, each major subsystem or component is developed as a distinct product by an interdisciplinary team. The team leader typically acts in the capacity of a "mini program manager" and assumes cost, schedule, and performance responsibility for a product. The team composition includes all members needed to make timely decisions, including customers and suppliers. This approach helps facilitate empowerment of the team for decision making at the lowest possible level. However, in the IPT approach special care must be taken by team members relative to contract issues when IPTs cross enterprise lines or involve fixed-price work. The assessment, measurement, analysis, and improvement of sustainability engineering processes also involve an IPT approach.

Team composition consists of members from all disciplines needed to support a concurrent engineering approach. This facilitates development of a system solution that is optimized for the entire life cycle including production and support. For example, a team may comprise members representing customer, design engineering, sustainability engineering, software engineering, systems engineering, manufacturing, test and evaluation, QA, reliability and maintainability, and logistics support. Members do not necessarily spend 100% of their time as part of one team and may belong to several teams.

With the IPT approach, the overall program may contain many product and process teams. For each product or process, which comprises both hardware and software components, the responsible team contains both hardware and software engineers. Hardware team membership includes engineers having each of the skills required by the product development phase. The largest benefit to this approach is that it enables decision making to occur in a timely manner. This is accomplished through the presence of all disciplines needed for resolution of issues and problems and the empowerment of decision making at the lowest possible level. The primary idea behind IPTs is to take advantage of team member expertise to produce a product correctly the first time. Concurrency in addressing production, test, and support shifts most engineering design changes to earlier phases of program development and reduces the overall number of changes incurred.

The IPT approach facilitates accomplishing all base practices contained in the understand customer needs and expectations PA (requirements development [RD]), due to the active participation of customer as a member of the team. This participation also provides an improvement in the overall efficiency of the customer's oversight and review process. For example, instead of the sequential preparation and review associated with the development of specifications, concurrent review and approval results in faster resolution of requirements conflicts and other requirements issues.

The IPT approach uses event-driven scheduling. This approach needs to be factored into the SEMP and other plans when defining procedures in support of the project planning (PP) and project monitoring and control (PMC) PAs. Event-driven scheduling relates program events to their accomplishments and corresponding accomplishment criteria. An event is considered complete only when all accomplishments are complete as defined by the corresponding criteria. This reduces risk by ensuring that product maturity is incrementally demonstrated prior to the start of subsequent activities.

9.2 Process Descriptions

In Sections 9.2.1 to 9.2.4 that follow, summary information is presented for each of the PAs of the sustainability maturity model (the CMMI model). The information follows the four-part format presented in Figure 9.1.

9.2.1 Engineering Category

The engineering category covers the development and maintenance activities that are shared across engineering disciplines (e.g., systems engineering and software engineering). There are six process areas in this category, and they have inherent interrelationships.

9.2.1.1 Requirements Management

Purpose: The purpose of requirements management (REQM) is to manage the requirements of a project's products and product components and identify inconsistencies between the requirements and the project's plans and work products.

Summary of BAs and subactivities (SAs):

BA 1: Manage requirements.
SA 1.1: Obtain an understanding of requirements.
SA 1.2: Obtain commitment to requirements.
SA 1.3: Manage requirements changes.
SA 1.4: Maintain bidirectional traceability of requirements.
SA 1.5: Identify inconsistencies between project work and requirements.

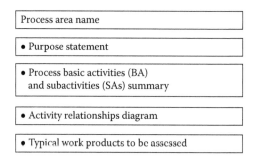

Figure 9.1 Process descriptions layout.

Activity relationships diagram:

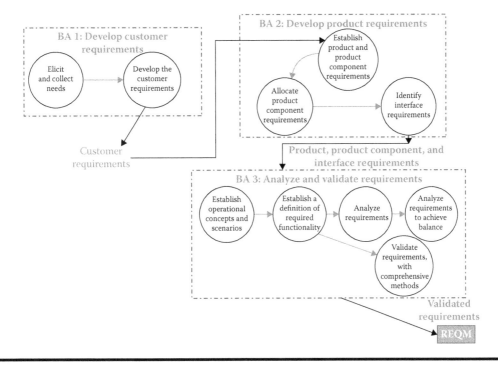

Figure 9.2 The REQM. (Software Engineering Institute 2006.)

Typical work products to be assessed:
1. Lists of criteria for distinguishing appropriate requirements providers
2. Criteria for evaluation and acceptance of requirements
3. Results of analysis against criteria
4. An agreed-on set of requirements
5. Requirements impact assessments
6. Documented commitments to requirements and requirements changed
7. Requirements status
8. Requirements database
9. Requirements traceability matrix
10. Requirements tracking system
11. Documentation of inconsistencies including sources, conditions, and rationale
12. Corrective actions

9.2.1.2 Requirements Development

Purpose: The purpose of RD is to produce and analyze customer, product, and product component requirements.

Summary of BAs and SAs:

BA 1: Develop customer requirements.

SA 1.1: Elicit needs.

SA 1.2: Develop customer requirements.

BA 2: Develop product requirements.
SA 2.1: Establish product and product component requirements.
SA 2.2: Allocate product component requirements.
SA 2.3: Identify interface requirements.
BA 3: Analyze and validate requirements.
SA 3.1: Establish operational concepts and scenarios.
SA 3.2: Establish a definition of required functionality.
SA 3.3: Analyze requirements.
SA 3.4: Analyze requirements to achieve balance.
SA 3.5: Validate requirements.
Activity relationships diagram:

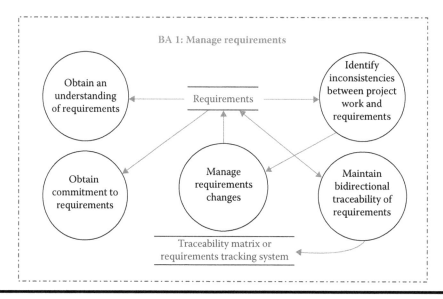

Figure 9.3 The RD. (Software Engineering Institute 2006.)

Typical work products to be assessed:
1. Customer requirements
2. Customer constraints on the conduct of verification
3. Customer constraints on the conduct of validation
4. Derived requirements
5. Product requirements
6. Product component requirements
7. Requirements allocation sheets
8. Provisional requirements allocations, design constraints, derived constraints
9. Relationships among derived requirements
10. Interface requirements
11. Operational concept
12. Product or product component installation; operational, maintenance, and support concepts
13. Disposal concepts use cases
14. Time line scenarios

15. New requirements
16. Functional architecture
17. Activity diagrams and use cases
18. Object-oriented analysis with services or methods identified
19. Requirements defects reports
20. Proposed requirements changed to resolve defects
21. Key requirements
22. Technical performance measures
23. Assessment of risks related to requirements
24. Record of analysis methods and results

9.2.1.3 Technical Solution

Purpose: The purpose of a technical solution (TS) is to design, develop, and implement solutions to requirements. Solutions, designs, and implementations encompass products, product components, and product-related processes either singly or in combination as appropriate.

Summary of BAs and SAs:

BA 1: Select product component solutions.
    SA 1.1: Develop alternative solutions and selection criteria.
    SA 1.2: Select product component solutions.

BA 2: Develop the design.
    SA 2.1: Design the product or product component.
    SA 2.2: Establish a technical data package.
    SA 2.3: Design interfaces using criteria.
    SA 2.4: Perform make, buy, or reuse analyses.

BA 3: Implement the product design.
    SA 3.1: Implement the design.
    SA 3.2: Develop product support documentation.

Activity relationships diagram:

Typical work products to be assessed:

1. Alternative solution-screening criteria
2. Evaluation reports of new technologies
3. Alternative solutions
4. Selection criteria and final selection
5. Evaluation reports of COTS products
6. Product component selection decisions and rationale
7. Documented relationships between requirements and product components
8. Documented solutions, evaluations, and rationale
9. Product architecture
10. Product component designs

9.2.1.4 Product Integration

Purpose: The purpose of product integration (PI) is to assemble the product from product components; ensure that the product, once integrated, functions properly; and deliver the product.

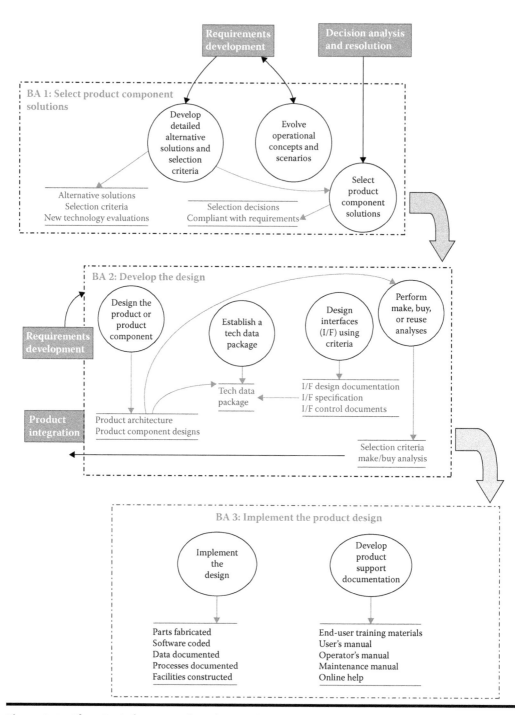

Figure 9.4   The TS. (Software Engineering Institute 2006.)

Summary of BAs and SAs:
 BA 1: Prepare for PI.
  SA 1.1: Determine integration sequence.
  SA 1.2: Establish the PI environment.
  SA 1.3: Establish PI procedures and criteria.
 BA 2: Ensure interface compatibility.
  SA 2.1: Review interface descriptions for completeness.
  SA 2.2: Manage interfaces.
 BA 3: Assemble product components and deliver the product.
  SA 3.1: Confirm readiness of product components for integration.
  SA 3.2: Assemble product components.
  SA 3.3: Evaluate assembled product components.
  SA 3.4: Package and deliver the product or product component.
Activity relationships diagram:

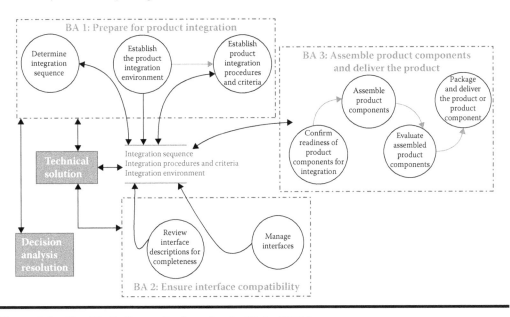

Figure 9.5 The PI. (Software Engineering Institute 2006.)

Typical work products to be assessed:
1. PI sequence
2. Rationale for selecting or rejecting integration sequences
3. Verified environment for PI
4. PI procedures
5. PI criteria
6. Categories of interfaces
7. Lists of interfaces per category
8. Table of relationships among product components and the external system environment (e.g., main power supply, fasteners, and computer bus system)
9. Table of relationships among different product components

10. List of agreed-on interfaces defined for each pair of product components, when applicable
11. Reports of interface control working group meetings
12. Action items for updating interfaces
13. Application program interfaces
14. Updated interface description or agreement
15. Acceptance documents for the received product components
16. Delivery receipts
17. Checked packing lists
18. Exception reports
19. Waivers

9.2.1.5 Work Product Verification

Purpose: The purpose of verification (VER) is to ensure that the selected work product meets its specified requirements.

Summary of BAs and SAs:

BA 1: Prepare for verification.

SA 1.1: Select work products for verification.

SA 1.2: Establish the verification environment.

SA 1.3: Establish verification procedures and criteria.

BA 2: Perform peer reviews.

SA 2.1: Prepare for peer reviews.

SA 2.2: Conduct peer reviews.

SA 2.3: Analyze peer review data.

BA 3: Verify selected work products.

SA 3.1: Perform verification.

SA 3.2: Analyze verification results.

Activity relationships diagram:

Typical work products to be assessed:

1. Lists of work products selected for verification
2. Verification methods for each selected work product
3. Verification environment
4. Verification procedures
5. Verification criteria
6. Peer review schedule
7. Peer review checklist
8. Entry and exit criteria for work products
9. Criteria for requiring another peer review
10. Peer review training material for selected work products
11. Peer review results
12. Peer review issues
13. Peer review data
14. Peer review action items
15. Verification results
16. Verification reports

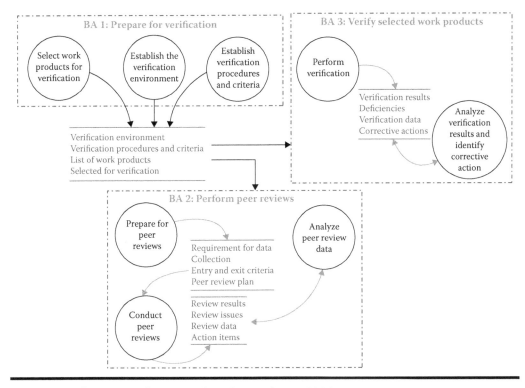

Figure 9.6   The VER. (Software Engineering Institute 2006.)

17. Demonstrations
18. As-run procedures log
19. Analysis report
20. Trouble reports
21. Change requests for verification methods, criteria, and environment

9.2.1.6 Work Product Validation

Purpose: The purpose of validation (VAL) is to demonstrate that a product or product component fulfills its intended purpose in its intended use when placed in its intended environment.

Summary of BAs and SAs:

BA 1: Prepare for validation.

SA 1.1: Select products for validation.

SA 1.2: Establish the validation environment.

SA 1.3: Establish validation procedures and criteria.

BA 2: Validate product or product components.

SA 2.1: Perform validation.

SA 2.2: Analyze validation results.

Activity relationships diagram:

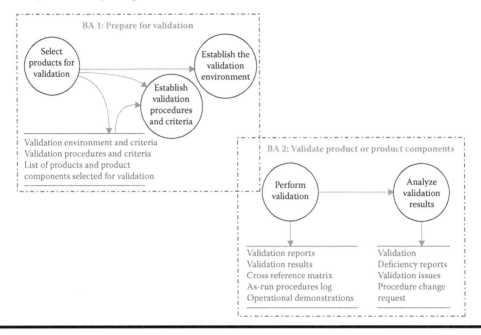

Figure 9.7   The VAL. (Software Engineering Institute 2006.)

Typical work products to be assessed:
1. Lists of work products selected for validation
2. Validation methods for each selected product or product component
3. Requirements for performing validation for each product or product component
4. Validation constraints for each product or product component
5. Validation environment
6. Validation procedures
7. Validation criteria
8. Test and evaluation procedures for maintenance, training, and support
9. Entry and exit criteria for work products
10. Validation results
11. Validation reports
12. Validation cross-reference matrix
13. As-run procedures log
14. Operational demonstrations
15. Validation deficiency reports
16. Validation issues
17. Change requests for procedures

9.2.2 Project Management Category

The project management category covers the activities related to planning, monitoring, and controlling the project.

9.2.2.1 Project Planning

Purpose: The purpose of project planning (PP) is to establish and maintain plans that define project activities.

Summary of BAs and SAs:

BA 1: Establish estimates.

　SA 1.1: Estimate the scope of the project.

　SA 1.2: Establish estimates of work product and task attributes.

　SA 1.3: Define project life cycle.

　SA 1.4: Determine estimates of effort and cost.

BA 2: Develop a project plan.

　SA 2.1: Establish the budget and schedule.

　SA 2.2: Identify project risks.

　SA 2.3: Plan for data management.

　SA 2.4: Plan for project resources.

　SA 2.5: Plan for needed knowledge and skills.

　SA 2.6: Plan for stakeholder involvement.

　SA 2.7: Establish the project plan.

BA 3: Obtain commitment to the plan.

　SA 3.1: Review plans that affect the project.

　SA 3.2: Reconcile work and resource levels.

　SA 3.3: Obtain plan commitment.

Activity relationships diagram:

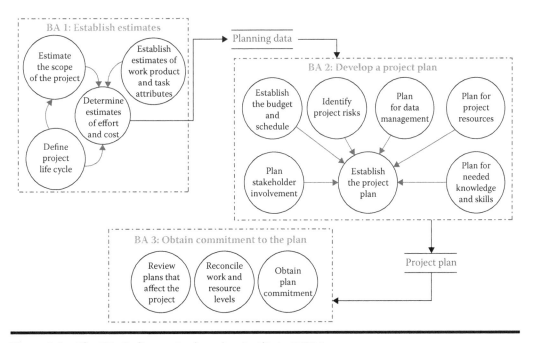

Figure 9.8　The PP. (Software Engineering Institute 2006.)

Typical work products to be assessed:
1. Task descriptions
2. Work-package descriptions
3. The WBS
4. Technical approach
5. Size and complexity of tasks and work products
6. Estimating models
7. Attribute estimates
8. Project life cycle phases
9. Estimation rationale
10. Project effort estimates
11. Project cost estimates
12. Project schedules
13. Schedule dependencies
14. Project budget
15. Identified risks
16. Risk impacts and probability of occurrence
17. Risk priorities
18. Data management plan
19. Master list of managed data
20. Data content and format description
21. Data requirements lists for acquirers and for suppliers
22. Privacy requirements; security requirements
23. Security procedures
24. Mechanism for data retrieval, reproduction, and distribution
25. Schedule for collection of project data
26. Listing of project data to be collected
27. The WBS work packages
28. The WBS task dictionary
29. Staffing requirements based on project size and scope
30. Critical facilities and equipment lists
31. Process and workflow definitions and diagrams
32. Program administration requirements list
33. Inventory of skills needs
34. Staffing and new hire plans
35. Database for skills and training
36. Stakeholder involvement plan
37. Overall project plan

9.2.2.2 Project Monitoring and Control

Purpose: The purpose of PMC is to provide an understanding of a project's progress so that appropriate corrective actions can be taken when the project's performance deviates significantly from the plan.

Summary of BAs and SAs:

BA 1: Monitor project against plan.

SA 1.1: Monitor PP parameters.

SA 1.2: Monitor commitments.

SA 1.3: Monitor project risks.

SA 1.4: Monitor data management.

SA 1.5: Monitor stakeholder involvement.

SA 1.6: Conduct progress reviews.

SA 1.7: Conduct milestone reviews.

BA 2: Manage corrective action to closure.

SA 2.1: Analyze issues.

SA 2.2: Take corrective action.

SA 2.3: Manage corrective action.

Activity relationships diagram:

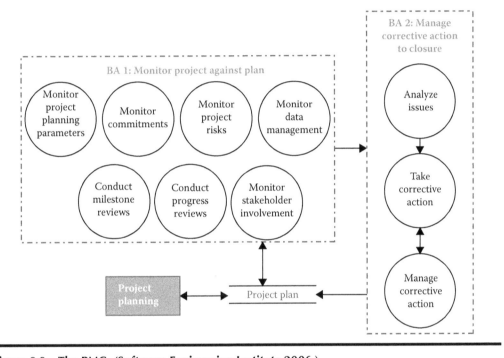

Figure 9.9   The PMC. (Software Engineering Institute 2006.)

Typical work products to be assessed:

1. Records of project performance
2. Records of significant deviations
3. Records of commitment reviews
4. Records of project risk monitoring
5. Records of data management
6. Records of stakeholder involvement
7. Documented project review results
8. Documented milestone review results
9. Lists of issues needing corrective actions
10. Corrective action plans
11. Corrective action results

9.2.2.3 Supplier Agreement Management

Purpose: The purpose of supplier agreement management (SAM) is to manage the acquisition of products from suppliers.

Summary of BAs and SAs:

BA 1: Establish supplier agreements.

SA 1.1: Determine acquisition type.

SA 1.2: Select suppliers.

SA 1.3: Establish supplier agreements.

BA 2: Satisfy supplier agreements.

SA 2.1: Execute the supplier agreement.

SA 2.2: Monitor selected supplier processes.

SA 2.3: Evaluate selected supplier work products.

SA 2.4: Accept the acquired product.

SA 2.5: Transition products.

Activity relationships diagram:

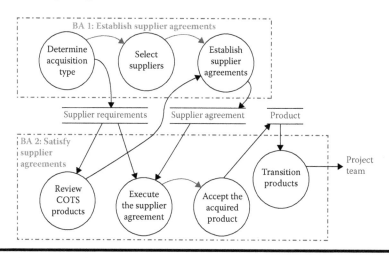

Figure 9.10 The SAM. (Software Engineering Institute 2006.)

Typical work products to be assessed:

1. List of acquisition types that will be used for all products and product components to be acquired
2. Market studies
3. List of candidate suppliers
4. Preferred supplier list
5. Trade study or other record of evaluation criteria, advantages and disadvantages of candidate suppliers, and rationale for selection of suppliers
6. Solicitation materials and requirements
7. The SOWs
8. Contracts
9. Memoranda of agreement
10. Licensing agreement
11. Supplier progress reports and performance measures

12. Supplier review materials and reports
13. Action items tracked to closure
14. Documentation of product and document deliveries
15. List of processes selected for monitoring or rationale for nonselection
16. Activity reports
17. Performance curves
18. Discrepancy reports
19. Lists of work products selected for monitoring or rationale for nonselection
20. Activity reports
21. Discrepancy reports
22. Acceptance test procedures
23. Acceptance test results
24. Discrepancy reports or corrective action plans
25. Transition plans
26. Training reports
27. Support and maintenance reports

9.2.2.4 Integrated Project Management

Purpose: The purpose of integrated project management (IPM) is to establish and manage a project and the involvement of relevant stakeholders according to an integrated and defined process that is tailored from the organization's set of standard processes.

Summary of BAs and SAs:

BA 1: Use the project's defined process.
   SA 1.1: Establish the project's defined process.
   SA 1.2: Use organizational process assets for planning project activities.
   SA 1.3: Establish the project's work environment.
   SA 1.4: Integrate plans.
   SA 1.5: Manage the project using integrated plans.
   SA 1.6: Contribute to organizational process assets.
BA 2: Coordinate and collaborate with relevant stakeholders.
   SA 2.1: Manage stakeholder involvement.
   SA 2.2: Manage dependencies.
   SA 2.3: Resolve coordination issues.
BA 3: Apply IPPD principles.
   SA 3.1: Establish the project's shared vision.
   SA 3.2: Establish the integrated team structure.
   SA 3.3: Allocate requirements to integrated teams.
   SA 3.4: Establish integrated teams.
   SA 3.5: Ensure collaboration among interfacing teams.

Activity relationships diagram:

Typical work products to be assessed:

1. Project's defined process
2. Project estimates
3. Project plans
4. Equipment and tools for the project
5. Installation, operation, and maintenance manuals for the project work environment

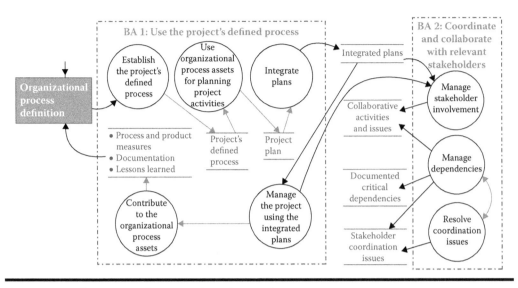

Figure 9.11   The IPM. (Software Engineering Institute 2006.)

6. User surveys and results
7. Usage, performance, and maintenance records
8. Support services for the project's work environment
9. Integrated plans
10. Work products created by performing the project's defined processes
11. Collected measures (actuals) and progress records or reports
12. Revised requirements, plans, and commitments
13. Proposed improvements to organizational process assets
14. Actual process and product measures collected from the project
15. Documentation of items such as process descriptions, plans, training modules, checklists, and lessons learned concerning processes
16. Process artifacts associated with tailoring and implementing the organization's set of standard processes on the project
17. Agendas and schedules for collaborative activities
18. Documented issues with items such as customer requirements, product and product component requirements, product architecture, and product design
19. Recommendations for resolving relevant stakeholder issues
20. Defects issues, and action items resulting from reviews with relevant stakeholders
21. Critical dependencies
22. Commitments to address critical dependencies
23. Status of critical dependencies
24. Relevant stakeholder coordination issues
25. Status of relevant stakeholder coordination issues

9.2.2.5 Risk Management

Purpose: The purpose of risk management (RSKM) is to identify potential problems before they occur so that risk-handling activities can be planned and invoked as needed across the life of the product or project to mitigate adverse impacts on achieving objectives.

Summary of BAs and SAs:
    BA 1: Prepare for RSKM.
        SA 1.1: Determine risk sources and categories.
        SA 1.2: Define risk parameters.
        SA 1.3: Establish a RSKM strategy.
    BA 2: Identify and analyze risks.
        SA 2.1: Identify risks.
        SA 2.2: Evaluate, categorize, and prioritize risks.
    BA 3: Mitigate risks.
        SA 3.1: Develop risk mitigation plans.
        SA 3.2: Implement risk mitigation plans.
Activity relationships diagram:

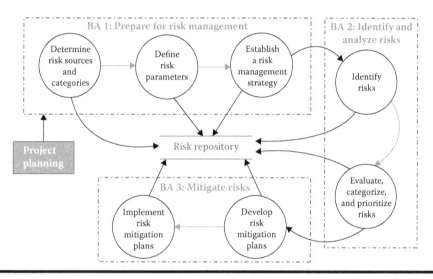

Figure 9.12 The RSKM. (Software Engineering Institute 2006.)

Typical work products to be assessed:
1. Risk source lists (external and internal)
2. Risk categories list
3. Risk evaluation, categorization, and prioritization criteria
4. RSKM requirements including control and approval levels, and risk assessment intervals
5. Project RSKM strategy
6. List of identified risks, including the context, conditions, and consequences
7. List of risks with a priority assigned to each risk
8. Documented handling options for each identified risk
9. Risk mitigation plans
10. Contingency plans
11. List of those responsible for tracking and addressing each risk
12. Updated lists of risk status
13. Updates assessments of risk likelihood, consequence, and thresholds
14. Updated lists of risk-handling options

15. Updated list of actions taken to handle risks
16. Updated risk mitigation plans

9.2.2.6 Quantitative Project Management

Purpose: The purpose of quantitative project management (QPM) is to quantitatively manage a project's defined process to achieve the project's established quality and process-performance objectives.

Summary of BAs and SAs:

BA 1: Quantitatively manage the project.

SA 1.1: Establish the project's objectives.

SA 1.2: Compose the defined process.

SA 1.3: Select the subprocesses that will be statistically managed.

SA 1.4: Manage project performance.

BA 2: Statistically manage subprocess performance.

SA 2.1: Select measures and analytic techniques.

SA 2.2: Apply statistical methods to understand variation.

SA 2.3: Monitor performance of the selected subprocesses.

SA 2.4: Record statistical management data.

Activity relationships diagram:

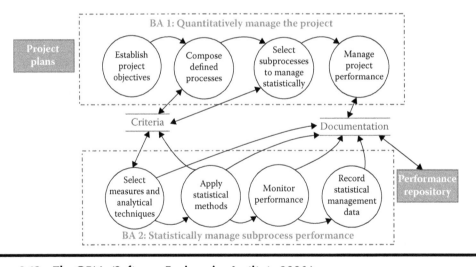

Figure 9.13   The QPM. (Software Engineering Institute 2006.)

Typical work products to be assessed:

1. The project's quality and process-performance objectives
2. Criteria used in identifying which subprocesses are valid candidates for inclusion in the project's defined process
3. Candidate subprocesses for inclusion in the project's defined process
4. Subprocesses to be included in the project's defined process
5. Identified risks when selected subprocesses lack a process-performance history

6. Quality and process-performance objectives that will be addressed by statistical management
7. Criteria used in selecting which subprocesses will be statistically managed
8. Subprocesses that will be statistically managed
9. Identified process and product attributes of the selected subprocesses that should be measured and controlled
10. Estimates (predictions) of achievement of the project's quality and process-performance objectives
11. Documentation of actions needed to address deficiencies in achieving the project's objectives
12. Definitions of measures and analytical techniques to be used in (or proposed for) statistically managing subprocesses
13. Operational definitions of the measures, their collection points in the subprocesses, and how the integrity of the measures will be determined
14. Traceability of measures back to the project's quality and process-performance objectives
15. Instrumented organizational support environment to support data collection
16. Collected measures
17. Natural bounds of process performance for each measured attribute of selected subprocesses
18. Process performance compared to the natural bounds of process performance for each measured attribute of each selected subprocess
19. Natural bounds of process performance for each selected subprocess compared to its established (derived) objectives
20. For each process, its process capability
21. For each subprocess, actions needed to address deficiencies in its process capability
22. Statistical and quality management data recorded in the organization's measurement repository

9.2.3 Support Category

The support category includes the activities that support product development and maintenance. The category addresses those processes that are targeted towards the project, and may address processes that apply more generally to the organization. For example, Process and Product Quality Assurance can be used with all the process areas to provide an evaluation of the work products described in all of the process areas.

9.2.3.1 Configuration Management

Purpose: The purpose of CM is to establish and maintain the integrity of work products using configuration identification, configuration control, configuration status accounting, and configuration audits.

Summary of BAs and SAs:

BA 1: Establish baselines.
   SA 1.1: Identify CIs.
   SA 1.2: Establish a CM system.
   SA 1.3: Create or release baselines.

BA 2: Track and control changes.
   SA 2.1: Track change requests.
   SA 2.2: Control CIs.

BA 3: Establish integrity.
    SA 3.1: Establish CM records.
    SA 3.2: Perform configuration audits.
Activity relationships diagram:

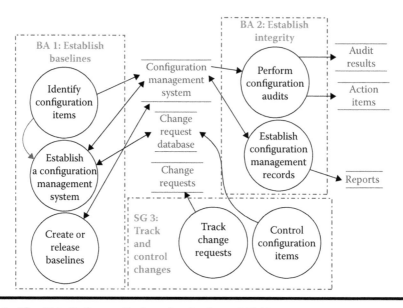

Figure 9.14　The CM. (Software Engineering Institute 2006.)

Typical work products to be assessed:
1. Identified CIs
2. The CM system with controlled work products
3. The CM access control procedures
4. Change request database
5. Baselines
6. Description of baselines
7. Change requests
8. Revision history of CIs
9. Archives of the baselines
10. Change log
11. Copies of change requests
12. Status of change requests
13. Differences between baselines
14. Configuration audit results
15. Action items from configuration audits

9.2.3.2 Process and Product Quality Assurance

Purpose: The purpose of process and product quality assurance (PPQA) is to provide staff and management with objective insight into processes and associated work products.

Summary of BAs and SAs:
　　BA 1: Align measurement and analysis (MA) activities.
　　　　SA 1.1: Establish measurement objectives.
　　　　SA 1.2: Specify measures.
　　　　SA 1.3: Specify data collection and storage procedures.
　　　　SA 1.4: Specify analysis procedures.
　　BA 2: Provide measurement results.
　　　　SA 2.1: Collect measurement data.
　　　　SA 2.2: Analyze measurement data.
　　　　SA 2.3: Store data and results.
　　　　SA 2.4: Communicate results.
Activity relationships diagram:

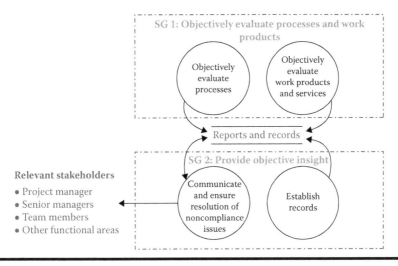

Figure 9.15　The PPQA. (Software Engineering Institute 2006.)

Typical work products to be assessed:
1. Evaluation reports
2. Noncompliance reports
3. Corrective actions
4. Corrective action reports
5. Quality trends
6. Evaluation logs
7. QA reports
8. Status reports of corrective actions
9. Reports of quality trends

9.2.3.3 Measurement and Analysis

Purpose: The purpose of MA is to develop and sustain a measurement capability that is used to support management information needs.

Summary of BAs and SAs:
    BA 1: Align MA activities.
        SA 1.1: Establish measurement objectives.
        SA 1.2: Specify measures.
        SA 1.3: Specify data collection and storage procedures.
        SA 1.4: Specify analysis procedures.

    BA 2: Provide measurement results.
        SA 2.1: Collect measurement data.
        SA 2.2: Analyze measurement data.
        SA 2.3: Store data and results.
        SA 2.4: Communicate results.
Activity relationships diagram:

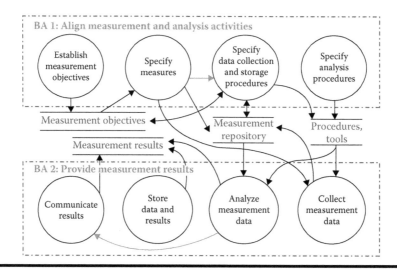

Figure 9.16 The MA. (Software Engineering Institute 2006.)

Typical work products to be assessed:
1. Measurement objectives
2. Specifications (description summaries) of base and derived measures
3. Data collection and storage procedures
4. Data collection tools
5. Analysis specifications and procedures
6. Data analysis tools
7. Base and derived measurement data sets
8. Results of data integrity tests
9. Analysis results and draft reports
10. Stored data inventory
11. Delivered reports and related analysis results
12. Contextual information or guidance to aid in the interpretation of analysis reports

9.2.3.4 Causal Analysis and Resolution

Purpose: The purpose of causal analysis and resolution (CAR) is to identify causes of defects and other problems and to take action to prevent them from occurring in the future.

Summary of BAs and SAs:

BA 1: Determine causes of defects.

SA 1.1: Select defect data for analysis.

SA 1.2: Analyze causes.

BA 2: Address causes of defects.

SA 2.1: Implement the action proposals.

SA 2.2: Evaluate the effect of changes.

SA 2.3: Record data.

Activity relationships diagram:

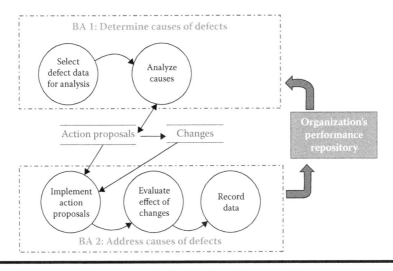

Figure 9.17   The CAR. (Software Engineering Institute 2006.)

Typical work products to be assessed:

1. Defect and problem data selected for further analysis
2. Action proposals
3. Action proposals selected for implementation
4. Improvement proposals
5. Measures of performance and performance change
6. The CAR records

9.2.3.5 Decision Analysis and Resolution

Purpose: The purpose of decision analysis and resolution (DAR) is to analyze possible decisions using a formal evaluation process that evaluates identified alternatives against established criteria.

Summary of BAs and SAs:

BA 1: Evaluate alternatives.

SA 1.1: Establish guidelines for decision analysis.

SA 1.2: Establish evaluation criteria.

SA 1.3: Identify alternative solutions.

SA 1.4: Select evaluation methods.

SA 1.5: Evaluate alternatives.

SA 1.6: Select solutions.

Activity relationships diagram:

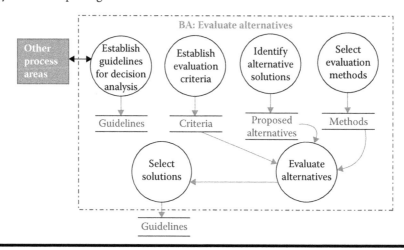

Figure 9.18 The DAR. (Software Engineering Institute 2006.)

Typical work products to be assessed:
1. Guidelines on when to apply formal evaluation process
2. Documented evaluation criteria
3. Rankings of criteria importance
4. Identified alternatives
5. Selected evaluation methods
6. Evaluation results
7. Recommended solutions to address significant issues

9.2.4 Process Management Category

The process management category contains the cross-project activities related to defining, planning, deploying, implementing, monitoring, controlling, appraising, measuring, and improving processes.

9.2.4.1 Organizational Process Focus

Purpose: The purpose of organizational process focus (OPF) is to plan, implement, and deploy organizational process improvements based on a thorough understanding of the current strengths and weaknesses of the organization's processes and process assets.

Summary of BAs and SAs:

BA 1: Determine process improvement opportunities.

SA 1.1: Establish organizational process needs.

SA 1.2: Appraise the organization's processes.

SA 1.3: Identify the organization's process improvements.

BA 2: Plan and implement process improvements.

SA 2.1: Establish process action plans.

SA 2.2: Implement process action plans.

BA 3: Deploy organizational process assets and incorporate lessons learned.

SA 3.1: Deploy organizational process assets.

SA 3.2: Deploy standard processes.

SA 3.3: Monitor implementation.

SA 3.4: Incorporate process-related experiences into organizational process assets.

Activity relationships diagram:

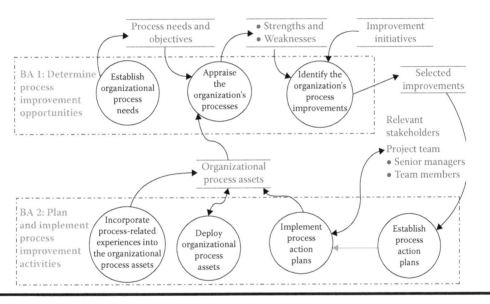

Figure 9.19   The OPF. (Software Engineering Institute 2006.)

Typical work products to be assessed:

1. Organization's process needs and objectives
2. Plans for the organization's process appraisals
3. Appraisal findings that address strengths and weaknesses of the organization's processes
4. Improvement recommendations for the organization's processes
5. Analysis of candidate process improvements
6. Identification of improvements for the organization's processes
7. Organization's approved process action plans
8. Commitments among various process action teams
9. Status and results of implementing process action plans
10. Plans for pilots
11. Plans for deploying organizational process assets and changes to them
12. Training materials for deploying organizational process assets and changes to them
13. Documentation of changes to organizational process assets
14. Support materials for deploying organizational process assets and changes to them

15. Organization's list of projects and status of process deployment on each project including existing and planned projects
16. Guidelines for deploying the organization's set of standard processes on new projects
17. Records of tailoring the organization's set of standard processes and implementing them on identified projects
18. Results of monitoring process implementation on projects
19. Status and results of process compliance evaluations
20. Results of reviewing selected process artifacts created as part of process tailoring and implementation
21. Process improvement proposals
22. Process lessons learned
23. Measurements on organizational process assets
24. Improvement recommendations for organizational process assets
25. Records of the organization's process improvement activities
26. Information on organizational process assets and improvements to them

9.2.4.2 Organizational Process Definition

Purpose: The purpose of organizational process definition (OPD) is to establish and maintain a usable set of organizational process assets and work environment standards.

Summary of BAs and SAs:

BA 1: Establish organizational process assets.

   SA 1.1: Establish standard processes.

   SA 1.2: Establish life cycle model descriptions.

   SA 1.3: Establish tailoring criteria and guidelines.

   SA 1.4: Establish the organization's measurement repository.

   SA 1.5: Establish the organization's process asset library.

   SA 1.6: Establish work environment standards.

Activity relationships diagram:

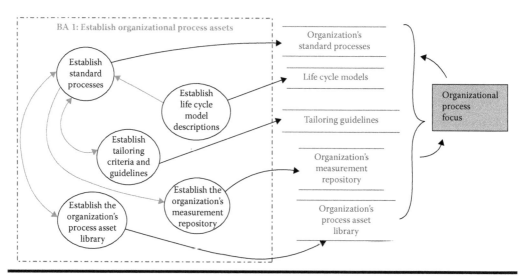

Figure 9.20   The OPD. (Software Engineering Institute 2006.)

Typical work products to be assessed:
1. Organization's set of standard processes
2. Descriptions of life cycle models
3. Tailoring guidelines for the organization's standard processes
4. Definition of the common set of product and process measures for the organization's set of standard processes
5. Design of the organization's measurement repository
6. Organization's measurement repository including structure and support environment
7. Organization's measurement data
8. Design of the organization's process asset library
9. Organization's process asset library
10. Selected items to be included in the organization's process asset library
11. Catalog of items in the organization's process asset library
12. Work environment standards

9.2.4.3 Organizational Training

Purpose: The purpose of organizational training (OT) is to develop skills and knowledge of people so that they can perform their roles effectively and efficiently.

Summary of BAs and SAs:

BA 1: Establish an organizational training capability.
    SA 1.1: Establish the strategic training needs.
    SA 1.2: Determine which training needs are the responsibilities of the organization.
    SA 1.3: Establish an organizational training tactical plan.
    SA 1.4: Establish training capability.

BA 2: Provide necessary training.
    SA 2.1: Deliver training.
    SA 2.2: Establish training records.
    SA 2.3: Assess training effectiveness.

Activity relationships diagram:

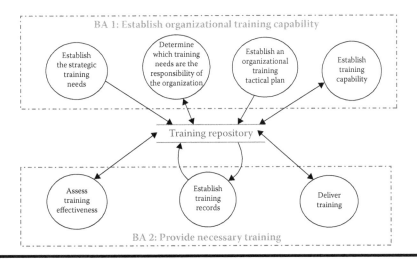

Figure 9.21 The OT. (Software Engineering Institute 2006.)

Typical work products to be assessed:
1. Training needs descriptions
2. Training needs assessment analysis
3. Common project and support group training needs
4. Training commitments
5. Organization's training tactical plan
6. Training materials and supporting artifacts
7. Delivered training courses
8. Training records
9. Training updates to the organizational repository
10. Training effectiveness surveys
11. Training program performance assessments
12. Instructor evaluation forms
13. Training examinations

9.2.4.4 Organizational Process Performance

Purpose: The purpose of organizational process performance (OPP) is to establish and maintain a quantitative understanding of the performance of the organization's set of standard processes in support of quality and process-performance objectives and to provide the process-performance data, baselines, and models to quantitatively manage the organization's projects.

Summary of BAs and SAs:

BA 1: Establish performance baselines and models.
  SA 1.1: Select processes.
  SA 1.2: Establish process-performance measures.
  SA 1.3: Establish quality and process-performance objectives.
  SA 1.4: Establish process-performance baselines.
  SA 1.5: Establish process-performance models.

Activity relationships diagram:

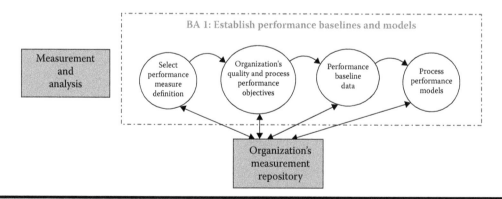

Figure 9.22 The OPP. (Software Engineering Institute 2006.)

Typical work products to be assessed:
1. Definitions for the selected measures of process performance
2. Organization's quality and process-performance objectives

3. Baseline data on the organization's process performance
4. Process-performance models

9.2.4.5 *Organizational Innovation and Deployment*

Purpose: The purpose of organizational innovation and deployment (OID) is to select and deploy incremental and innovative improvements that measurably improve the organization's processes and technologies. Such improvements support the organization's quality and process-performance objectives as derived from the organization's business objectives.

Summary of BAs and SAs:

BA 1: Select improvements.

    SA 1.1: Collect and analyze improvement proposals.

    SA 1.2: Identify and analyze innovations.

    SA 1.3: Pilot improvements.

    SA 1.4: Select improvements for deployment.

BA 2: Deploy improvements.

    SA 2.1: Plan the deployment.

    SA 2.2: Manage the deployment.

    SA 2.3: Measure improvement effects.

Activity relationships diagram:

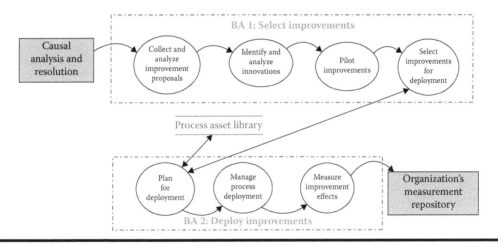

Figure 9.23   The OID. (Software Engineering Institute 2006.)

Typical work products to be assessed:

1. Analyzed process and technology improvement proposals
2. Candidate innovative improvements
3. Analysis of proposed innovative improvements
4. Pilot evaluation reports
5. Documented lessons learned from pilots
6. Process and technology improvements selected for deployment
7. Deployment plan for selected process and technology improvements
8. Updated training to reflect deployed process and technology improvements

9. Documented results of process and technology improvement deployment activities
10. Revised process and technology improvement measures, objectives, priorities, and deployment plans
11. Documented measures of the effects resulting from deployed process and technology improvements

9.3 Generic Activities Applicable to All Processes

This section describes all the generic goals and generic practices of the sustainability maturity model that directly address process institutionalization for maturity levels 2 (well defined) and 3 (planned and tracked). Requirements for the application of generic activities are defined for each stage of organizational maturity when an assessment is performed by a certified CMMI assessor. The definitions presented here are intended for a general understanding of the types of activities associated with institutionalization of processes described in Section 9.2.

Institutionalization is an important concept in process improvement. When mentioned in the BAs and SA descriptions, institutionalization implies that a process is ingrained in the way work is performed and there is commitment and consistency to performing the process.

There are 10 generic activities expected to be applied to each of the processes, either individually or collectively. They are summarized here and more fully defined in Sections 9.3.1 through 9.3.10:

1. Establish an organizational policy applicable to a process.
2. Plan development, deployment, use, and improvement of the process.
3. Provide resources for management of the process.
4. Assign responsibility for maintenance of the process.
5. Train people in understanding and using the process.
6. Manage configurations of the process components.
7. Identify and involve relevant stakeholders.
8. Monitor and control the process.
9. Objectively evaluate adherence to the policy for using the process.
10. Review the status of process adherence with higher-level management.

9.3.1 Establish an Organizational Policy

Establish and maintain an organizational policy for planning and performing the process.

The purpose of this generic practice is to define organizational expectations for the process and make these expectations visible to those in the organization who are affected. In general, senior management is responsible for establishing and communicating guiding principles, direction, and expectations for the organization.

9.3.2 Plan Development, Deployment, Use, and Improvement of the Process

Establish and maintain the plan for performing the process. The purpose of this generic practice is to determine what is needed to perform the process and to achieve the established objectives, prepare a plan for performing the process, prepare a process description, and get agreement on the plan from relevant stakeholders.

9.3.3 *Provide Resources for Management of the Process*

Provide adequate resources for performing the process, developing work products, and providing services of the process. The purpose of this generic practice is to ensure that resources necessary to perform the process as defined by the plan are available when they are needed. Resources include adequate funding, appropriate physical facilities, skilled people, and appropriate tools.

9.3.4 *Assign Responsibility for Maintenance of the Process*

Assign responsibility and authority for performing the process, developing work products, and providing services of the process. The purpose of this generic practice is to ensure that there is accountability for performing the process and achieving the specified results throughout the life of the process. The people assigned must have appropriate authority to perform the assigned responsibilities.

9.3.5 *Train People*

Train the people performing or supporting the process as needed. The purpose of this generic practice is to ensure that people have the necessary skills and expertise to perform or support the process.

9.3.6 *Manage Configurations*

Place designated work products of the process under appropriate levels of control. The purpose of this generic practice is to establish and maintain the integrity of the designated work products of the process (or their descriptions) throughout their useful life.

9.3.7 *Identify and Involve Relevant Stakeholders*

Identify and involve relevant stakeholders of the process as planned. The purpose of this generic practice is to establish and maintain the expected involvement of stakeholders during the execution of the process. Involve relevant stakeholders as described in an appropriate plan for stakeholder involvement.

9.3.8 *Monitor and Control the Process*

Monitor and control the process against the plan for performing the process and take appropriate corrective action. The purpose of this generic practice is to perform direct day-to-day monitoring and controlling of the process. Appropriate visibility into the process is maintained so that appropriate corrective action can be taken when necessary. Monitoring and controlling the process involves measuring appropriate attributes of the process or work products produced by the process.

9.3.9 *Objectively Evaluate Adherence to the Policy for Using the Process*

Objectively evaluate adherence of the process against its process description, standards, and procedures, and address noncompliance. The purpose of this generic practice is to provide credible assurance that the process is implemented as planned and adheres to its process description, standards, and procedures. This generic practice is implemented, in part, by evaluating selected work products of the process.

9.3.10 Review the Status of Process Adherence with Higher-Level Management

Review the activities, status, and results of the process with higher-level management and resolve issues. The purpose of this generic practice is to provide higher-level management with appropriate visibility into the process.

References

Chrissis, M. B., M. Konrad, and S. Shrum. 2011. *CMMI® Guidelines for Process Integration and Product Improvement*. 3rd ed. ISBN 978-0-321-71150-2, March 2011. Upper Saddle River, NJ: Addison-Wesley.

Software Engineering Institute (SEI). 2006. "Capability Maturity Model Integrated-Development® (CMMI-DEV®), Version 1.2." Technical Report CMU/SEI-2006-TR-008. Pittsburgh, PA: Carnegie Mellon University.

Index